Pedagogical Revelations and Emerging Trends

Edited By

Dr. C. Sheeba Joice and Dr. M. Selvi

Pedagogical Revelations and Emerging Trends

About the Conference

INTERNATIONAL CONFERENCE ON EDUCATING THE POST MILLENNIALS (ICEM'24)

ICEM'24 was the second edition organized by Saveetha Teaching Learning Centre (STLC), Saveetha Engineering College, India held on May 17, 2024.. This signature programme is where the STLC connected the academic experts, budding educators and eminent members of the teaching-learning community. This confluence explored and enhanced the potential of Engineering Educators and Learners. ICEM'24 provided a global platform to identify best practices in teaching and applaud the evolutionary aspects involved in reaching the zenith.

ICEM'24 consisted of two keynote sessions and paper presentations under two tracks namely:

TRACK 1: TECHNOLOGY ORIENTED STUDENT- CENTRIC TEACHING-LEARNING

- Engineering Education for Sustainable Development
- Integrating STEM in Curriculum
- First-Year Engineering Laboratory
- Experiential Learning
- Innovative Teaching Tools
- Problem/Project-Based Learning
- Critical Thinking and Decision Making in Engineering
- Effective Assessment of Teaching-Learning Strategies
- Evolving Engineering Pedagogy for Implementing NEP 2020

TRACK 2: INDUSTRIAL COLLABORATION IN TEACHING LEARNING

- Progressing to Industry 5.0
- Role of Industrial Experts in Curriculum Design
- Impact of Virtual Internship
- Design Thinking
- Entrepreneurship and Engineering Innovation
- Teaching for the Attainment of 4Cs
- Engineering Leadership
- Quality Assurance in Engineering Education
- E- Learning Eco System

Proceedings of the International Conference on Educating the Post Millennials (ICEM'24)

Pedagogical Revelations and Emerging Trends

Edited by

Dr. Sheeba Joice. C

Dean,
Saveetha Teaching Learning Center,
Professor and Deputy Head
Department of ECE, Saveetha Engineering College.

Dr. M. Selvi

Professor and Deputy Head,
Department of Electronics and Communication Engineering
Saveetha Engineering College

CRC Press
Taylor & Francis Group
Boca Raton London New York

CRC Press is an imprint of the
Taylor & Francis Group, an **informa** business

First edition published 2025
by CRC Press
4 Park Square, Milton Park, Abingdon, Oxon, OX14 4RN

and by CRC Press
2385 NW Executive Center Drive, Suite 320, Boca Raton FL 33431

CRC Press is an imprint of Informa UK Limited

British Library Cataloguing-in-Publication Data
A catalogue record for this book is available from the British Library

ISBN: 9781032960029 (pbk)
ISBN: 9781032960012 (hbk)
ISBN: 9781003587538 (ebk)

DOI: 10.1201/9781003587538

Typeset in Sabon LT Std
by HBK Digital

Contents

List of Figures

List of Tables

Preface

We feel greatly honoured to have been assigned with the job of organizing the second edition of International Conference on Educating the Post Millennials (ICEM'24) on May 17, 2024 at Saveetha Engineering College. This international conference is a platform that brings the brightest minds across the globe to share their ideas and insights on the recent innovations and emerging technologies involved in the domain of pedagogy. With an aim to promote collaboration and foster innovation, this conference promises to be a melting pot of ideas and knowledge sharing.

We would like to extend our gratitude to the management for permitting us to organize the conference, and for conceptualizing and executing this event with utmost dedication and perseverance. We sincerely thank our beloved Founder-President Dr. N. M. VEERAIYAN and our Director Dr. S. RAJESH for their significant support to make this conference a great success. We record our heartfelt thanks to our Principal Dr. V. VIJAYA CHAMUNDEESWARI and Vice Principal Dr. R. SENTHIL KUMAR for their continuous support and guidance in organizing the conference. We thank the team members of Saveetha Teaching Learning Centre and all our staff, learners and participants for making this conference a successful one.

We would also like to express our gratefulness to the Indo Universal Collaboration for Engineering Education (IUCEE) for inspiring us to conduct path-breaking events on teaching-learning domain.

This second edition connected the academic experts, budding educators and eminent members of the teaching-learning community. This confluence set an arena for enhancing the potential of Engineering Educators and Learners. Dr. K.N. Subramanya, an eminent educator and Principal, RV College of Engineering, Bengaluru inaugurated ICEM'24. He addressed the gathering on the methods of quality assurance in engineering education during this digital era. Besides, ICEM'24 witnessed an interactive keynote session by Educational Psychologist Dr. Prathiba Nagabhushan, based at St. Mary MacKillop College in Canberra highlighting the need for educators to adapt their pedagogical approaches for post-millennial learners.

In order to facilitate participants from other countries, ICEM'24 was a hybrid programme that hosted paper presentation session in both virtual and face-to face mode. Several research works on digital fluency, experiential learning, emotional empowerment, ecological pedagogy and curriculum enhancement were presented.

Out of 150 submissions, 50 research papers were selected for presentation. All papers that were presented in the conference has been peer-reviewed and approved for publication in the Taylor & Francis Conference Proceedings Series titled Pedagogical Revelations.

Hence, ICEM'24 witnessed the works of passionate educators and learners who were instrumental in changing the paradigm of teaching domain.

ABOUT THE INSTITUTION

Saveetha Engineering College, started in the year 2001 by the Saveetha Medical & Educational Trust, which is committed to develop this college into a renowned institution for Engineering education and research. The Saveetha Group of Institution is headed by Dr. N. M. Veeraiyan, a committed and dedicated Medical Professional.

The campus of Saveetha Engineering College is spread over a 120 acres of scenic beauty, facing Chembarambakkam lake on the Chennai- Bangalore National Highway (NH4). Located at about 8 km from Poonamalee township, the college buildings are laid out amidst a serene environment. The Institution offers UG, PG and Doctoral Programmes through departments such as AGRI, BIOMED, CIVIL, CHEM, CSE, CSE (Cyber Security), CSE (IoT), EEE, EIE, ECE, MED ELEC, AI&DS, AI&ML, IT and MECH.

The college consists of a spacious canteen and well equipped laboratories with the state of the art computers equipment and gadgets. Separate rest rooms for boys and girls are provided in the college campus. The campus also houses well-ventilated hostels for boys and girls. In all, the learners are provided with an atmosphere conducive to pursue their studies freely.

The motto of the Institution is "BE THE BEST" which motivates the learners to aim for nothing less than the best in every aspect of their academics and profession.

Vision and mission

Our vision: to be and to be recognized for setting the standard of excellence in engineering education and high quality research in science and technology.

Our mission: to promote academic excellence; widen intellectual horizon; inculcate self- discipline and high ideals for the total personality of the individual.

ABOUT SAVEETHA TEACHING LEARNING CENTRE (STLC)

Saveetha Teaching Learning Centre, is a pioneer initiative of Saveetha Engineering College to promote excellence in teaching and learning among academicians and avid learners.

Mission of STLC

To provide holistic approach in graduating teacher centric educators to learner centric educators to build a brand of students who'll enjoy learning and establish themselves to constructively contribute in the field of Engineering.

Objectives of STLC

- To become facilitators to encourage self-learning
- To keep oneself updated to equip the learners
- To be trained to utilize pedagogical tools
- Motivated to create self-learners
- Highly committed to the profession
- Should be able to encourage learners to become inquisitive and innovative
- Study and develop customized modules to train the faculty
- To provide platform to exhibit their learning

The Saveetha Teaching Learning Centre received the Outstanding Teaching Centre Award at the ICTIEE 2023, organized by the Indo Universal Collaboration for Engineering Education (IUCEE) from January 6-8, 2023 at Vidyavardhaka College of Engineering in Mysuru for its exceptional efforts in adopting teaching-learning methodologies and organizing programs to equip faculty members with necessary skills for excellence in teaching practices.

FOUNDER PRESIDENT'S MESSAGE

I'm delighted to extend a warm welcome to the International Conference on Educating the Post Millennials (ICEM)'24, meticulously organized by the Saveetha Teaching Learning Centre (STLC) at Saveetha Engineering College.

The core objective of engineering education is to instill a sense of curiosity in learners, inspiring them to explore and pursue sustainable living practices. ICEM'24 promises to be a distinctive conference, bringing together academicians and dedicated researchers committed to advancing the teaching-learning process. This remarkable gathering of insightful minds will serve as an exceptional platform for sharing innovative ideas and engaging in thought-provoking discussions across a wide spectrum of educational topics, reflecting the latest trends in the field.

The conference has attracted exceptional research contributions spanning various domains of education and related themes, reflecting the fulfillment of its objectives through the reception of real-time research endeavors. I am certain that attendees will greatly appreciate the conference venue, Saveetha Engineering College, known for its environmentally conscious campus and state-of-the-art facilities.

My heartfelt congratulations go out to all the faculty members and learners whose efforts have contributed to the success of this conference. I wish you all a day filled with fruitful discussions and stimulating exchanges of knowledge, as together, we strive to contribute towards building a stronger India.

Once again, my sincere congratulations to the organizers, accompanied by my best wishes for the event's success.

Dr. N. M. Veeraiyan
Founder President

DIRECTOR'S MESSAGE

I feel honored and immensely pleased to extend a warm welcome to all participants, inviting them to share their teaching experience and innovative ideas on educational development with us.

Engineering challenges are often multifaceted, presenting myriad solutions, ranging from good to bad and everything in between. The true art lies in crafting a solution that stands out. It's a creative endeavor, fueled by imagination, intuition, and purposeful decision-making.

The International Conference on Educating the Post Millennials (ICEM'24), spearheaded by Saveetha Teaching Learning Centre (STLC), stands as a pioneering event aimed at providing a fertile ground for nurturing imagination and fostering innovative ideas to shape the future.

In an era where educators and learners alike are embracing Industry 5.0 and flexible learning approaches, ICEM'24 presents an unparalleled opportunity and platform for both novice and seasoned academicians to exchange views, insights, and experiences on cutting-edge teaching and learning methodologies implemented worldwide.

With the recent overhaul of the education landscape through the National Education Policy (NEP 2020), our institution is at the forefront of adapting to these changes, preparing industry-ready professionals to serve society. I am confident that ICEM'24 will further galvanize our efforts and inspire us to continuously strive for excellence.

My heartfelt congratulations to the organizers of ICEM'24 for orchestrating this successful conference.

I eagerly anticipate reconnecting with you all at ICEM'24, where we can continue to share and explore even more innovative ideas together.

Dr. S. Rajesh

Director

PRINCIPAL'S MESSAGE

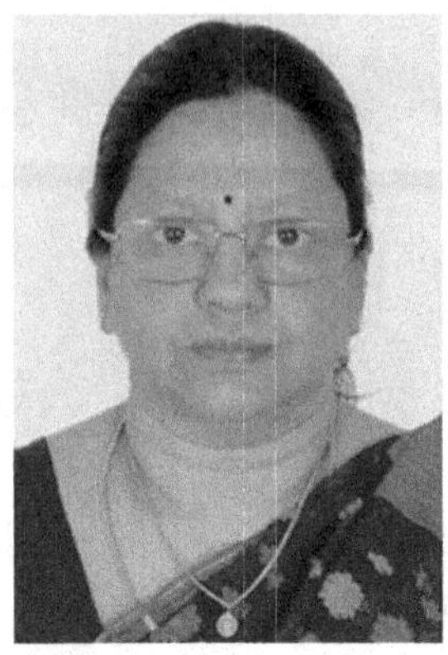

I extend a warm welcome to all, inviting you to join us in shaping a prosperous future filled with both joy and responsibility. It's with great pleasure that I share this message to inspire the faculty, students, and participants of the upcoming International Conference on Educating the Post Millennials (ICEM'24).

The venue of ICEM'24, Saveetha Engineering College with her exemplifying and highly qualified academic facilitators along with state-of-the-art facilities upholds the motto "Be the Best". Our mission extends beyond mere instruction as we aim to nurture individuals to confront both current and future challenges. Engaging in peer-learning and discussions at esteemed conferences like ICEM'24 offers invaluable opportunities for collaboration and networking, benefiting both academicians and learners alike.

Saveetha Teaching Learning Centre (STLC) spearheads innovative teaching methodologies, providing a framework for academicians to implement creative ideas in the classroom. In today's dynamic world, marked by technological advancements and global integration of knowledge, the demand for novelty and creativity is paramount. It is essential to educate learners to become industry-ready professionals, and ICEM'24 serves as a crucial step towards achieving this goal.

I extend my congratulations to the ICEM'24 team for organizing this intellectually enriching event. Best wishes for your future endeavors.

Dr. V. Vijaya Chamundeeswari

Principal

VICE PRINCIPAL'S MESSAGE

In the words of the renowned quote, "Education is not the filling of a pail, but the lighting of a fire," implying that teaching is not just about imparting knowledge, but igniting a passion for learning. As educators, it is our responsibility to inspire and empower our students to pursue their dreams and navigate the challenges they encounter in the world.

I firmly believe that the International Conference on Educating the Post Millennials (ICEM'24), organized by the Saveetha Teaching Learning Centre (STLC), will stand as a significant event, motivating numerous members of the teaching community to explore innovative tools and technologies to enhance their teaching practices.

Over the years, our institution has been recognized for its adaptability to various innovative teaching methods. The success of these experimental approaches has bolstered our confidence to share our experiences and engage in collaborative learning with peers during this conference. With this optimism, I envision a fruitful event and look forward to many more editions in the future.

Once again, I extend my congratulations to the organizing team of ICEM'24 and wish them continued success in their future endeavors.

Dr. R. Senthil Kumar

Vice Principal

CONFERENCE CHAIR'S MESSAGE

"Continuous effort is the key to unlocking great achievements", encapsulates the essence of our journey. The establishment and progressive evolution of Saveetha Teaching Learning Centre, initiated by our management, stand testament to this principle. I extend my heartfelt gratitude to our management and fellow faculty members for their collaborative efforts in this venture, alongside the commendable adaptability showcased by our learners in embracing diverse teaching methodologies over the years.

The path leading to the organization of the International Conference on Educating Post Millennials (ICEM'24) has been marked by incremental progress rather than sudden leaps. It has involved numerous internal active learning initiatives, brainstorming sessions, and peer discussions. We've received more than 100 original research papers focusing on contemporary and indigenous teaching methods.

I extend my sincere appreciation to our management for their unwavering financial, infrastructural and moral support in bringing this conference to fruition. Additionally, I acknowledge the invaluable assistance provided by the Indo-Universal Collaboration for Engineering Education (IUCEE), which has played a significant role in gathering consortium members for the program.

Lifelong learning is an ongoing, self-motivated pursuit of knowledge.

I extend my best wishes to all participants for a successful learning experience during ICEM'24.

Dr. C. Sheeba Joice

Organizing Chair

ORGANIZING CHAIR'S MESSAGE

Saveetha Teaching Learning Centre (STLC) has emerged as a robust and progressive entity, making significant strides since its establishment in 2019. As part of our ongoing academic endeavors, we have orchestrated the International Conference on Educating Post Millennials (ICEM'24), scheduled for May 17, 2024.

The response from the academic community has been overwhelmingly positive, with research papers pouring in from various corners of the globe. Each submission underwent rigorous scrutiny by our internal review team to ensure the dissemination of high-quality research. Consequently, 50 papers have been selected for presentation. These research articles have been peer-reviewed for prospective publication in the Scopus-indexed Taylor & Francis Conference Proceedings titled 'Pedagogical Revelations'. I firmly believe that these research contributions have enriched the knowledge base and understanding within the teaching-learning community.

I extend my heartfelt gratitude to the dignitaries, delegates, faculty members, and learners whose enthusiastic contributions have rendered this conference a resounding success. Moreover, the unwavering financial and technical support extended by our management has been instrumental in achieving this milestone.

Looking ahead, I eagerly anticipate our future engagements and the forthcoming editions of the conference.

Dr. M. Selvi
Organizing Chair

Details of Programme Committee

Dr. Krishna Vedula, Executive Director, IUCEE
Dr. Prathiba Nagabhushan, Faculty, Australian Catholic University, Canberra
Dr. Rashpinder Kaur, Assitant Professor, Chitkara University
Dr. Ruhi Bakhare, Head Research Cell & Assistant Professor, Dr. Ambedkar Institute of Management Studies & Research, Deekshabhoomi
Dr. Jeyamala Chandrasekaran, Associate Professor, Thiagarajar College of Engineering, Madurai
Dr. D. Anitha, Assistant Professor, Thiagarajar College of Engineering, Madurai
Dr. K. Rajeswari, Associate Professor, Thiagarajar College of Engineering, Madurai
Dr. K. Velmurugan, Assistant Professor, Anurag University
Dr. Prakash Hegade, Assistant Professor, KLE Technological University, Hubli
Dr. R. Kaladevi, Associate Professor, Panimalar Engineering College

Chief Patron

Dr. N. M. Veeraiyan,
Founder President & Chancellor,
Saveetha Institute of Medical and Technical Sciences.

Patrons

Dr. V. Saveetha Rajesh,
Director,
NMV University.

Dr. S. Rajesh,
Director,
Saveetha Engineering College.

Co-Patrons

Dr. V. Vijaya Chamundeeswari,
Principal,
Saveetha Engineering College.
Organizing Committee

Conference Chair

Dr. Sheeba Joice. C
Dean,
Saveetha Teaching Learning Center,
Professor and Deputy Head
Department of ECE, Saveetha Engineering College.

Organizing Chair

Dr. M. Selvi,
Professor & Deputy Head/ECE,
Saveetha Engineering College.

STLC Team

Dr. Sharmila Queenthy,
Associate Professor, SEC
Dr. S. Padma,
Assistant Professor, SEC
Dr. Sravya. P.R.
Assistant Professor, SEC

Dr. K. Aruna Devi,
Professor, SEC
Dr. S. Imran Hussain,
Assistant Professor, SEC
Mr. R. Kannan,
Assistant Professor, SEC
Dr. G. Nagappan,
Professor & Head/CSE, SEC
Mr. Jerome Nithin Gladson,
Assistant Professor, SEC
Mr. S. Joyal Isac,
Assistant Professor, SEC
Ms. K. Padma Priya,
Assistant Professor, SEC

Chief Guest

Dr. K.N. Subramanya, an eminent educator and Principal, RV College of Engineering, Bengaluru.

Key Note Speakers

Dr. K.N. Subramanya, an eminent educator and Principal, RV College of Engineering, Bengaluru

Dr. Prathiba Nagabhushan, Educational Psychologist based at St. Mary MacKillop College in Canberra

About the Editors

Dr. Sheeba Joice. C. Dean, Saveetha Teaching Learning Center, Professor and Deputy Head, Department of ECE, Saveetha Engineering College. She received her B.E.(EEE) Degree from Madras University, Chennai, India, in 1998 and her M.E. (Applied Electronics) from Anna University, Chennai, India, in 2007. She received her Doctorate from Anna University, in the area of Embedded Control of Drives in 2012. She is the Associate Dean of Saveetha Teaching Learning Centre at Saveetha Engineering College, Chennai. She is serves as a Professor and Deputy Head in the Department of Electronics and Communication Engineering. She was an Executive Committee Member IETE – Chennai Centre and Life Member ISTE and IETE. Her areas of interest in research includes Embedded Systems, Image Processing and Pedagogical innovations. She has been resource person for different Faculty Development Programmes. She has more than 15 publications to her credit.

Dr. M. Selvi currently serves as the Professor and Deputy Head in the Department of Electronics and Communication Engineering, Saveetha Engineering College, Anna University Chennai. She holds a Bachelor of Engineering in Electronics and Communication Engineering from Government College of Engineering, Tirunelveli, Tamil Nadu along with an M.E (communication Systems) from National Engineering College, Kovilpatti, Tamil Nadu. She received her Ph.D. from Anna University, Chennai by 2014 in the field of Free Space Optical communication. Accumulating over 25 years of experience in Teaching, Industry, and Research, she has established herself as a seasoned professional. She served as the Executive Committee Member and Honorary Secretary of IETE during 2020. Her research interest includes IoT, Assessment Methodology and Online content development.

1 The AI tutor: Revolutionizing education through personalized learning

Sornavalli G.[a], *Sreeshma M. Nair*[b], *Srija Vardharajan*[c] *and Srujana Srinivasan*[d]
Department of Information Technology, Sri Sivasubramaniya Nadar College of Engineering, Tamil Nadu, India

Abstract

This paper is a transformative tool built to harness the power of artificial intelligence (AI) to deliver personalized and adaptive learning experiences. AI tutor offers tailored support across diverse subjects, to diverse learners. Its key features include – a recommendation algorithm for content suggestions based on learning styles, text summarization features for swift information synthesis, quiz generation tools for adaptive quizzes, optical character recognition (OCR) technology and image processing techniques to enable the conversion of handwritten to digital text, and scheduling algorithms for robust time management – offering a dynamic approach to digitized learning. The motivation behind the AI tutor system is rooted in a commitment to revolutionize education. Conventional learning methods often present limitations in adapting to the unique learning styles and preferences of individual students. Thus, the need for an innovative educational platform that addresses these challenges becomes apparent. This paper explores the multifaceted functionalities and innovative features of AI tutor, aimed at meeting the escalating demand for accessible and effective learning platforms.

Keywords: Artificial intelligence, e-learning, image processing, accessibility, constraint satisfaction problem techniques, natural language processing

Introduction

In today's dynamic landscape of modern education, applications for artificial intelligence (AI)-based tutoring are essential resources that are expanding the possibilities for individualized education. These applications have developed into essential parts of modern educational systems, surpassing simple technological innovations thanks to a multitude of developments that have fueled their growth.

The innovative quiz maker system developed by Gabajiwala et al. (2022) uses natural language processing (NLP) techniques to automatically create quizzes from textual content. This allows educators to assess students' comprehension and critical thinking abilities in an easy-to-use and efficient manner. Building on this framework, Sung et al. (2007) and Le et al. (2013) carried out an extensive analysis of AI-supported tutoring methods. These systems provide individualized recommendations and content suggestions based on user interactions and learning patterns, analyzed by sophisticated algorithms that encourage learners to become more engaged and proficient.

Furthermore, new opportunities for content creation and pathway optimization have been made possible by the use of AI technology into tutoring programmes. This is demonstrated by the AI-powered learning system created by Diwan et al. (2023), which uses ongoing learner behavior analysis to create personalized learning pathways for each student's changing needs and preferences, these applications guarantee that learning stays effective, relevant, and engaging.

Meanwhile in the field of group learning, Maravanyika et al. (2017) shows how these platforms can help students connect with peers who have similar interests and aspirations, which can promote peer support and knowledge exchange. Moreover, Kerres and Buntins (2020) explore the intersection of AI and recommender systems, evaluating how well recommender systems powered by AI can direct students to the best possible learning materials. These technologies enable students to confidently traverse challenging educational environments.

The research by AlShaikh and Hewahi (2021) draw attention to the variety of methodologies and technological advancements that these systems are based upon – highlighting the diverse applications of AI in education amongst different learner classes, and Ahmad et al. (2021) investigate the use of AI in tutoring specifically, highlighting how it has the potential to transform conventional teaching strategies and improve student outcomes. Gromyko et al. (2017) conducted a similarly significant study in which they investigated artificial intelligence's potential as a teaching aid for human cognition.

Pardamean et al.'s (2022) study looks into AI-based learning style prediction in online primary education, showing how AI algorithms may modify course

[a]sornavallig@ssn.edu.in, [b]sreeshma2110158@ssn.edu.in, [c]srija2110108@ssn.edu.in, [d]srujana2110214@ssn.edu.in

DOI: 10.1201/9781003587538-1

material to fit each student's unique learning preferences. Troussas et al. (2020) also investigated personalized tutoring using a stereotype student model with a hybrid learning style tool, emphasizing the possibility of AI to tailor learning experiences according to unique user traits. In a similar vein, Truong (2016) emphasize the significance of accommodating a variety of learning preferences and demands when discussing the integration of learning styles and adaptive e-learning systems.

Sreelakshmi et al.'s (2019) study presents a chatbot for education that can generate quizzes and answer questions, demonstrating the potential of conversational agents powered by artificial intelligence to enhance student learning. Additionally, Kukulska-Hulme and Shield (2008) gave an overview of mobile-assisted language learning and talked about how AI can be used to create individualized language learning experiences. Their study shows how AI tools may improve student participation and offer individualized support instantly.

In conclusion, AI-based tutoring programmes are a game-changer in education, providing students with individualized and flexible help. We developed our own AI-powered tutoring system because we believe that certain issues and constraints must be addressed as this field of study moves further.

Literature review

Artificial intelligence has completely transformed education. Sung et al. (2007) automatic quiz generation system for English text demonstrated the progress made in creating quizzes using AI. An extensive evaluation of text summarizing approaches was carried out by Abualigah et al. (2020), who emphasized the significance of these strategies in educational settings. In a similar vein, Gambhir and Gupta (2017) reviewed the most recent advancements in automatic text summarization and emphasized its importance in the synthesis of instructional content. Furthermore, T-BERTSum, a topic-aware text summary method based on BERT, was proposed by Ma et al. (2022).

An intelligent personal assistant created for task and time management was demonstrated by Myers et al. (2007), while Meng et al. (2022) showed how AI can be used to introduce an adaptive recommender based framework for teaching and learning.

Alam et al. (2023) further highlighted how AI can be used to build intelligent tutoring systems that improve classroom experiences. According to Ahmad et al. (2021), it has the ability to completely transform conventional teaching techniques. In their discussion of AI as a companion for human intellect, Gromyko et al. (2017) demonstrated the cooperative potential of AI in boosting human capacities. Gupta and Lehal (2010) concentrated on AI-driven strategies that seek to distill volumes of instructional material into concise summaries.

The combination of learning styles and adaptive e-learning systems was studied by Troussas et al. (2020), who highlighted the possible synergy between AI and pedagogical theories. On that note, Pardamean et al. (2022) suggested AI-based learning style prediction for elementary education. Furthermore, Fernandopulle et al. (2021) presented EduHelp, an online tutoring platform that uses AI to offer individualized learning experiences in a variety of educational contexts. In order to help content authors in educational contexts, Dang et al. (2022) investigated the usage of automatic text summaries. Diwan et al. (2023) illustrated learning pathway augmentation and AI-based learning material creation.

After looking at these bodies of work, we have identified shortcomings in the current AI-supported learning environment. A significant deficiency concerns the small number of systems that include thorough evaluations of students' cognitive preferences, even though some provide adaptive learning routes based on data. In order to close this gap, our solution incorporates the VARK quiz into our coaching software, offering personalized learning experiences that are tailored to the individual learning styles and aptitudes.

Additionally, a lot of the research that has been done ignores how crucial accessibility is in education, especially for students from a variety of backgrounds. Although AI-driven tutoring programmes have demonstrated efficiency, their uptake in underprivileged populations is still not fully understood. Our research explores this gap by making sure that our software is made to fit learners with different skill levels and degrees of technical literacy thanks to its abundance of resources as well as user-friendly interface.

There is a clear lack of information in the literature about how to support students in effectively managing their time. Our scheduler provides students the freedom to set realistic goals. The text summarizer's compression ratio deviates from conventional strategies that outline a desired word count for the material to be condensed into. There is room for future study to improve AI tutor's features, such as investigating how to use augmented reality to develop engaging learning environments to improve retention and knowledge transfer by creating realistic scenarios.

It is imperative to recognize the priceless contributions made by the body of literature that has already been reviewed for this study. Through a process of information synthesis and accumulation, we have developed a thorough grasp of the opportunities and problems associated with intelligent tutoring systems.

Methodology

System overview

The AI tutor system comprises several modules, each serving a specific function to optimize the learning process as depicted in Figure 1.1. The users' learning preferences, identified through the VARK quiz module, inform subsequent interactions within the system. The text summarization module allows to extract key information and eliminating redundant details, enabling users to grasp complex concepts quickly and efficiently. Complementing the text summarization module is the quiz generation module. Powered by semantic analysis, this module automatically generates quizzes tailored to reinforce the key concepts covered in the summaries. The handwritten-to-text module adds another layer of functionality to the AI tutor system. By incorporating optical character recognition (OCR) technology, this module allows users to upload handwritten notes and convert them into digital text. Finally, the scheduling module assists users in effectively managing their study sessions by considering parameters such as study duration, break times, and preferred intervals.

Personalized time scheduler

The AI tutor's personalized scheduler employs constraint satisfaction problem (CSP) solving techniques to tailor solutions to individual scheduling needs, focusing on creating optimal study and leisure balance. By utilizing CSP approaches, the system systematically explores various combinations of tasks within given constraints, such as preferred study times and breaks, to construct personalized schedules. This method ensured that schedules not only adhere to user-defined preferences and requirements but also promote effective study habits and adequate leisure time, resulting in schedules that are well-suited to individual users' preferences and constraints accounting for leisure time and effective study schedules.

Users possess the capability to input their intended duration of study, including the commencement time and break intervals, specifying the number and length of breaks desired. Subsequently, the system generates a customized schedule tailored to these preferences, delineating distinct time slots to accommodate these requirements. This functionality streamlines the process of schedule creation, alleviating users from the burden of manual scheduling by automating the task seamlessly.

Handwritten to text converter

This tool empowers users with the ability to effortlessly convert handwritten notes into digital text. Users can conveniently upload photos of their written notes, either individually or in batches, providing flexibility in input methods. Leveraging the PyTesseract module, our system seamlessly processes the uploaded images, accurately converting them into editable text format as shown in Figure 1.2.

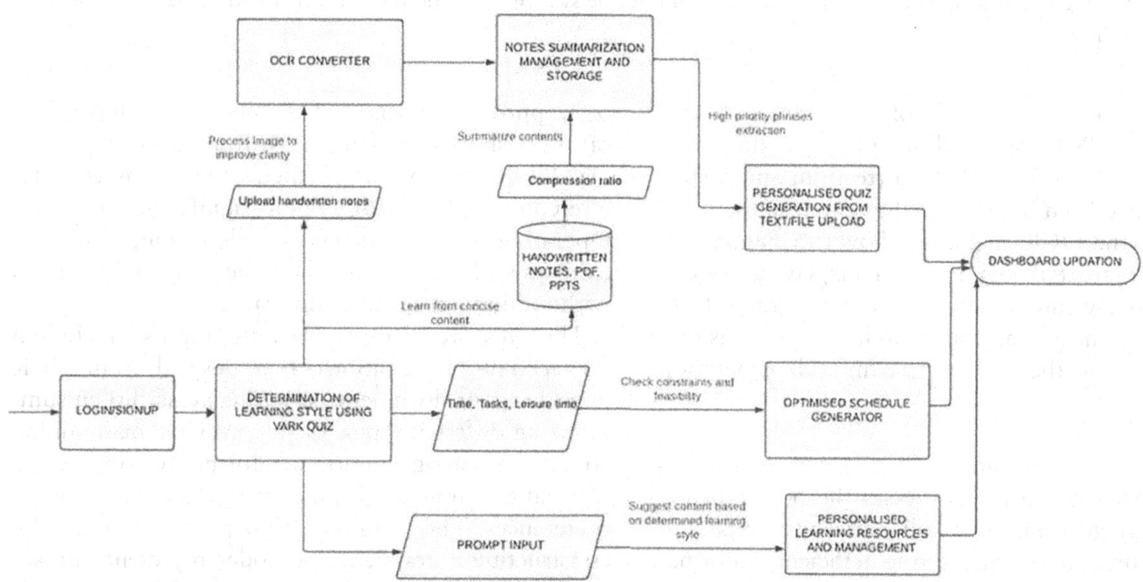

Figure 1.1 System architecture diagram depicting the various module
Source: Author

Figure 1.2 Flowchart illustrating the image processing techniques employed by the system
Source: Author

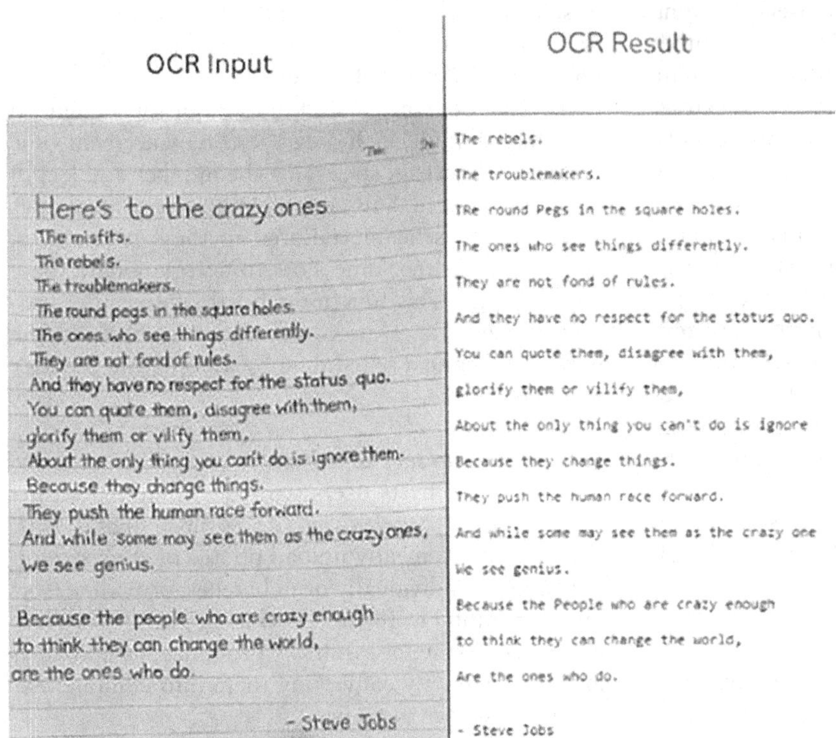

Figure 1.3 Figure depicts the handwritten notes input into the system (left) and the text generated by the system (right)
Source: Author

To improve the accuracy of the extracted text, robust Image Processing techniques – including binarization, noise removal, skew correction, and a spell checker, have been implemented to enhance the functioning of the OCR engine as shown in Figure 1.3. With the extracted text now on display, users can choose to view summaries of content generated from the OCR engine, or can choose to have questions prepared based on the content, via the quiz generation module.

Notes summarizer feature

Our summarizer feature harnesses the power of the TextRank algorithm to distill extensive texts into concise summaries, facilitating efficient information consumption. While previous research focuses on summary levels ranging from full text, to selected (central) sentences, down to a collection of keywords, our approach consists of a summary compression ratio (as shown in Figure 1.4). Rooted in graph theory, TextRank represents sentences or words as nodes interconnected by edges that signify relationships. Operating as an unsupervised algorithm, TextRank exhibits scalability to handle large texts and operates without domain-specific constraints.

This feature accepts various inputs, including PowerPoint presentations, text derived from OCR processing, or documents, ensuring versatility in summarizing different types of content while maintaining context, retaining important information, and ensuring that the summary is tailor-made to meet the user's preferences. The summarization process initiates by constructing a graph, where nodes represent phrases or words, and edges are established based on their semantic relationships, often quantified through measures like Jaccard similarity. Edge weights are assigned

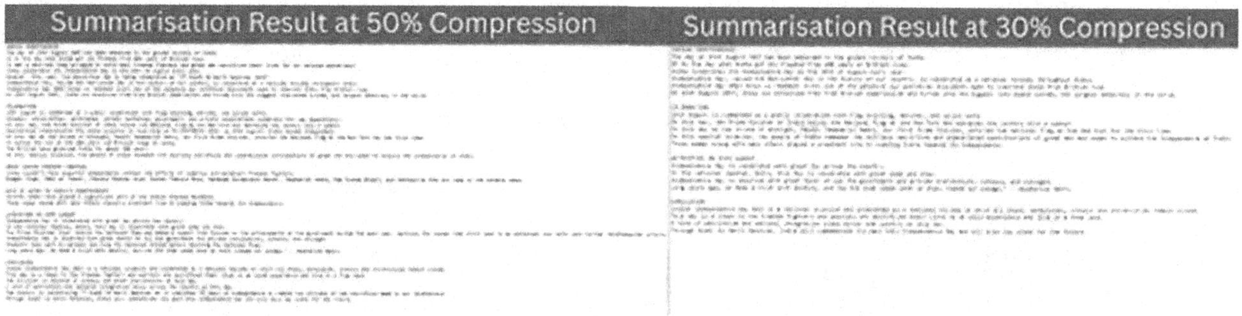

Figure 1.4 Figure depicts the summarized notes produced by the system at 50% compression (left) and 30% compression (right)
Source: Author

Table 1.1 Evaluation of our OCR by comparing PyTesseract with EasyOCR.

OCR tool	Mean accuracy
PyTesseract	0.91473
EasyOCR	0.71632

Source: Author

to reflect the strength of these relationships, aiding in the identification of crucial information. Subsequently, the algorithm executes a random walk, with transition probabilities determined by edge weights, while assigning importance scores to nodes based on their significance.

These scores heavily influence the final rankings, guiding the selection of top-ranked sentences or units to craft the summary. By adopting this extractive summarization approach, the Summa summarizer ensures that the generated summaries encapsulate the most pertinent and informative aspects of the original text, facilitating efficient comprehension and knowledge retention.

Quiz generation feature

The quiz generation module adopts a systematic methodology to dynamically generate quizzes from input text, with a focus on multiple-choice questions (MCQs) that maintain relevance to the actual answer and uphold contextual integrity. Following tokenization into sentences, keyword extraction prioritizes nouns and utilizes normalized Levenshtein similarity to filter phrases, resulting in the identification of the top 50 noun phrases based on length. Subsequently, a sentence-to-vector model determines the eligibility of words for MCQ formulation. Concurrently, each extracted keyword is linked to a text snippet containing the first three sentences containing that keyword, ensuring contextual relevance. This mapping is stored within the keyword sentence mapping repository.

Table 1.2 Evaluation of our summarizer by comparing Summa summarizer and BERT summarizer.

Metric	Summa summarizer	BERT summarizer
F-measure	0.75	0.60
Recall	0.80	0.65
Precision	0.70	0.55

Source: Author

During question generation, contextual information from the text snippet and its corresponding answer is leveraged to formulate MCQs, ensuring alignment with the input text and facilitating effective assessment. This comprehensive approach ensures the creation of quizzes that are both pertinent to the content and conducive to meaningful learning outcomes.

Performance evaluation

OCR

Our OCR model represents a state-of-the-art solution designed to accurately extract text from images, while Easy OCR serves as a benchmark for comparison. It is clearly highlighted in Table 1.1 that our OCR model has higher efficiency.

Summarizer

In assessing the efficacy of our summarizer module, we conducted a comparative exploration between two prominent algorithms: Summa and BERT. BERT, a state-of-the-art natural language processing model, offers advanced capabilities in semantic understanding and contextual analysis.

In analyzing the results from the above comparison in Table 1.2, several key insights emerge. Firstly, Summa consistently demonstrates higher values of precision, recall, and F-measure compared to BERT across multiple (10) evaluations. This indicates that Summa generally achieves better performance in

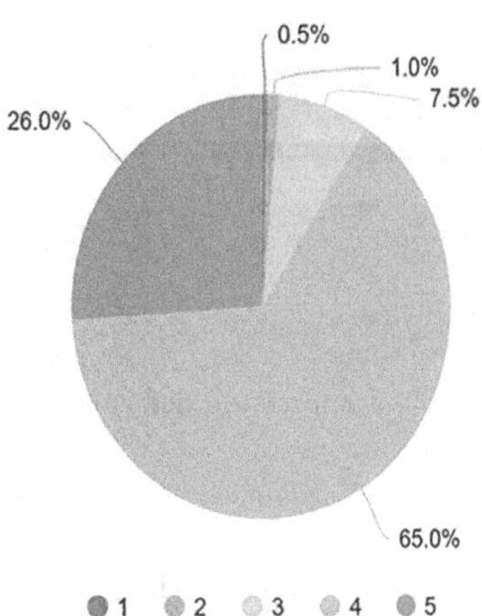

Figure 1.5 Survey on the likelihood of recommendation of the app
Source: Author

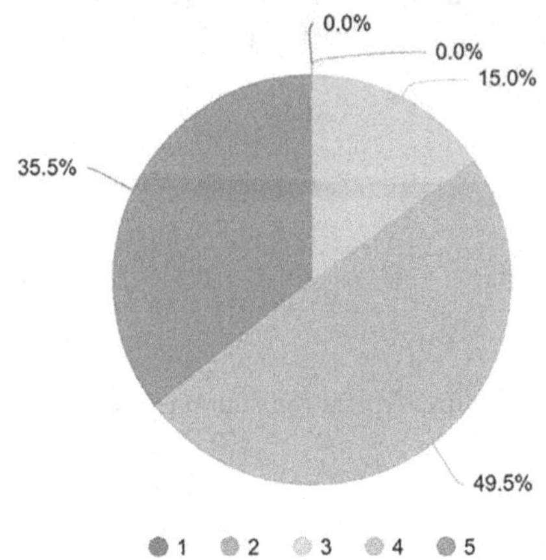

Figure 1.6 Survey on the impact assessment on individual learning
Source: Author

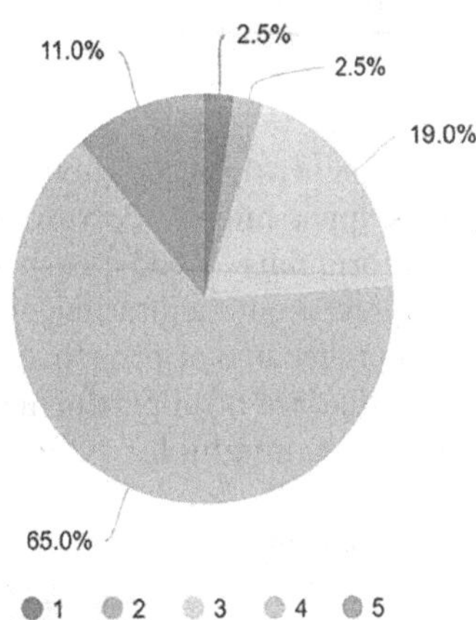

Figure 1.7 Survey on the accuracy in adaptation of the app to user learning style
Source: Author

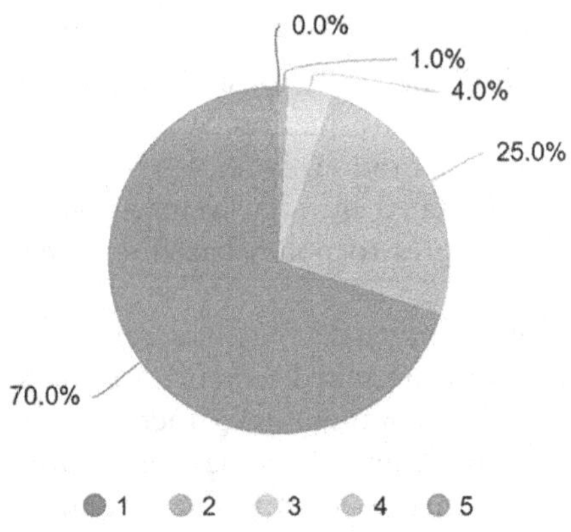

Figure 1.8 Survey on the quality of content fetched by the app on various topics
Source: Author

terms of producing summaries that capture relevant information while minimizing redundancy and omission of important details. A higher precision score signifies that a larger proportion of the information in the generated summary is relevant and accurate compared to the reference summary. Similarly, a higher recall score suggests that Summa tends to capture a greater portion of the relevant information present in the reference summary.

Results

To delve into the effectiveness of AI Tutoring, we conducted a study to understand its impact on their learning outcomes and overall academic performance. We focused this study mainly on school students between classes 9 and 12 and undergraduate college students. Through the feedback collected from a diverse sample of 200 students, the analytics provided below, provide insights into the efficacy of AI-powered educational tools. The 200 participants were comprised of 42 school students studying in classes 9–12th and the rest 158 were undergraduate college students.

The high percentage of respondents choosing 4 – likely in Figure 1.5 suggests a strong preference for recommendation of application to peers, indicating the good performance of the app

The high percentage of respondents choosing 4 – good and 5 – very good in Figure 1.6, suggests that the app has been successful in facilitating self-paced education, indicating users are very likely to use the app.

The high percentage of respondents choosing 4 – good in Figure 1.7, suggests that the app has been able to be quite accurate in tailoring content

The high percentage of respondents choosing 5 – very good in Figure 1.8, suggests that the app has been able to fetch appropriate learning materials.

Conclusions

In conclusion, the emergence of AI-driven tutoring systems signifies a profound shift in educational paradigms, offering bespoke support and personalized learning pathways adaptable to diverse student profiles. Our comprehensive examination of AI tutor's sophisticated features underscores its potential to transcend traditional pedagogical constraints by adeptly catering to individual learning modalities and preferences. Through meticulous analysis involving participants across school and undergraduate levels, our study unequivocally demonstrates the efficacy of AI tutoring in enhancing learning outcomes and academic performance. The amassed feedback corroborates its capacity to engage learners, nurture critical thinking aptitudes, and dynamically accommodate evolving educational exigencies. To propel this transformative trajectory forward, sustained efforts in refining AI tutor and fostering collaborative endeavors within the realm of AI-enabled education are imperative. Such endeavors are pivotal in ensuring accessibility, inclusivity, and ongoing relevance, thereby fostering a more equitable and empowered educational landscape poised for sustained advancement.

References

Sung, L., Lin, Y., and Chen, M. C. (2007). An automatic quiz generation system for English text. *Seventh IEEE Internat. Conf. Adv. Learn. Technol*, pp. 196–197. https://doi.org/10.1109/icalt.2007.56.

Gabajiwala, E., Mehta, P., Singh, R., and Koshy, R. (2022). Quiz maker: Automatic quiz generation from text using NLP. *Lec Notes Elec. Engg.*, 523–533. https://doi.org/10.1007/978-981-19-5037-7\textunderscore37.

Abualigah, L., Bashabsheh, M. Q., Alabool, H., and Shehab, M. (2020). Text summarization: A brief review. In: Abd Elaziz, M., Al-qaness, M., Ewees, A., Dahou, A. (eds.) Recent Advances in NLP: The Case of Arabic Language. *Stud. Comput. Intell.*, 874. Springer, Cham (https://doi.org/10.1007/978-3-030-34614-0\textunderscore1.

Gambhir, M. and Gupta, V. (2017). Recent automatic text summarization techniques: A survey. *Artif. Intell. Rev.*, 47, 1–66. https://doi.org/10.1007/s10462-016-9475-9.

Ma, T., Pan, Q., Rong, H., Qian, Y., Tian, Y., and Al-Nabhan, N. (2022). T-BERTSum: Topic-aware text summarization based on BERT. *IEEE Trans. Comput. Soc. Sys.*, 9(3), 879–890. doi: 10.1109/TCSS.2021.3088506.

Myers, K., Berry, P., Blythe, J., Conley, K., Gervasio, M., McGuinness, D. L., Morley, D., Pfeffer, A., Pollack, M., and Tambe, M. (2007). An intelligent personal assistant for task and time management. *AI Mag.*, 28(2), 47. https://doi.org/10.1609/aimag.v28i2.2039.

Kerres, M. and Buntins, K. (2020). Recommender in AI-enhanced learning: An assessment from the perspective of instructional design. *Open Edu. Stud.*, 2(1), 101–111. https://doi.org/10.1515/edu-2020-0119.

Meng, N., Dhimolea, T. K., and Ali, Z. (2022). AI-enhanced education: Teaching and learning reimagined. In: Albert, M. V., Lin, L., Spector, M. J., Dunn, L. S. (eds.) Bridging Human Intelligence and Artificial Intelligence. *Edu. Comm. Technol. Iss. Innov*. Springer, Cham, pp 107–124. https://doi.org/10.1007/978-3-030-84729-6\textunderscore7.

Maravanyika, M., Dlodlo, N., and Jere, N. (2017). An adaptive recommender-system based framework for personalized teaching and learning on e-learning platforms. 2017 IST-Africa Week Conf. (IST-Africa), pp. 1–9. https://doi.org/10.23919/istafrica.2017.8102297.

Le, N., Strickroth, S., Groß, S., and Pinkwart, N. (2013). A review of AI-supported tutoring approaches for learning programming. Stud. Comput. Intell., 267–279. https://doi.org/10.1007/978-3-319-00293-4\textunderscore20.

AlShaikh, F. and Hewahi, N. M. (2021). AI and machine learning techniques in the development of intelligent tutoring system: A review. *2021 Internat. Conf. Innov. Intell. Informat. Comput. Technol. (3ICT)*, pp. 403-410. https://doi.org/10.1109/3ict53449.2021.9582029.

Alam, A. (2023). Harnessing the power of AI to create intelligent tutoring systems for enhanced classroom experience and improved learning outcomes. *Lec. Notes Data Engg. Comm. Technol.*, 571–591. https://doi.org/10.1007/978-981-99-1767-9\textunderscore42.

Ahmad, S. F., Rahmat, M. K., Mubarik, M. S., Alam, M. M., and Hyder, S. I. (2021). Artificial intelligence and its role in education. *Sustainability*, 13(22), 12902. https://doi.org/10.3390/su132212902.

Gromyko, V., Kazaryan, V. P., Vasilyev, N., Simakin, A. G., and Anosov, S. (2017). Artificial intelligence as tutoring partner for human intellect. *Adv. Intell. Sys. Comput.*, 238–247. https://doi.org/10.1007/978-3-319-67349-3\textunderscore22.

Gupta, V. and Lehal, G. S. (2010). A survey of text summarization extractive techniques. *J. Emerg. Technol. Web Intell.*, 2(3), pp. 258–268. https://doi.org/10.4304/jetwi.2.3.258-268.

Dang, H., Benharrak, K., Lehmann, F., and Buschek, D. (2022). Beyond text generation: Supporting writers with continuous automatic text summaries. *Proc. 35th Ann. ACM Sympos. User Interface Softw. Technol*, pp. 1–13 https://doi.org/10.1145/3526113.3545672.

Fernandopulle, S. R., Warnasooriya, W., Jayasinghe, J., Theeraj, S., Samarakoon, U., and Kumari, S. (2021). EduHelp – An online tutoring application. *2021 3rd Internat. Conf. Adv. Comput. (ICAC)*. IEEE, pp. 377–382. https://doi.org/10.1109/icac54203.2021.9671110.

Kukulska-Hulme, A. and Shield, L. (2008). An overview of mobile assisted language learning: From content delivery to supported collaboration and interaction. *ReCALL*, 20(3), 271–289. https://doi.org/10.1017/s0958344008000335.

Sreelakshmi, A. S., Abhinaya, S., Nair, A., and Nirmala, S. J. (2019). A question answering and quiz generation chatbot for education. 2019 Grace Hopper Celebration India (GHCI) 1–6. https://doi.org/10.1109/ghci47972.2019.9071832.

Diwan, C., Srinivasa, S., Suri, G., Agarwal, S., and Prasad, R. (2023). AI-based learning content generation and learning pathway augmentation to increase learner engagement. *Comp. Edu. Art. Intell.*, 4, 100110. https://doi.org/10.1016/j.caeai.2022.100110.

Pardamean, B., Suparyanto, T., Cenggoro, T. W., Sudigyo, D., and Anugrahana, A. (2022). AI-based learning style prediction in online learning for primary education. *IEEE Acc.*, 10, 35725–35735. https://doi.org/10.1109/access.2022.3160177.

Troussas, C., Chrysafiadi, K., and Virvou, M. (2020). Personalized tutoring through a stereotype student model incorporating a hybrid learning style instrument. *Edu. Inform. Technol.*, 26(2), 2295–2307. https://doi.org/10.1007/s10639-020-10366-2.

Truong, H. M. (2016). Integrating learning styles and adaptive e-learning system: Current developments, problems and opportunities. *Comp. Human Behav.*, 55, 1185–1193. https://doi.org/10.1016/j.chb.2015.02.014.

2 Strategies for effectively facilitate online courses and leverage technology for teaching educators

N. Ashokkumar[1,a], *P. Nagarajan*[2,b], *M. Bharathi*[1,c] *and N. Swetha*[3,d]

[1]Department of ECE, Mohan Babu University, (Erstwhile Sree Vidyanikethan Engineering College), Tirupati, Andhra Pradesh. India

[2]Department of ECE, SRM Institute of Science and Technology, Vadapalani Campus, Chennai, Tamil Nadu, India

[3]Department of ECE, Sethu Institute of Technology, Virudhunagar, Madurai, Tamil Nadu, India

Abstract

This study explores and evaluates strategies for training educators to proficiently facilitate online courses while leveraging technology for enhanced teaching experiences. The higher number of enrolments in the different online courses throughout the world indicates the demand for online learning aspects. The literature review section provides information about the challenges as well as strategies that would provide an effective learning approach for educators. On the other hand, this section elaborates on the way of technological integration in online education. In the methodology section, a primary research approach was used that included positivism research philosophy. On the other hand, deductive approaches help in the existing approach testing. In that case, primary data collection used where online survey by using Google form have been seen. In that case, 70 participants contributed to the survey where their valuable information helped in the statistical data analysis. By using IBM SPSS, statistical analysis has been done where descriptive analysis as well as linear regression analysis and correlation analysis are performed. In the discussion, different strategies are discussed that should be used to increase the effectiveness of their learning style that helps in the online courses. In the end, it can be concluded that effective approaches to support educators in mitigating the evolving landscape of online teaching in the education industry.

Keywords: Online learning, adaptive learning, online teaching, online courses, educators

Introduction

The developments in technology have brought about revolutionary changes in education that change the ways of delivered as well as consumed education over the last 20 years. In that case, the development of the internet not only brought creation as well as advancement in the learning aspects but also helped to increase the effectiveness among educators. Here, different kinds of advanced technology provide growth in the education field. In terms of traditional educational strategies, this learning model relies on physical buildings (Drgoňa et al. 2020). On the other hand, due to the accessible as well as flexible nature of the learning system, the online learning system not only reduces the geographical barriers but also fulfills the needs of a variety of learners. Due to the increases in the number of digital technologies as well as flawless internet connectivity and having access to lifetime learning opportunities, the increasing percentages behind the acceptance of online courses have been seen in this 21st century.

Figure 2.1 graphical image provides information about the demand as well as acceptance of online learning courses throughout the world. In 2021, over 20 million new learners joined Coursera's online learning platform, marking significant growth in student registrations compared to the 3 years before the pandemic (weforum.org, 2021). Despite the advantages of online learning, there are several challenges that educators face when transitioning from traditional to online education. These challenges are related to proficiency in technology as well as delivering effective as well as detailed content and the ability to foster virtual communities among educators. By adopting different types of pedagogical approaches, educators can mitigate the lack of face-to-face interactions that not only increase the engagement of students but also improve learning in the education industry (Singh et al., 2021). In addition, the COVID-19 epidemic not only increased the percentages of the adoption of online education but also made it a mainstream educational approach. The purpose of the study is to observe, evaluate and recommend successful strategies for preparing educators to effectively manage online courses while using technology to expand the learning involvement.

The objectives of the study are

To recognize the challenges encountered by instructors in facilitating online courses.

[a]ashoknoc@gmail.com, [b]nagarajan.pandiyan@gmail.com, [c]bharathi891@gmail.com, [d]swethanagarajan234@gmail.com

DOI: 10.1201/9781003587538-2

More learners are accessing online learning

The demand for online learning on Coursera continues to outpace pre-pandemic levels.

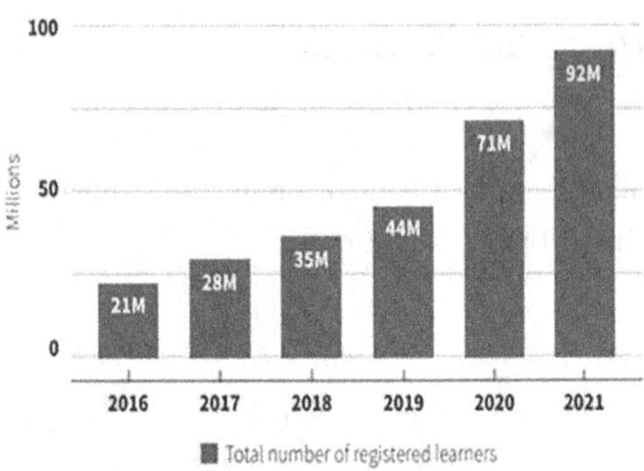

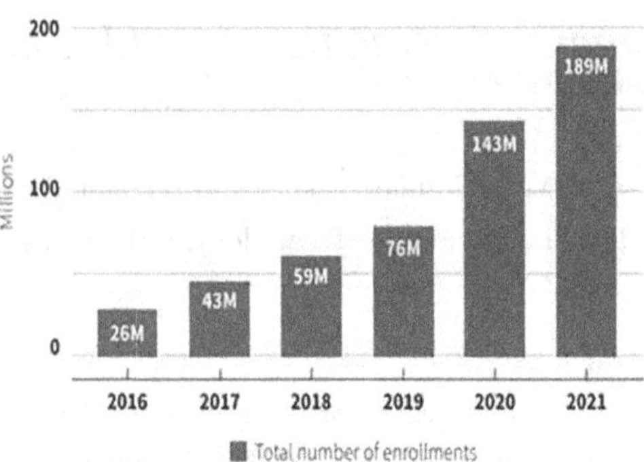

Figure 2.1 The demand for online learning after Covid 19
Source: weforum.org, 2021

To examine pedagogical strategies that contributes to effective online instruction.
To explore the integration of technology to enhance online course facilitation in the education industry.
To analyze existing professional development programs for educators in online settings.

The research queries are

What are the challenges faced by educators in facilitating online courses?
What pedagogical strategies contribute to effective online teaching?
In what way technology can be effectively integrated to enhance online course facilitation?
What professional development initiatives exist for educators in online settings?

Hypotheses of the study

H1: Online courses are more effective when instructors receive training, in strategies specifically tailored for online environments.
H2: By incorporating technology into training programs for educators it greatly improves their capacity to effectively utilize tools to create interactive and captivating courses.

Literature review

The challenges faced by educators in facilitating online courses

In this 21st century, educators face different types of challenges while facilitating online courses for students. In that case, lack of in-person communication not only creates communication gaps between educators and students but also reduces the effectiveness of learning aspects (Cilliers et al., 2022). On the other hand, different types of learning styles among students brought complications that created negative impacts on the students' performances. In that case, educators need to adopt different instructional approaches to cater to varied preferences. Different types of technological barriers related to unequal access to resources and technical issues can bring disparities that affect the engagement of students as a result, there is disruption seen in the learning experience.

Figure 2.2 provides information about the challenges that faced educators during facilitating online courses. In terms of challenges that are faced by educators is time management. In that case, students and educators may struggle with time management in online courses which brings burnout among both participants. In the virtual world, developing effective assessments and providing timely feedback are complex undertakings that demand careful consideration for educators that improve the learning experiences (Christopoulos et al., 2020). Digital literacy is essential but a lack of skills among educators hinders the utilization of online platforms proficiently. Throughout an online course, sustaining student engagement can be challenging. In that case, by using different interactive and collaborative strategies by educators, this kind of challenge can be mitigated to increase the participation of the students.

Pedagogical strategies contribute to effective online teaching

The accomplishment of online learning – the use of teaching methods that are specifically designed for the digital learning environment. As per the study by

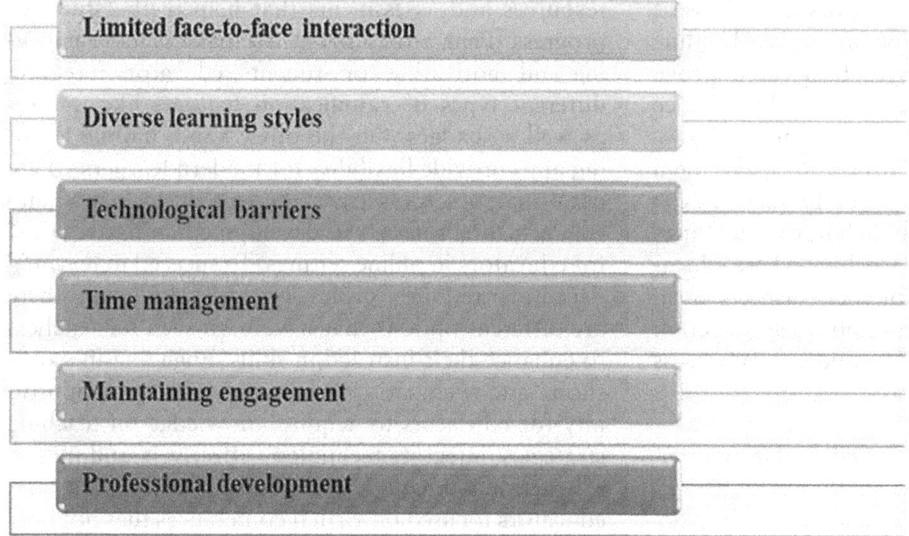

Figure 2.2 Challenges faced by educators during online courses
Source: Christopoulos et al., 2020

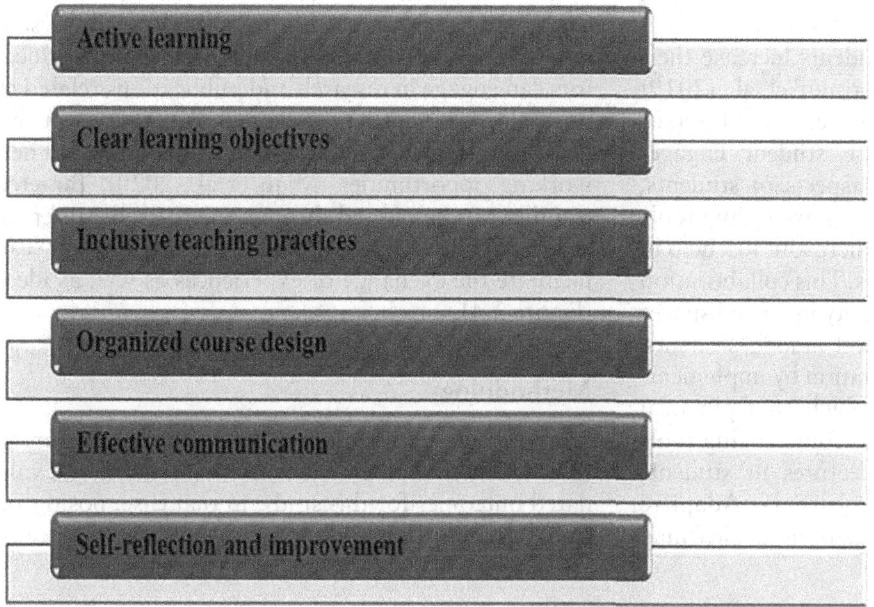

Figure 2.3 Strategies that brings effective online learning system
Source: Bradley, 2021

Stone and Springer (2019), the number of student participants increased due to the engaging activities offered by the educators. In that case, different types of collaborative projects as well as the use of discussion forums and virtual promotions not only promote active learning but also increase the students' participation. On the other hand, Saiyad et al. (2020) stated that clear learning objectives by educators also bring effective online teaching. Due to clear communication, students can understand the expectations of the course in a positive manner. Similarly, organized course design also increases the effectiveness of online teaching. To improve the overall learning experience, educators should use consistent formatting and organization that fulfill the students' demands.

Figure 2.3 offers information that helps to increase the effectiveness of online learning systems which increases the participation in online courses. In the financial year 2020, the total enrolment of the online courses is 143 million where different types of online platforms help in this aspect. Educators should assess the student's progress by offering an unbiased online teaching approach. In that case, feedback can improve the students' weaknesses. Educators should enhance experiences by incorporating multimedia, interactive content and educational platforms to make them more engaging. On the other hand, Bradley (2021) stated that different types of collaborative activities and discussions improve online teaching where educators can offer different kinds of social aspects positively. Here, educators should promote self-reflection strategies and using these strategy students can assess their learning progress.

Technology can be effectively unified to improve online course facilitation

To improve the excellence of online course instruction it is crucial to integrate technology. As per the review of Turnbull et al. (2020), educators need to utilize a learning management system (LMS) that helps to centralize course materials as well as assignments and communication between students and educators. Here, this course management helps in the grade-tracking aspects that help the students increase their strength. On the other hand, Serrano et al. (2019) stated that interactive virtual classes and different multimedia content can increase student engagement and improve the learning aspects of students. Similarly, different types of collaborative online tools that are Google Workspace and Microsoft 365 help in real-time collaboration on projects. This collaboration not only fosters a sense of community in virtual spaces but also increases the interest level among students. Students get hands-on experimentation by implementing virtual labs and simulations which offer practical experience. Different types of web conferencing tools help educators provide virtual lectures to students which reduces the demographic barriers. Adaptive learning platforms provide customized learning resources and assessments that help in the student's progress (Peng et al., 2019). To make it more engaging and motivating for students, educators integrate different types of gamification features like quizzes as well as badges. On the other hand, mobile learning apps provide flexibility for modern learners where data analytics tools track student performance and engagement. Professional development initiatives exist for educators in online settings. To succeed in teaching educators requires professional development, there are different options that have been seen for teachers to enhance their instruction skills. Engaging in workshops and webinars presents a wonderful opportunity for educators to acquire knowledge on teaching strategies, integrate technology effectively and master instructional design (Xie and Rice, 2021). In that case, educators focused on virtual conferences that not only help to learn from experts but also exchange practices that improve engagement in the online learning aspects. On the other hand, Okolie et al. (2020) stated that different types of mentorship programs, institutional training initiatives, and qualified growth grants can greatly assistance educators. To stay conversant with the modern trends in online education, educators can engage in research and publications related to this field. By joining professional associations in this field, educators can get access to conferences and networking opportunities (Mian et al., 2020). Targeted training can be offered through micro-credential programs, while cross-institutional collaboration can facilitate the exchange of experiences as well as ideas (Figure 2.4).

Methodology

In terms of methodology, this study used primary research methodology which offered reliable and validated outcomes for this study. In that case, positivism research philosophy due to its quantitative nature of

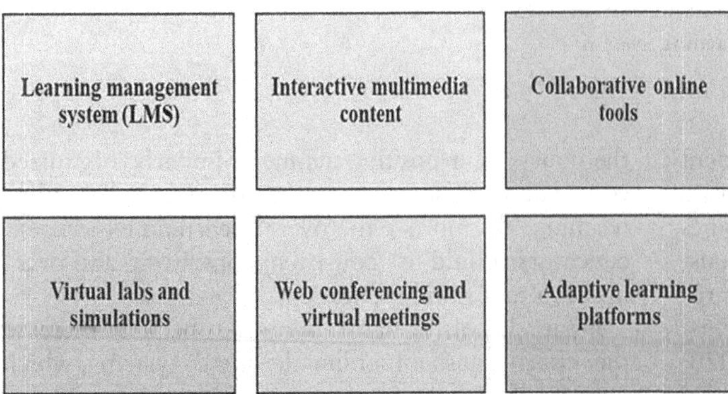

Figure 2.4 Technology can be effectively unified to improve online course facilitation
Source: Serrano et al., 2019

data. Here, by analyzing all the data associated with the topic, this study offered empirical research results where educators get proper strategies to improve the engagement of students for the online learning aspects. On the other hand, a deductive research approach was also used for this study where this approach examined different theories or existing knowledge that help in hypothesis testing aspects (Casula et al., 2021). This testing not only helps to understand the resulting association between autonomous variables and reliant on variables. Instead, the descriptive research design utilized for this study plays a vibrant part in understanding the distribution of variables associated with the research. In this research, the primary data collection process was used. As per the investigation of Laato and Tregel (2023), primary data collection refers to the data collection process where all the data is collected from non-existing sources. This study used an online survey where a total of 70 participants who are educators in different educational institutes provided their valuable experiences. In that case, the quantitative data analysis method was used for the investigation of all the collected data. In this statistical investigation, the IBM SPSS statistical tool is used that offers descriptive analysis as well as linear regression analysis and Pearson correlation analysis.

Findings and analysis and demographic analysis

Age (Table 2.1, Figure 2.5)

Gender

According to the survey findings 42.9% of the participants were male while 50.0% were female. A small percentage of 7.1% identified themselves as "Other". These results give us a picture of the gender composition, within the surveyed group offering insights, into the diverse range of individuals who took part in the study (Table 2.2, Figure 2.6).

Educational qualification

The survey participants have a range of backgrounds. Specifically, 22.9% completed their education 7.1% have obtained a graduation degree 48.6% have pursued postgraduate qualifications and 21.4% hold a Ph.D. This information effectively illustrates the profiles within the surveyed group (Table 2.3, Figure 2.7).

Regression analysis

Hypothesis 1

The regression model investigates how educators' readiness for teaching (dependent variable) is influenced by their participation in professional development workshops (independent variable). The R-squared worth of 0.760 proposes that the model can reason for 76 in a hundred of the disparity in preparedness. Furthermore, the standardized coefficient of 0.872 indicates a relationship. The higher value in Durbin-Watson suggested that there is no autocorrelation. Therefore, participating in development workshops creates positive impacts on online teaching (Table 2.4).

Hypothesis 2

Hypothesis 2 is aligned with the regression analysis, which examines the influence of educators'

Table 2.1 Age.

		Frequency	Percent	Valid percent	Cumulative percent
Valid	1	35	50.	50.0	50.0
	2	38	11.4	11.4	61.4
	3	11	15.7	15.7	77.1
	4	16	22.9	22.9	100.0
Total		70	100.0	100.0	

Source: Author

Table 2.2 Gender.

		Frequency	Percent	Valid percent	Cumulative percent
Valid	1	30	42.9	42.9	42.9
	2	35	50.0	50.0	
	3	5	7.1	7.1	92.9
Total		70	100.0	100.0	100

Source: Author

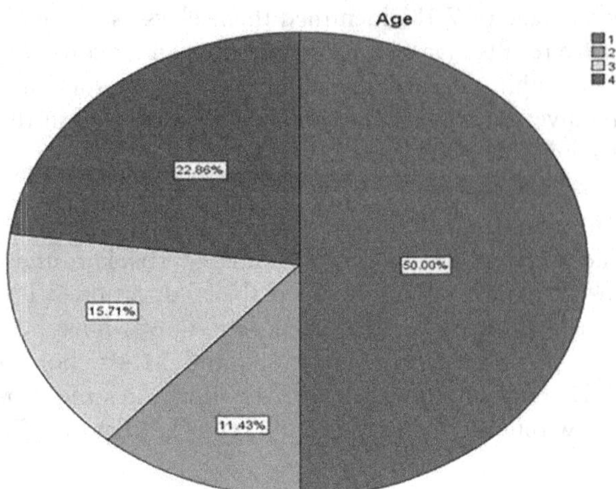

Figure 2.5 Age distribution
Source: Author

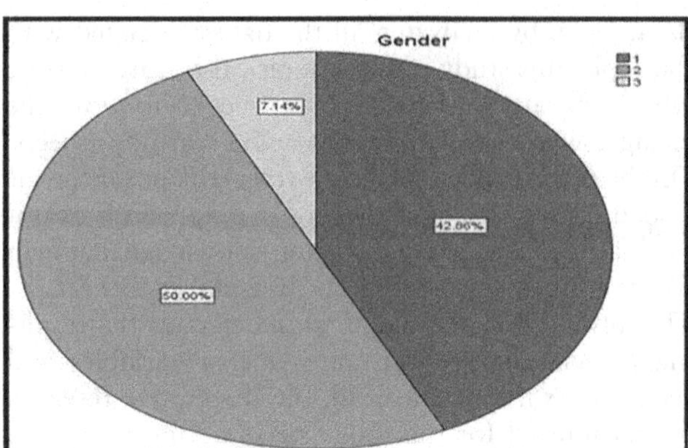

Figure 2.6 Gender distribution
Source: Author

Table 2.3 Educational qualification.

		Frequency	Percent	Valid percent	Cumulative percent
Valid	1	16	22.9	22.9	22.9
	2	5	7.1	7.1	30.0
	3	34	48.6	48.6	78.6
	4	15	21.4	21.4	100.0
Total		70	100.0	100.0	

Source: Author

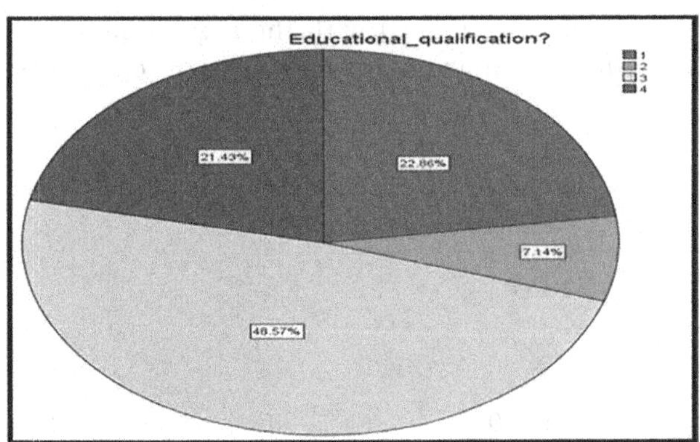

Figure 2.7 Educational qualification distribution
Source: Author

engagement in collaborative learning communities (IV2) on their preparedness for online teaching (DV). Here, the R-value 0.340 shows a positive relation between these two variables. The value of coefficient is 0.340 indicates a moderate positive relationship between engagement in collaborative learning communities and educator preparedness for online education. This finding shows the importance of collaborative practices in enhancing skills for educators that help in online teaching (Table 2.5).

Correlation test

Table 2.6 offers information about the relationship amongst different variables. In this case, DV1 makes positive relationships with IV1 as well as IV2 where both participation in professional development workshops and engagement in collaborative learning communities are positively correlated with effective online teaching strategies for educators.

Discussion

The results of the statistical analysis provide valuable information for this study. In that case, professional development for workshops improves the teaching aspects in the online context. On the other hand, ongoing training not only enhances skills as well as confidence of educators in different educational centers in the online atmosphere but also helps to improve the performances of pupils (Hwang et al., 2021). Here, professional training also increases the teaching ability of teachers that help to mitigate all the issues

Table 2.4 Linear regression investigation for hypothesis 1 (Source: IBM SPSS)
Model summary[b].

Model	R	R-square	Adjusted R-square	Std. error of the estimate	Durbin-Watson
1	0.872[a]	0.760	0.757	1.46598	0.149

[a]Predictors: (Constant), IV1_Participation in professional development workshops
[b]Dependent variable: DV_Educator preparedness for online teaching

ANOVA[a]

Model		Sum of square	df	Mean square	F	Sig.
Valid	Regression	463.632	1	463.632	215.732	0.000[b]
	Residual	146.140	68	2.149		
	Total	609.771	69			

[a]Dependent variable: DV_Educator preparedness for online teaching
[b]Predictors: (Constant), IV1_Participation in professional development workshops

Coefficents[a]

Model	Unstandardized coefficients		Standardized coefficients	t	Sig.
	B	Std. error	Beta		
1 (Constant)	1.156	0.431		2.681	0.009
Residual IV1_Participation in professional development workshops	0.823	0.056	0.872	14.686	0.000

[a]Dependent variable: DV_Educator preparedness for online teaching
Source: Author

Table 2.5 Linear regression investigation for hypothesis 2.
Model Summary[b]

Model	R	R-square	Adjusted R-square	Std. error of the estimate	Durbin-Watson
1	0.340[a]	0.116	0.103	2.81562	0.229

[a]Predictors: (Constant), IV2_Engagement in collaborative learning communities
[b]Dependent variable: DV_Educator preparedness for online teaching

ANOVA[a]

Model		Sum of square	df	Mean square	F	Sig.
Valid	Regression	70.688	1	70.688	8.917	0.004[b]
	Residual	539.083	68	7.928		
	Total	609.771	69			

[a]Dependent variable: DV_Educator preparedness for online teaching
[b]Predictors: (Constant), IV2_Engagement in collaborative learning communities

Coefficents[a]

Model	Unstandardized coefficients		Standardized coefficients	t	Sig.
	B	Std. error	Beta		
1 (Constant)	4.427	0.907		4.881	0.000
Residual IV2_Engagement in collaborative learning communities	0.527	0.177	0.340	2.985	0.004

[a]Dependent variable: DV_Educator preparedness for online teaching
Source: Author

Table 2.6 Pearson relationship test.

Correlations				
		DV_Educator_ Preparedness for online teaching	IV1_Participation in professional development workshops	IV2 Engagement in collaborative learning communities
DV_Educator_ Preparedness for online teaching	Pearson connection Sig (2-tailed) N	1 70	0.872** 0.000 70	0.340** 0.000 70
IV1 _Participation in professional development workshops	Person connection Sig (2-tailed) N	0.872* 0.000 70	1 70	0.617** 0.000 70
IV2 Engagement in collaborative learning communities	Person connection Sig (2-tailed) N	0.340** 0.004 70	0.617** 0.000 70	1 70

**Correlation is significant at the 0.001 level (2-tailed)

Source: Author

that arise in online education. Similarly, knowledge sharing as well as resource exchange brings effective online teaching strategies where collaborative e-learning platforms take major roles. On the other hand, attractive digital content not only increases attraction among students but also increases motivation among students to join online learning systems. Sustaining collaborative learning communities is an effective strategy that improves the online learning approach among educators that would improve academic performances of students in the future (Al-Kumaim et al., 2021). Amalgamation of technology in education also fulfills the needs of education where different types of learning applications or platforms provide reduction of demographic barriers that build a strong relationship between students and educators.

Conclusions

In the end, it can be concluded that in remote education, educators are faced with different types of issues. These issues took place due to the less amount of knowledge about the technologies as well as lack of skills. On the other hand, limited face to face interaction is another challenge faced by educators that reduces the effectiveness of the online courses. In the online learning aspects, time management is another issue that hampers the students' performance where clear objectives help the educators as well as students to mitigate these issues. Effective communication helps the students to make positive communication where educators bring interest to participants in online courses that improve academic performances of the students. In this modern era, integration of technologies in the education system offers effective intervention that helps educators in this digital age.

References

Al-Kumaim, N. H., Alhazmi, A. K., Mohammed, F., Gazem, N. A., Shabbir, M. S., and Fazea, Y. (2021). Exploring the impact of the COVID-19 pandemic on university students' learning life: An integrated conceptual motivational model for sustainable and healthy online learning. *Sustainability*, 13(5), 2546. https://www.mdpi.com/2071-1050/13/5/2546/pdf.

Bradley, V. M. (2021). Learning management system (LMS) use with online instruction. *Internat. J. Technol. Edu.*, 4(1), 68–92. https://files.eric.ed.gov/fulltext/EJ1286531.pdf.

Casula, M., Rangarajan, N., and Shields, P. (2021). The potential of working hypotheses for deductive exploratory research. *Qual. Quant.*, 55(5), 1703–1725. https://link.springer.com/article/10.1007/s11135-020-01072-9.

Christopoulos, A., Pellas, N., and Laakso, M. J. (2020). A learning analytics theoretical framework for STEM education virtual reality applications. *Edu. Sci.*, 10(11), 317. https://www.mdpi.com/2227-7102/10/11/317/pdf.

Cilliers, J., Fleisch, B., Kotze, J., Mohohlwane, N., Taylor, S., and Thulare, T. (2022). Can virtual replace in-person coaching? Experimental evidence on teacher professional development and student learning. *J. Dev. Econ.*, 155, 102815. https://www.sciencedirect.com/science/article/am/pii/S0304387821001668.

Drgoňa, J., Arroyo, J., Figueroa, I. C., Blum, D., Arendt, K., Kim, D., and Helsen, L. (2020). All you need to know about model predictive control for buildings. *Ann. Rev. Con.*, 50, 190–232. https://www.sciencedirect.com/science/article/pii/S1367578820300584.

Hwang, G. J., Wang, S. Y., and Lai, C. L. (2021). Effects of a social regulation-based online learning framework on students' learning achievements and behaviors in mathematics. *Comp. Edu.*, 160, 104031.

Laato, S. and Tregel, T. (2023). Into the Unown: Improving location-based gamified crowdsourcing solutions for geo data gathering. *Entertain. Comput.*, 46, 100575. https://www.sciencedirect.com/science/article/pii/S1875952123000307.

Mian, S. H., Salah, B., Ameen, W., Moiduddin, K., and Alkhalefah, H. (2020). Adapting universities for sustainability education in industry 4.0: Channel of challenges and opportunities. *Sustainability*, 12(15), 6100. https://www.mdpi.com/2071-1050/12/15/6100/pdf.

Okolie, U. C., Nwajiuba, C. A., Binuomote, M. O., Ehiobuche, C., Igu, N. C. N., and Ajoke, O. S. (2020). Career training with mentoring programs in higher education: facilitating career development and employability of graduates. *Edu. Training*, 62(3), 214–234. https://www.researchgate.net/profile/Ugochukwu-Okolie-2/publication/340165436_Career_training_with_mentoring_programs_in_higher_education_Facilitating_career_development_and_employability_of_graduates/links/5f0d7314299bf15bd70b1922/Career-training-with-mentoring-programs-in-higher-education-Facilitating-career-development-and-employability-of-graduates.pdf.

Peng, H., Ma, S., and Spector, J. M. (2019). Personalized adaptive learning: an emerging pedagogical approach enabled by a smart learning environment. *Smart Learn. Environ.*, 6(1), 1–14. https://link.springer.com/article/10.1186/s40561-019-0089-y.

3 Investigating innovative approaches to learning – Gradient descent and regression for millennials

Varsha S.[1,a], *Kinshuk Agarwal*[1,b], *Kushagra Awasthi*[1,c], *Prapulla S. B.*[1,d], *Deepamala N.*[1,e], *Prakash R.*[2,f] *and Jayalatha*[2,g]

[1]Department of CSE, RV College of Engineering, Bengaluru, Karnataka, India

[2]Department of Mathematics, RV College of Engineering, Bengaluru, Karnataka, India

Abstract

In contemporary education, the effective conveyance of complex mathematical concepts, such as gradient descent and regression, remains a significant challenge. This paper presents an innovative approach to tackle this issue through the application of design thinking principles in educational content development. This paper outlines the design process, highlighting key decisions made at each stage and the rationale behind them. We also discuss the implementation of the website and its reception among students through preliminary user feedback. The website features interactive simulations, animated video and curated content based on student reviews, designed to foster active engagement and comprehension. Furthermore, we reflect on the potential implications of our approach for improving the teaching and learning of mathematical concepts beyond gradient descent and regression. Implementing concise design thinking methods into educational content development, our project offers a novel perspective on addressing the challenges associated with teaching complex mathematical concepts.

Keywords: Regression, design thinking, simulations, active engagement

Introduction

Design thinking provides a comprehensive problem-solving approach, focusing on empathy, ideation, prototyping, and testing. We aimed to create a user-friendly platform for math education by understanding student needs, prioritizing interactivity, and fostering engagement. By integrating simulations, videos, and curated content, we transform passive learning into an immersive journey. This introduction sets the stage for discussing our design process, implementation, and potential impacts on mathematical education.

Empathy phase

Identification of stakeholders

1. Students: Students who have learnt concepts of gradient descent and regression as a part of their curriculum. Students who implement these concepts and those who are yet to learn this concept will be the end users.
2. Teachers: Teachers having thorough knowledge of the subject and particular teaching experience in this concept. Also includes experts on the subject who have insights about the industry requirements.
3. Enthusiasts: AIML has become the go-to trend in the tech industry – with many people wanting to get into the field. Mastering statistics and in particular regression analysis is the start.
4. Tutors: Tutors who have taught numerous students, have worked on the implementation and have a very clear idea about the areas of difficulty.

Questionnaire for student

1. Rate your understanding of ideas related to gradient descent and regression.
2. What were the difficulties faced in understanding these concepts?
3. Do you prefer access to educational content outside the classroom?
4. What resources or methods do you find most effective in understanding these topics?
5. Would you prefer online lectures or interactive videos to gain more insights on these concepts?
6. Do you feel that this is vital for your field of study or future career?
7. Are there any advanced techniques related to gradient descent that you want to explore in future?

[a]varshas.cs22@rvce.edu.in, [b]kinshukagarwal.cs22@rvce.edu.in, [c]kawasthi.cs22@rvce.edu.in, [d]prapullasb@rvce.edu.in, [e]deepamalan@rvce.edu.in, [f]prakashr@rvce.edu.in, [g]jayalathag@rvce.edu.in

DOI: 10.1201/9781003587538-3

Student responses

We conducted around 50 interviews with students, covering various branches, to gauge their understanding and identify areas of improvement. Most students are keen to learn more about statistics, with many favoring online video tutorials. Challenges include grasping formulas and limited course-specific resources. We gaged for student interest levels and for common study resources.

Ideation phase

Problem statement: Students think it's difficult and challenging to understand the concepts of statistics, as they're not able to relate to its real world examples, and aren't able to visualize its working clearly due to the lack of in-depth knowledge, lack of proper learning resources and application-based resources.

How might we (HMW) questions

1. Design an interactive platform for gradient descent and regression info?
2. Develop an engaging gamified tool for mastering gradient descent and regression?
3. Design real-world examples and visuals for better understanding of gradient descent and regression?
4. Use VR or AR for practical learning?
5. Integrate real-world data sets to demonstrate statistical applications?
6. Incorporate gamification to boost statistical learning participation?

Brainstorming

Brainstorming: This consists of having a constructive conversation for a session, improvising based on the ideas that were raised. Using problem statements and previously created how might we questions served as a guide to keep us focused on the task and prevent us from steering too far from the original issues.

To organize ideas, we made a mind-map to collect our ideas, shown in Figure 3.1.

Affinity map depicted in Figure 3.2 helped us to identify patterns, themes and connections among the ideas, gave deeper understanding and guided decision-making.

The customer journey map depicted in Figure 3.3 helped us to visualize and understand the student's experience of working with our product.

We plan to integrate visualization tools, parameter tuning games, and Python coding exercises to teach core concepts like gradient descent and regression. Neumann et al (2011) discuss how integrating technology in statistics lectures enhances student learning by simulating concepts, creating interactive

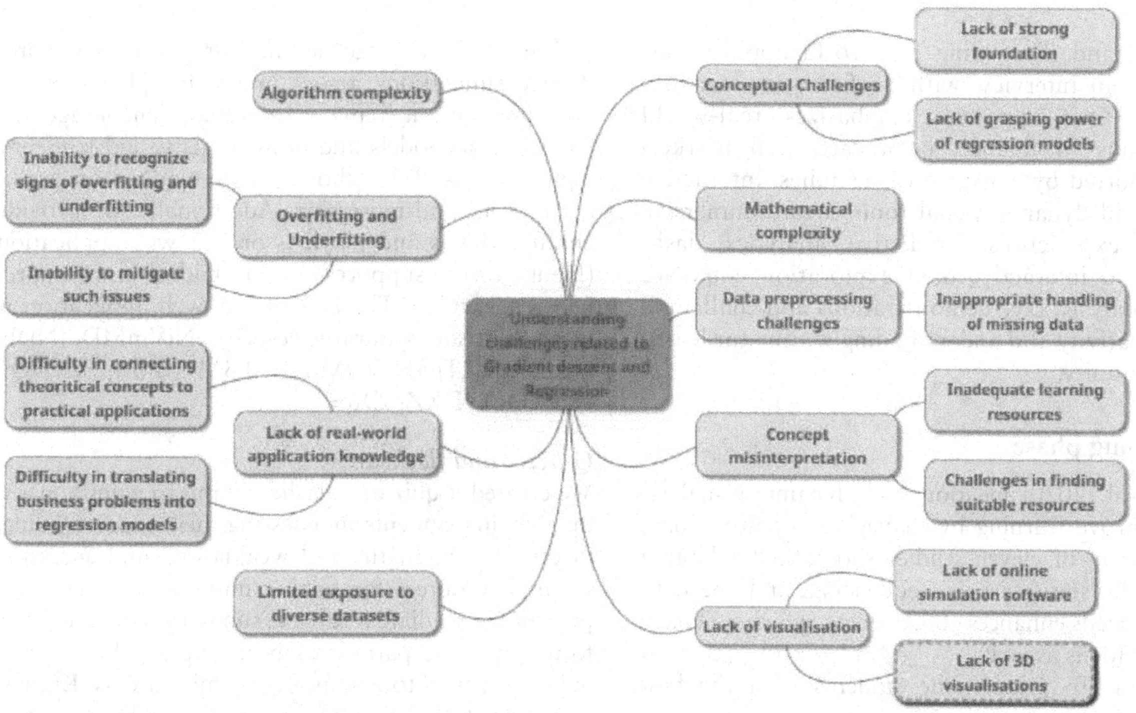

Figure 3.1 Mind map: Challenges faced by students
Source: Author

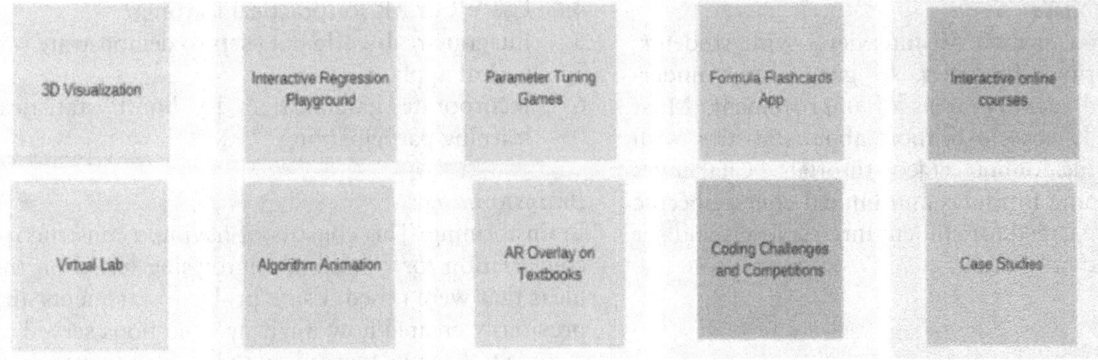

Figure 3.2 Affinity Map
Source: Author

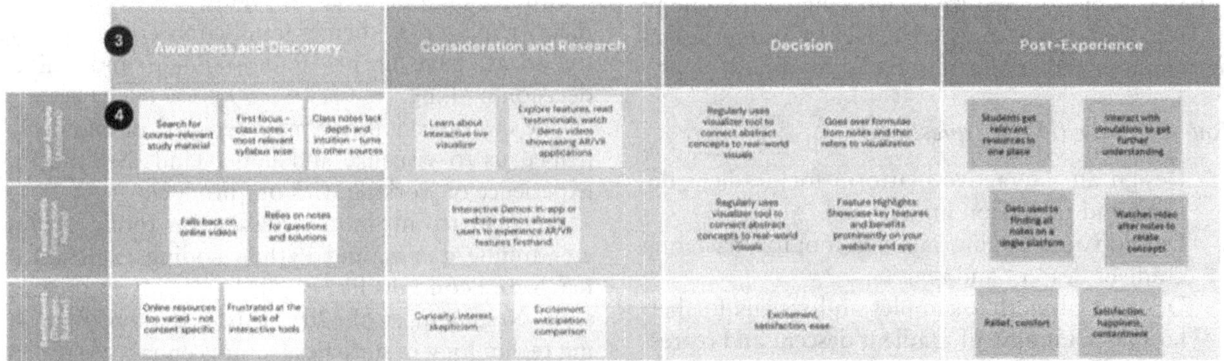

Figure 3.3 Customer Journey Map
Source: Author

exercises, and illustrating real-world applications, based on an interview with 38 first year statistics students. Our approach emphasizes real-world applications in finance, healthcare, and marketing, supported by engaging case studies. Interactive quizzes and dynamic visual tools ensure immersive learning experiences. Predictive analytics dashboards and interactive data exploration empower deeper analysis. Live demonstrations and challenges foster creativity and understanding within our learning community.

Prototyping phase

Rekik et al (2020) mention how dynamic visualizations improve learning by managing cognitive load. Their review of eleven studies shows that adapting designs like static visuals, sequencing, and speed to learners' needs enhances effectiveness, which will have to be our focus as well.

Features: To provide the students with the best learning experience we have built an application with numerous features. We offer animated videos, flashcards, and quizzes to enhance student understanding and assess knowledge depth. Our platform features diverse simulations across interactive platforms for experimentation. Teplá et al (2022) encourage the inclusion of models and animations to aid students. Figures 3.5 and 3.6 showcase simulations on gradient descent and regression. Additionally, we provide curated videos and playlists on our web application (Figure 3.4) to support concept building and comprehensive learning. The link to our web application is https://statista.mydurable.com/?pt=NjVmMDdiNDFjNzdlN2JkZTg4NjM0MjE2OjE3MTM0MjU0NDQuNzU5OnByZXZpZXc=

Quizzes and flashcards

We created a quiz and flashcard app to gamify learning, helping students in assessing their understanding, identifying strengths and weaknesses, and accessing tailored resources for further improvement. The app provides immediate feedback, summaries, and reviews, fostering active participation and personalized learning experiences to enhance academic success. Kumari and Yadav (2018) provide an overview of linear regression. It also includes step-by-step instructions for performing linear regression calculations in SPSS and

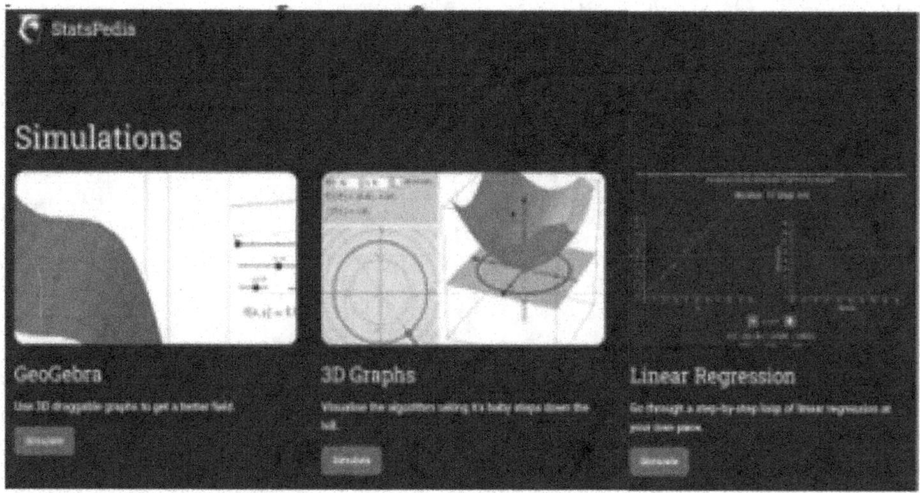

Figure 3.4 Website for statistics
Source: Author

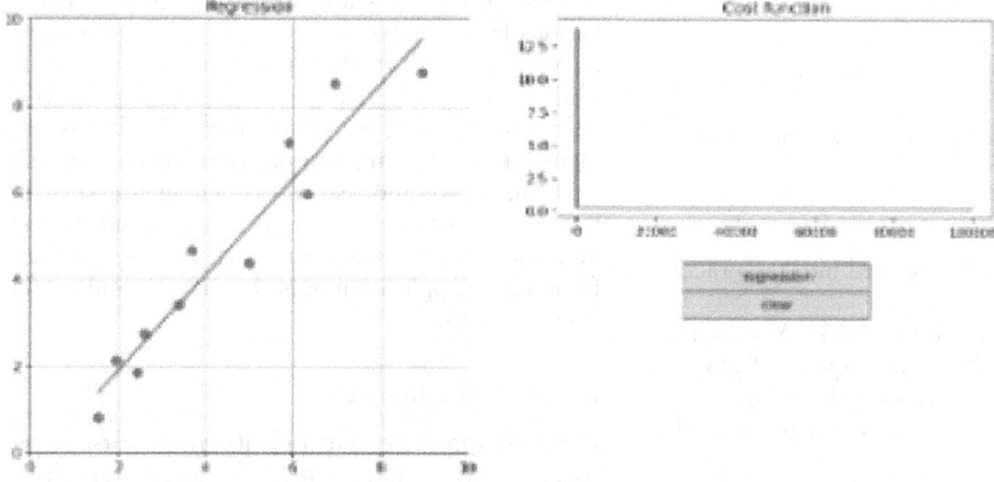

Figure 3.5 Regression Simulation
Source: Author

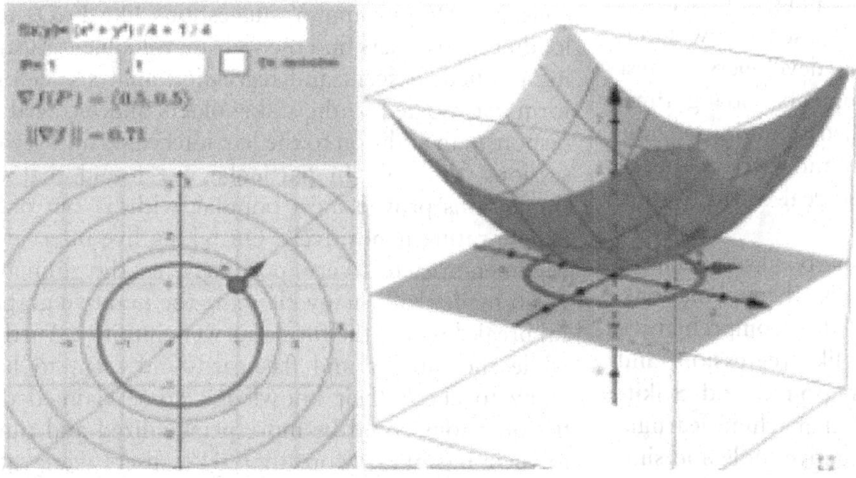

Figure 3.6 Gradient Descent Simulation
Source: Author

Table 3.1 Responses from stakeholders after experiencing the model

Feature	Strongly agree	Agree	Neutral	Disagree	Strongly disagree
Overall experience	25	31	17	4	2
Helpfulness of resources	27	32	15	2	3
Helpfulness of sims	65	-	13	-	1
Depth of coverage	32	33	11	2	1
Satisfaction	37	29	8	4	1

Source: Author

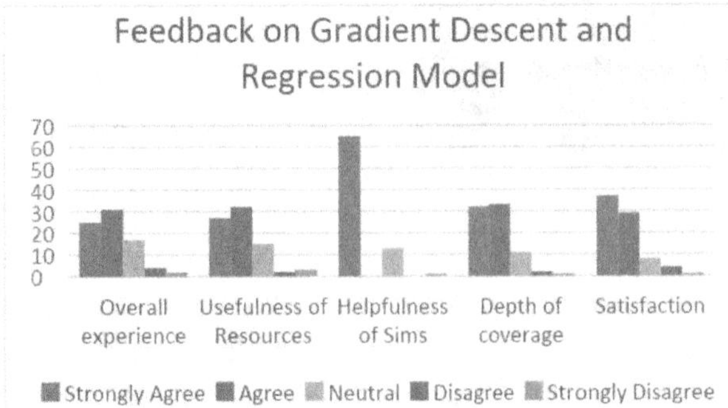

Figure 3.7 Feedback on gradient descent and regression model
Source: Author

Excel. These points can be factored into the practical evaluation.

Tapkir (2023), Ganie et al (2023), Manorathna (2020) and Saleh (2022) discuss aspects of gradient descent and relevant optimisation algorithms - these topics are crucial, and can be tested using coding tests.

Finding resources for theory

YouTube hosts lots of educational resources, particularly in fields such as machine learning, data science, and AR/VR. Lacoche et al (2022) discuss how simulations aid learning by allowing developers to test AR applications in VR environments, saving time, resources, and providing controlled conditions. These simulations can evaluate usability and user experience effectively, potentially matching real-world testing results.

We searched for channels or playlists such as "3Blue1Brown" for intuitive visual explanations, "StatQuest with Josh Starmer" for comprehensive coverage of statistical concepts like regression, and simulators like TensorFlow Playground and Scikit-learn for hands-on exploration of machine learning tasks. These resources offer interactive tools and simulations that facilitate learning through experimentation with different scenarios and edge cases.

Animated videos

We have included animated videos to explain linear regression and curve fitting concepts. Kuhlmann et al (2024) discuss how effective STEM learning with instructional videos depends on cognitive engagement, influenced by prior knowledge. Digital trace data showed that aligning engagement strategies with prior knowledge improves learning. These videos cater to various learning styles, providing clarity and engagement. They cover fundamental concepts with visualizations, real-world examples, quizzes, and interactive elements, fostering active learning. The videos also highlight practical applications of regression analysis, bridging theory and real-world outcomes across different domains like finance, healthcare, and marketing. This comprehensive approach prepares students for effective application in both academic and professional settings.

Results and outcomes

Our web app, tested on students with diverse expertise, received positive feedback for its coverage, interactive features, and animated videos. Users suggest adding quizzes, detailed analysis, and practical ML code examples while expanding coverage to related concepts. Despite minor suggestions like FAQs and loading signs, users find the website valuable for learning gradient descent and regression analysis. Google form was floated to the stakeholders and received 79 responses in addition to the live interviews and review sessions we took. In particular, we found that the simulations proved most popular, with 37 out of 79 people rating it positively. On taking live interviews, beginners reacted very positively to the animated video made, with many claiming the iterative graphical breakdown of the formulae were intuitive. Further, while the quizzes and flashcards did seem to help many to check their knowledge, some claimed that the questions could be more personalized and tuned to their mistakes. More than 41% users were satisfied with the depth of coverage of content. Few users were particularly happy with the progressive nature of the site – with a gradual increase in difficulty level and more focus on application and the coding aspect. Overall, our keen focus on simulations and visuals proved positive. Table 3.1 below shows our responses.

Feedback and conclusions

As the final phase, we mainly focused on two aspects: usability and effectiveness. For effectiveness, we observed real time users as they interacted with our material, with a key focus on user-friendliness and intuitionism. By analyzing user behavior, feedback, and pain points during usability testing sessions, we identified areas of improvement in terms of interface design, navigation flow, and overall user experience. Secondly, effectiveness testing allowed us to assess the extent to which the platform achieves its intended learning outcomes measuring learning outcomes before and after using the platform. This was done through pre- and post-tests, quizzes, surveys, and assessments to gauge knowledge retention, comprehension, and application of learned concepts. Table 3.1 depicts the responses from the stakeholders after experiencing the model. Figure 3.7 shows the graph for the feedback given for the model after using this model.

Design thinking proved instrumental in our project by enabling us to empathize with stakeholders, identify common hurdles in understanding complex statistical concepts, and innovating solutions such as interactive simulations and engaging videos, resulting in a comprehensive learning platform fostering enhanced understanding and engagement. This approach ensures continuous improvement, equipping students for academic and career success.

References

Richards, D., Dobriban, E., and Rebeschini, P. (n.d.). Comparing classes of estimators: When does gradient descent beat ridge regression in linear models? arXiv:2108.11872v2, 12 Jun 2022, pp. 1–3.

E. Lacoche, E. Villain, and A. Foulonneau, "Evaluating usability and user experience of AR applications in VR simulation," Frontiers in Virtual Reality, vol. 3, Jul. 2022, pp. 1–3.

Q. Zheng and J. Lafferty, "A Convergent Gradient Descent Algorithm for Rank Minimization and Semidefinite Programming from Random Linear Measurements," *arXiv preprint arXiv:1506.06081*, Jun. 2017, pp. 1–3.

S. Ruder, "An overview of gradient descent optimization algorithms," *arXiv:1609.04747 [cs.LG]*, Sep. 2016, revised Jun. 2017. https://doi.org/10.48550/arXiv.1609.04747

Weihs, C. and Ickstadt, K. (2018). Data science: The impact of statistics. *Internat. J. Data Sci. Analyt.*, 6(1), pp. 1–3. https://doi.org/10.1007/s41060-018-0102-5.

Sarstedt, M. and Mooi, E. (2014). Regression analysis. A concise guide to market research, pp. 1–3. https://doi.org/10.1007/978-3-642-53965-7_7.

Saleh, H. and Layous, J. (2022). Machine learning - Regression. https://doi.org/10.13140/RG.2.2.35768.67842.

Khasanov, D., Tojiyev, M., and Primqulov, O. (2021). Gradient descent in machine learning. pp. 1–3. https://doi.org/10.1109/ICISCT52966.2021.9670169.

Ganie, A. and Dadvandipour, S. (2023). From big data to smart data: A sample gradient descent approach for machine learning. *J. Big Data*, 10(1), pp. 1–3. https://doi.org/10.1186/s40537-023-00839-9.

Dogo, E., Afolabi, O., Nwulu, N., Twala, B., and Aigbavboa, C. (2018). A comparative analysis of gradient descent-based optimization algorithms on convolutional neural networks. https://doi.org/10.1109/CTEMS.2018.8769211.

Manorathna, R. (2020). Linear regression with gradient descent.

Iqbal, M. A. (2021). Application of regression techniques with their advantages and disadvantages. *J. XYZ*, 4(1), pp. 11–17.

Tapkir, A. (2023). A comprehensive overview of gradient descent and its optimization algorithms. *IARJSET*, 10(1), pp. 1–3. https://doi.org/10.17148/IARJSET.2023.101106.

Niu, L. (2018). A review of the application of logistic regression in educational research: Common issues, implications, and suggestions. *Edu. Rev.*, 72(1), pp. 1–27. https://doi.org/10.1080/00131911.2018.1483892.

Kumari, K. and Yadav, S. (2018). Linear regression analysis study. *J. Prac. Cardiovas. Sci.*, 4(1), pp. 33. https://doi.org/10.4103/jpcs.jpcs_8_18.

Agarwal, S. and Ijmtst, Editor. (2021). Use of statistics in research. *Internat. J. Modern Trends Sci. Technol.*, 7(1), pp. 98–103 . https://doi.org/10.46501/IJMTST0711017

D. L. Neumann, M. M. Neumann, and M. Hood, "Evaluating computer-based simulations, multimedia and animations that help integrate blended learning with lectures in first year statistics," Australasian Journal of Educational Technology, vol. 27, no. 2, pp. 274–289, 2011.

Rekik G, Belkhir Y, Jarraya M, Bouzid MA, Chen YS, Kuo CD. Uncovering the Role of Different Instructional Designs When Learning Tactical Scenes of Play through Dynamic Visualizations: A Systematic Review. Int J Environ Res Public Health. 2020 Dec 31;18(1):256. doi: 10.3390/ijerph18010256.

M. Teplá, P. Teplý, and P. Šmejkal, "Influence of 3D models and animations on students in natural subjects," International Journal of STEM Education, vol. 9, no. 65, pp. 1–20, Oct. 2022.

Kuhlmann, S. L., Plumley, R., Evans, Z., Bernacki, M. L., Greene, J. A., Hogan, K. A., Berro, M., Gates, K., & Panter, A. (2024). Students' active cognitive engagement with instructional videos predicts STEM learning. *Computers & Education, 216,* 1–20. https://doi.org/10.1016/j.compedu.2024.105050

4 Evaluating the impact of learning analytics on assessments within a learning management system: A case study

Sheeba Joice C.[1,a], *Selvi. M.*[1,b] *and Ramyapandian Vijayakanthan*[2,c]

[1]Department of ECE, Saveetha Teaching Learning Centre, Saveetha Engineering College, Chennai, Tamil Nadu, India

[2]Department of Computer and Information Sciences, Towson University, Maryland, USA

Abstract

Assessing learners' abilities is crucial, demanding suitable evaluation methods and well-designed questions. The effectiveness of assessment methods in distinguishing students' abilities is crucial for educational outcomes. However, designing test items that accurately reflect learners' knowledge levels and identify performance variations remains challenging. This study focuses on evaluating the test items used in the "Digital Principles and System Design" course for 53 learners from the Department of Electronics and Communication Engineering. The assessment was carried out through learning management system (LMS) (Moodle). Notable performance variations among test items, are categorized as easy, moderate, and tough. Furthermore, an overall analysis of the questions highlights the classification of the question paper, with a Cronbach's alpha value of 0.7 indicating acceptable reliability. This case study aims to identify potential areas for enhancing the quality and effectiveness of question papers in accurately assessing learners' abilities.

Keywords: Learning analytics, learning enhancement, facilitation index, discrimination index, Cronbach alpha

Introduction

In the realm of modern education, the integration of learning management systems (LMSs) has redefined the dynamics of teaching and learning. The advent of technology has enabled educators to harness a wealth of learner data, thereby paving the way for the emergence of learning analytics – a pivotal tool that offers insights into the learning process. In this context, the present paper delves into a critical aspect of the educational landscape: the impact of learning analytics on assessments within a LMS. Through a detailed case study, we would like to emphasize on the significance of this relationship and its implications for educational efficacy. Assessments play a crucial role in measuring students' understanding, only quality assessments can help to enhance the quality in education, by identifying the learners' strengths, and addressing their learning needs. Reliability and validity, are two fundamental attributes of assessments, which stand as pivotal determinants of their effectiveness. As educators continuously strive to design assessments that accurately gauge students' performance, the application of learning analytics brings forth the potential to uncover valuable insights and optimize the assessment process. In this paper special focus is on the impact of learning analytics and assessments, within the realm of a LMS. Through a comprehensive case study, it is aimed to ascertain the extent to which learning analytics influences the reliability and validity of assessments. By delving into this area, it is seeked to contribute to the broader discourse on educational effectiveness. Novel approaches are implemented to elevate the quality and precision of assessments through the integration of learning analytics.

This paper is sequenced with five sections, of which the section II gives a detailed literature review, section III describes the evaluation of impact of assessment questions in terms of facilitation value and discrimination index. Section IV deals with reliability and validity analysis from the obtained parameter values followed by the conclusions in the section V.

Review of related work

A review of literature is presented in this section, which exhibits the quantum of work that is going on to interpret the performance of learners in their assessment. Analyzing student responses to assessments is vital for assessment quality and teaching improvement. Many educators lack familiarity with formal statistical methods, hindering informed judgment. The authors have studied various data presentation formats for assessment analysis and suggest simple, effective visual tools for educators. A staff survey assessed academics' response to these formats, revealing potential for aiding teaching and learning through data reflection. Formal reflection on learner responses by academic developers can guide educators towards employing specific items for diagnostic and formative assessments (Crisp and Palmer, 2007).

[a]sheebajoice@saveetha.ac.in, [b]selvim@saveetha.ac.in, [c]rvijay1@students.towson.edu

DOI: 10.1201/9781003587538-4

Educators face challenges teaching engineering to millennial. An inventive approach, multiple interactive learning algorithm (MILA), was implemented for second-year Electronics and Communication Engineering undergraduates. Each one-hour session was divided into three segments with focused activities, including revision before summative assessment. Performance over a semester via internal assessments and end semester examinations showed enthusiastic participation, improved comprehension, and better results (84.03% vs. 76.26%) compared to conventional teaching methods (Selvi and Sheeba Joice, 2022).

In the data science era, higher education institutions (HEIs) recognize the significance of assessing students' learning outcomes using available resources. Enhancing teaching and learning often hinges on appraisal processes. Learning Management Systems (LMS) hold valuable data, yet require processing in line with quality assurance and accreditation principles. This paper introduces a framework for extending open-source Moodle LMS, aiding academic program assessment and student outcomes. The framework comprises a customized Moodle system for course and student outcome calculation, coupled with a backend system for program-level outcome evaluation and report generation (Muhammad Abu et al., 2022).

As learning analytics and learning design advance, their convergence gains significance for scholarly exploration. This article summarizes key findings from an extensive analysis of empirical evidence regarding learning analytics' use in learning design. It also examines how learning analytics inform design decisions and their contextual applications. The study surveyed seven academic databases, including 43 papers. Results reveal consistent design and learning patterns emerging from the interplay of analytics and design. The review highlights the need for a framework integrating design data based on analytics and theory, and emphasizes documenting educators' design choices' effects on learning activities and outcomes (Katerina and Giannakos, 2019).

Analytics in higher education is an emerging field with evolving terminology. Existing literature shows overlapping definitions for similar terms and distinct definitions for related terms. This paper aims to delineate and categorize diverse analytics types discussed in academia and practice, followed by proposing a conceptual framework illustrating their interrelationships. A standardized set of definitions for commonly used analytics terms in academia is also provided (van Barneveld et al., 2012).

This discussion introduces course signals, an early intervention solution for educators. Course signals employs predictive models, considering various factors such as grades, demographics, academic history, and engagement with the blackboard vista system. It enables personalized, real-time feedback delivered via emails and color-coded indicators. The system's mechanism, outcomes, and perceptions of both faculty and students are elaborated upon (Kimberly and Pistilli, 2012).

Computer-aided LMSs have extensive applications in higher education, particularly with the transition from traditional to online learning. While LMSs provide tools for knowledge sharing and assessment, exploring their efficacy in students' engagement and progress tracking is vital. This study examines the "Moodle" LMS and evaluates the effect of "Moodle quizzes" on knowledge improvement and assessment in a civil engineering course at an Australian university. The course includes a database of 62 formative and 61 summative quiz questions, incorporating multimedia elements. Across four course groups of 169 students, quizzes assess competencies at different stages using automated grading and psychometric analysis, including facility and discrimination indices from the Moodle quiz framework (Sithara et al., 2019).

With LMSs becoming widespread in education, data on students' online behavior has grown, enabling performance predictions. Disparate outcomes due to LMS variations make overarching conclusions challenging. The authors have outlined theoretical foundations and common predictor variables. Analyzing 17 blended courses (4,989 students) through Moodle LMS, multi-level and standard regression was used to forecast performance. Despite single-institution data, predictive models vary significantly, indicating limited transferability. LMS data's minimal impact on early intervention or assessment grades underscores the need for diverse data sources and specific theoretical bases for robust predictive models (Rianne et al., 2017).

The adoption of e-learning platforms has made education more accessible and interactive, yet increasing users and data challenges arise. This study assesses novel technology's impact on e-learning, methodically evaluating existing systems, categorization, challenges, and trends. A proposed big data-based framework addresses growing e-learning needs, enabling adaptable course delivery and personalized learning using data-driven insights (Mengchi and·Yu, 2023).

Over the last 10 years, LMS have rapidly gained attraction. Educational institutions have invested substantial funds to implement these systems, aiming to enhance education quality and boost enrollment via distance and blended learning. Nevertheless, assessing the influence of LMS on learner performance has become a prominent research area, often relying on subjective survey data, which may be unreliable. To address this, a learning analytics tool was created. Its application on data from two courses at Mbeya University of Science and Technology established that factors like discussion engagement and peer interaction significantly impacted academic success. However, time spent on LMS, downloads, and login frequency did not significantly affect learning outcomes (Imani and Mtebe, 2017).

Assessing course learning outcomes is a fundamental gauge of learner's success. Learning analytics informs learners, educators, and institutions about progress, but integrated tools for predicting student achievement against specific outcomes are lacking. A framework for a learning analytics tool within Moodle is proposed which will assess learning outcomes. Our approach directly measures outcomes through LMS activities, analyzes results, and suggests improvements for upcoming semesters (Sahar et al., 2016).

Learning management systems (LMS) have become increasingly popular, particularly due to the COVID-19 pandemic, because of their efficiency. Online exams conducted within LMS are used to assess student comprehension. This paper presents a blockchain-based framework designed for secure and reliable exam administration and evaluation. By employing cryptographic hashing and proof of stake, the framework ensures the integrity and security of data. A module integrated with Moodle LMS showcases this approach, storing exam data directly on the blockchain, thus enhancing security compared to traditional centralized methods. The results confirm blockchain's accuracy and reliability in protecting exam data, ensuring trustworthy and precise assessment of students' academic performance (Mohamed et al., 2017).

The aim was to explore how "ideal questions" (with good difficulty and discrimination indices) might be influenced by non-functioning distractors (NF-Ds). A cross-sectional study occurred at Fatima Jinnah Dental College, Karachi, involving 102 1st year dental students. The analysis centered on the first-semester physiology paper with 50 one-best multiple choice questions (MCQs) and 5 options each. Items with suitable difficulty and discrimination indices were identified, while effective distractors were defined as those chosen by ≥5% of students (Mozaffer and Jaleel, 2012).

E-learning is increasingly prevalent in higher education, either as a supplement or replacement for traditional teaching. This study explores the impact of Moodle e-learning platform usage on student performance and satisfaction in a University of Ljubljana public administration program. Empirical results show positive correlations between platform use, learner performance, and satisfaction, offering insights for decision-makers aiming to improve success of learners and their contentment via e-learning (Lan et al., 2015). The effectiveness of the content in the online learning id generally evaluated using MCQ. Each question referred as test item influences the mark scored in the test which further reflects the students learning and the online content. In this work an attempt has been made to analyze the test item for its reliability and validity. This analysis would help the facilitators to improvise the learning content and the level of questions used in the MCQ test.

Evaluation of impact of assessment

A reliability and validity assessment were conducted for the "Digital Principles and System Design" course, catering to 2nd year Electronics and Communication Engineering learners. The initial internal assessment comprised part A, consisting of twenty questions, facilitated through the open access learning management system, Moodle. The questions spanned topics within each of the course's five units, resulting in a question bank of over 100 items. Random selections from this pool formed the assessment quiz, ensuring unique sets for learners seated next to each other. Continuous assessment test-I (CIA I) comprised three parts: part A, with twenty MCQ, part B, featuring five two-mark questions, and part C, involving two descriptive/numerical questions. This paper specifically scrutinizes the validity and reliability of the MCQ marks. The validity analysis employs facilitation value and discrimination index calculations in line with outcome-based education guidelines.

A. Facilitation value

The facility value indicates the difficulty level of a question, represented as the percentage of students who answer it correctly. A high facility value (over 80%) suggests that the question is easy. This can be due to:

(a) The question assessing a fundamental concept that students are expected to understand before finishing the unit.
(b) Poorly constructed distractors, making the correct answer obvious.

Conversely, a low facility value implies that the question is difficult, which can be due to:

(a) The question being misleading, confusing, or poorly constructed.

(b) The question assessing a concept that was not well-explained or not aligned with the learning outcomes.

The facilitation value (FV) is calculated, using the formula,

$$Fl = \frac{\Sigma \text{ Sum of Mark of Upper Count} + \Sigma \text{ Sum of Mark of Lower Count}}{(2 \times \text{Group count}) \times \text{Maximum mark of the test}}$$

Dividing the class into upper and lower groups helps determine the facilitation value, which indicates the question's ease. Students with the highest grades are placed in the upper group, while those with the lowest grades are placed in the lower group. If all the learners have scored full marks or all have given the correct answer, then the FV is 1. And if all the learners have failed to give the correct answers to the question, then the FV is equal to 0. The nature of the question whether it's easy, moderate or tough can be identified, based on the FV. FV=1 implies, the question is very easy and FV=0, implies question is very tough. The optimum FV in the range of 0.5–0.7 is expected for any test item.

B. Discrimination index

The discrimination index (DI) is to indicate how good the question is used to separate the good performers and the poor performers. Discrimination index is calculated using the formula,

$$DI = \frac{\Sigma \text{ Sum of Mark of Upper Count} + \Sigma \text{ Sum of Mark of Lower Count}}{\text{Group count} \times \text{Maximum mark of the test}}$$

The interpretation of discrimination index is as follows:

50 and above: Indicates very good discrimination, meaning the question effectively distinguishes between students who understand the material and those who do not.

30–50: Indicates adequate discrimination, meaning the question moderately distinguishes between higher and lower-performing students.

20–29: Indicates weak discrimination, meaning the question only slightly differentiates between students who understand the material and those who do not.

0–19: Indicates very weak discrimination, meaning the question does not effectively distinguish between higher and lower-performing students.

Negative values: Suggest that the question is probably invalid, as it may be misleading or poorly constructed, causing higher-performing students to answer incorrectly more often than lower-performing students.

This index will help the educators to distinguish the academically good and academically poor learners. With this index, special attention can be given to the learners, based on their intellectual ability.

C. Cronbach's alpha

Cronbach's alpha (α) coefficient measures the internal consistency or reliability of a set of survey items. This statistic helps determine whether a collection of items consistently measures the same characteristic. Cronbach's alpha quantifies the level of agreement on a standardized scale from 0 to 1, with higher values indicating greater agreement between items.

The reliability of the test is evaluated from the Cronbach alpha value which is calculated using the formula given below:

$$\alpha = \left(\frac{k}{k-1}\right)\left(\frac{s_y^2 - \Sigma s_i^2}{s_y^2}\right),$$

where, k represents the number of items in the measure,

S_i^2 represents the variance of each test item.

S_y^2 represents the variance of all items together.

The Cronbach alpha scores obtained for the test items can be interpreted by referring to the values tabulated in Table 4.1.

Reliability and validity analysis

Table 4.2 gives FV value and the DI value for the part A questions of CIA I test for the course digital principles and system design, a course taken by 2nd year ECE students. From Table 4.2, it is observed that out of 20 questions, two questions are tough (T), six questions are easy (E) and remaining twelve questions are at moderate level (M) of toughness. This is inferred based on the FV value, since the FV for six questions are greater than 0.6 and for two questions the FV is near

Table 4.1 Interpretation of Cronbach alpha score (Atina, 2017).

Cronbach's alpha score	Level of reliability
0.0–0.20	Poor reliability
>0.20–0.40	Fair reliability
>0.40–0.60	Satisfactory reliability
>0.60–0.80	Good reliability
>0.80–1.00	Very good reliability

Source: Author

Table 4.2 Question wise FV, DI values.

Q. No.	FV	DI	Variance	Type
1	0.65625	0.4375	0.211765	E
2	0.65625	0.3125	0.203137	E
3	0.6875	0.625	0.193725	E
4	0.5625	0.875	0.217195	E
5	0.375	0.25	0.248571	T
6	0.53125	0.3125	0.254902	M
7	0.5	0	0.251508	M
8	0.4375	0.375	0.238431	M
9	0.4375	0.5	0.251429	M
10	0.5	0.375	0.255102	M
11	0.3125	0.375	0.219608	T
12	0.5	0.5	0.253394	M
13	0.5	0.375	0.255102	M
14	0.5625	0.375	0.237245	M
15	0.53125	0.8125	0.250196	M
16	0.71875	0.3125	0.226667	E
17	0.78125	0.1875	0.241327	E
18	0.46875	0.3125	0.255319	M
19	0.53125	0.5625	0.25	M
20	0.5625	0.5	0.248227	M
	Total marks variance		**14.08446**	

Source: Author

to 0.3. Medium toughness questions have FV around 0.5. It is as per the requirement of outcome-based education (OBE), 10% of the question has higher toughness level that is attempted by high performer students. Both these tough questions have low DI value, so these questions are not properly attempted by upper group. Hence, upper and lower group were not discriminated with the help of two tough questions. At the same time, two medium level questions were able to distinguish the upper and lower group.

The Cronbach alpha value for the 20 Quiz question is obtained as 0.7 which represents that the question paper is **RELIABLE.** The learning analysis reveals that the educators have designed a reliable question paper, and still there is enough scope to improve the Cronbach value, that is to design a very reliable assessment.

Conclusions

This analysis focuses on evaluating assessments administered through Moodle, where questions are generated randomly. The intent is to provide faculty with insights on designing questions that maintain reliability even in random paper generation. The Cronbach's alpha value gauges question paper quality, aiding faculty in enhancing standards. Learner comprehension is evident through FV and DI. The faculty members can improvise the questions used for summative assessments through LMS that enhances the learning ability of the students. Each test is expected to have 10% of the questions that can be answered by upper group of the class and 50% of the questions that can be answered by all categories of the students. Such questions can be taken from the essential and basic concept of the syllabus of the course. These metrics inform assessment refinement, ultimately elevating learning outcomes. In essence, this evaluation examines the impact of assessments on enhancing the learning process.

References

Crisp, G. T. and Palmer, E. J. (2007). Engaging academics with a simplified analysis of their multiple-choice question (MCQ) assessment results. *J. Univ. Teach. Learn. Prac.*, 4(2), 31–50.

Selvi, M. and Sheeba Joice, C. (n.d.). Performance analysis of conventional and innovative teaching learning methodologies in engineering. *J. Engg. Edu. Transform.*, 36 (special issue), 110–114.

Muhammad Abu A., El-khalili, N., Al-sheikh Hasan, M., and Banna, A. A. (2022). Extending learning management system for learning analytics. *IEEE Internat. Conf. Busin. Analyt. Technol. Sec.* DOI: 10.1109/ICBATS54253.2022.9759070,1-6 .

Katerina, M. and Giannakos, M. (2019). Learning analytics for learning design: A systematic literature review of analytics-driven design to enhance learning. *IEEE Trans. Learn. Technol.*, 12(4), 516–534.

Van Barneveld, A., Arnold, K. E., and Campbell, J. P. (2012). Analytics in higher education: Establishing a common language. *Edu. Learn. Initiative,* 1–11.

Kimberly, E. A. and Pistilli, M. D. (2012). Course signals at purdue: Using learning analytics to increase student success. *Proc. 2nd Internat. Conf. Learn. Analyt. Knowl.*, 267–270.

Sithara H. P. W. G., Ayres, J. R., Behrend, M. B., and Smith, E. J. (2019). Optimising Moodle quizzes for online assessments. *Internat. J. STEM Edu.*, 6(1), 1–14.

Rianne, C., Snijders, C., Kleingeld, Ad, and Matzat, U. (2017). Predicting student performance from LMS data: A comparison of 17 blended courses using Moodle LMS. *IEEE Trans. Learn. Technol.*, 10(1), 17–29.

Mengchi, L. and Yu, D. (2023). Towards intelligent E-learning systems. *Edu. Inform. Technol.*, 7845–7876.

Imani, M. and Mtebe, J. S. (2017). Using learning analytics to predict students' performance in Moodle learning management system: A case of Mbeya University of science and technology. *Elec. J. Inform. Sys. Dev. Countries*, 1–13.

Sahar, Y., Kadry, S., Sicilia, M.-A. (2016). A framework for learning analytics in Moodle for assessing course outcomes. *Proc. IEEE Global Engg. Edu. Conf.*, 261–266.

Mohamed, A., Idrees, A. M., and Shokry, M. (2024) . A proposed model for improving the reliability of online exam results using blockchain. *IEEE Access*, vol. 12, pp. 7719–7733, 2024

Mozaffer, R. H. and Jaleel, F. (2012). Analysis of one-best MCQs: the difficulty index, discrimination index and distractor efficiency. *J. Pakistan Med. Assoc.*, 62(2), 142–147.

Lan, U., Aristovnik, A., Tomaževič, N., and Keržič, D. (2015). Analysis of selected aspects of students' performance and satisfaction in a Moodle-based e-learning system environment. *Eurasia J. Math. Sci. Technol. Edu.*, 11(6), 1495–1505.

Atina, A. (2017). Improvement of quality, interest, critical, and analytical thinking ability of students through the application of research based learning (RBL) in introduction to stochastic processes subject. *Internat. Elec. J. Math. Edu.*, 12(2), 167–191.

5 Course management e-learning evaluation to predict students' achievement factors using a novel normalized discounted cumulative gain (NNDCG) technique in comparison with Silhouette score cluster model with improved retention rate

Parthasarathy V.[a] *and Devi T.*[b]

[1]Research Scholar, Saveetha University, Chennai, Tamil Nadu, India

[2]Department of Computer Science and Engineering, Saveetha School of Engineering, Saveetha Institute of Medical and Technical Sciences, Saveetha University, Chennai-602105, Tamil Nadu, India

Abstract

Aim: Research work concentrates on improving e-learning retention by predicting student achievements with novel normalized discounted cumulative gain (NNDCG), comparing it with Silhouette score cluster (SSC) in course management, for enhanced engagement and success strategies. **Materials and methods:** Predicting students' achievement factor to improve retention rate using novel NDCG (N=30) whose size of sample and SSC whose size of sample (N=30), G power 80%. **Results:** Novel NDCG obtains retention rate as 90.33% whose value becomes more than SSC whose retention rate becomes 59.47 and between the novel NDCG algorithm as well as SSC gets p=0.003 for difference value (independent sample t-test p<0.05). **Conclusion:** Novel NDCG obtains finer retention rate with 90.33% in predicting students' achievement factor than SSC model's retention rate of 59.47%.

Keywords: Achievement factor, e-learning, engagement, machine learning, normalized discounted cumulative gain, Silhouette score cluster, strategies, retention rate

Introduction

One of the significant subsets in case of artificial intelligence (AI) is machine learning (ML) which helps in the learning process for a computer and make decisions by avoiding the need of programming in an explicit manner. Involvement from various algorithms help in learning patterns from data, enabling systems to improve their performance over time (Rozić et al., n.d.; Lee et al., 2018). Machine learning finds applications in diverse fields, from predicting outcomes to image recognition, and is integral to the development of intelligent systems that adapt and evolve based on experience (Kant et al., 2021). Traditional methods employed for e-learning evaluation often encounter limitations in accurately predicting students' achievements (Teng et al., 2023). These conventional approaches may struggle in adjusting with dynamic and diversity of environments such as e-learning, where factors such as individual learning styles, varying engagement levels, and evolving course content pose challenges for assessment (Reschly, 2020). The application of these insights holds the focus on improving retention rates, acknowledges the importance of student engagement, personalized learning experiences, and the adaptability of the educational platform, optimize e-learning strategies, foster a more inclusive learning environment, and contribute to the overall success of students in diverse educational settings (Lykourentzou et al., 2009).

Over the past 5 years, a growing scholarly interest has been observed in students' achievement in e-learning (Aljohani et al., 2019). IEEE explorer has become a repository for over 7882 articles, reflecting the increasing attention given to this specific topic (Kotsiantis et al., 2004). Similarly, Science Direct has made a substantial contribution with 7832 articles and Springer has contributed with 5605 articles highlighting the growing significance of this subject within academic discourse (Aydoğdu, 2019). A noteworthy contribution within this body of research is the introduction of a pioneering framework proposed by Coman et al. (2020) cited by 1526. This framework represents a significant leap in predicting students' achievement factors in the context of e-learning, particularly due to the NDCG technique's unique ability to evaluate and rank the relevance of information (Rajabalee and Santally, 2020).

[a]192111029.sse@saveetha.com, [b]devit.sse@saveetha.com

DOI: 10.1201/9781003587538-5

Existing algorithms for predicting the student's achievement factor have a retention rate of 59.47% and those in literature help in prediction of the student's achievement factor with higher retention rate. The novel normalized discounted cumulative gain (NNDCG) becomes specifically tailored for the prediction of students' achievement factor with higher retention rate (Shah Hussain and Khan, 2021). This algorithm significantly increases retention rates in e-learning. By prioritizing student engagement, personalized learning experiences, and the adaptability of the educational platform, the algorithm aims to address existing challenges (Calli et al., 2013). With a focus on optimizing e-learning strategies, it anticipates fostering a more inclusive and effective learning environment, ultimately contributing to a substantial improvement in retention rates (Gray and DiLoreto, 2016).

Materials and methods

The research recently occurred in the lab of augmented reality located in SSE, Chennai with systems comprising of more computational capacity. This in turn resulted in accurate values computation and the groups used where two in number with size of sample as 30 (Castejon et al., 2021). G power becomes 80% whereas alpha takes 0.05 as well as with beta 0.8 including confidence interval 95% (Azzi et al., 2019).

The dataset utilized in this study is named "Success rate data", a CSV file comprises 5075 data. This dataset was obtained from Kaggle and focuses on students' performance in both e-learning and conventional learning settings. The attributes within the dataset delineate various aspects of student engagement, encompassing attendance in actual and e-learning classes, examination performance, attention rates, and related metrics. In this research, an algorithm is developed and implemented using the dataset. This algorithm aims to analyze and predict student performance, leveraging the wealth of information available in the dataset.

Google Colab serves as an internet-connected tool for data science and ML. Like Jupyter Notebooks, it offers a Python programming environment accessible via a web browser. Utilizing Google's cloud infrastructure, it provides high-performance computing without local installations. Integration with Google drive simplifies dataset access, and users can create visualizations using libraries like Matplotlib and Seaborn, as well as work with ML frameworks such as TensorFlow. Its collaborative features allow real-time sharing and concurrent work on notebooks, beneficial for education, research, and development projects.

Normalized discounted cumulative gain (NDCG)

Normalized discounted cumulative gain (NDCG) method proposed is an innovative strategy specifically tailored to assess student achievement and retention rates as sample preparation group 1 in the realm of e-learning. Its primary focus lies in evaluating normalized cumulative gains, providing a holistic assessment of the relevance and significance of diverse factors

Table 5.1 Pseudo-code for normalized discounted cumulative gain.

Input: Success rate dataset

Output: Retention rate

Step 1: Import libraries
The necessary libraries are imported – pandas for data manipulation, numpy for numerical operations, and matplotlib for plotting.

Step 2: Read dataset
The script reads a dataset from a CSV file specified by "file_path" and stores it in a pandas DataFrame named "data".

Step 3: Define final grades
Extracts the "final_grade" column from the dataset and converts it to a list.

Step 4: Ideal ranking
Creates an ideal ranking by sorting the final grades in descending order.

Step 5: Define discounted cumulative gain (DCG) function.
Defines a function to calculate the discounted cumulative gain (DCG) at a specified position "k" in the ranking.

Step 6: Define normalized discounted cumulative gain (NDCG) function and calculate NDCG score
Defines a function to calculate the normalized discounted cumulative gain (NDCG) specified position "k" by comparing the actual ranking with the ideal ranking and calculate the NDCG score

Source: Author

influencing student success. Diverging from traditional methods, NDCG takes into account the weighted importance of these factors in a normalized manner, aiming to furnish a detailed evaluation. This approach is intended to contribute to more effective strategies in enhancing student retention within online educational environments. Table 5.1 represents the pseudo code of NDCG.

Silhouette score cluster model (SSC)

The Silhouette score cluster (SSC) model serves as a benchmark in the analysis of student achievement and retention rates in e-learning. It offers a quantitative measure as a sample preparation group 2 to assess the quality of clustering in student data. Unlike the NNDCG technique, which focuses on evaluating normalized cumulative gains, the Silhouette score determines how well students are grouped based on similarity within clusters. This method aids in understanding the homogeneity of clusters, enabling comparisons to identify patterns and potentially improve retention strategies in e-learning environments. Table 5.2 represents the pseudo code of SSC.

Statistical analysis

IBM SPSS utilization is better for analysis in a statistical way and is utilized for analysis of standard error mean including mean along with value of deviation. In this study, the algorithms NNDCG and SSC are compared and data obtained from multiple iterations independent sample T-test analysis is performed (Azzi et al., 2019). In this study resources as dependent values. Input text data which is novel, session id as independent variables and analysis of T-test is conducted in this research work (Solé-Beteta et al., 2022).

Results

Analysis in a statistical way denotes that NNDCG group 1 (90.33%) has more mean retention than the SSC group 2 (59.47%) for predicting students' achievement factor.

Table 5.1 provides pseudo-code for NNDCG and SSC. It begins with library imports, data reading, and extraction of "final_grade" values. The code then computes NDCG scores, visualizing the alignment of the original ranking with the ideal ranking for evaluating the system's performance.

Table 5.2 outlines code for clustering involving library imports, dataset reading, and feature selection. It iterates through various cluster sizes, employs K-means clustering, computes Silhouette scores, and plots them against cluster numbers to determine an optimal solution for the dataset.

Table 5.3 provides a comparison of retention rate between the NDCG algorithm and various iterations of the SSC.

Table 5.4 present group statistics results for testing independent samples statistically between NDCG and SSC methods. NDCG demonstrates a mean retention of 90.33% and a standard deviation of 5.281, while SSC shows a mean retention of 59.47% as well as standard deviation with 9.043. Utilizing T-test, the standard error mean for NDCG (0.964) is compared with SSC (1.651).

Table 5.5 illustrates the application of the tests, setting the significance at ($p=0.003$, $p<0.05$) where confidence interval 95% is obtained. After applying SPSS calculations to NDCG.

Figure 5.1 is presented as a bar graph comparing the mean retention of NDCG. The y-axis represents mean retention, while the x-axis displays error bars

Table 5.2 Pseudo-code for Silhouette score cluster.

Input: Success rate dataset

Output: Retention rate

Step 1: Import libraries
The necessary libraries are imported – pandas for data manipulation, numpy for numerical operations, and matplotlib for plotting.

Step 2: Read dataset
The script reads a dataset from a CSV file specified by "file_path" and stores it in a pandas DataFrame named "data".

Step 3: Select relevant columns for clustering
Selects relevant columns ("retention_rate","forum participation","assignments completed" and "final_grade") from the dataset for clustering.

Step 4: Calculate Silhouette scores for different cluster sizes
In this loop, the script iterates through different cluster sizes, starting from 2 up to the total number of data points. For each cluster size, K-means clustering is performed on the selected features, and the Silhouette score is calculated and appended to the list of Silhouette scores.

Source: Author

Table 5.3 The raw data with a sample size of N=30 for both normalized discounted cumulative gain and Silhouette score cluster.

S. No.	NDCG access time (ms)	SSC access time (ms)
1	69	79
2	81	80
3	85	55
4	85	55
5	87	76
6	88	70
7	87	60
8	89	65
9	90	60
10	90	62
11	91	60
12	91	61
13	91	62
14	91	62
15	92	60
16	92	56
17	92	60
18	92	59
19	92	60
20	93	60
21	93	59
22	93	55
23	93	53
24	94	52
25	94	50
26	95	48
27	95	45
28	95	45
29	95	45
30	95	70

The retention rate is calculated for each of the 30 iterations. In comparison to the Silhouette score cluster, normalized cumulative gain technique demonstrates higher retention rate.

Source: Author

for both NDCG techniques, including their ±2 standard deviations and 95% confidence intervals.

Figure 5.2 is presented as a bar graph comparing the mean retention of SSC. The y-axis represents mean retention, while the x-axis displays error bars for SSC techniques, including their ±2 standard deviations and 95% confidence intervals.

Figure 5.3 compares the mean retention of NNDCG with SSC, indicating NDCG's higher mean retention of 90.33% versus SSC's 59.47%. NDCG also exhibits a lower standard deviation. x-axis: NDCG versus SSC algorithm, y-axis: retention value for mean, with error bar ±2 SD.

Discussion

After investigation of results got from T-test, determination of significance is done. 0.003 becomes the value which is smaller compared to the 0.05 which is significant and novel DCG retention rate is 90.33% with better result as well as the retention rate for SSC model becomes 59.47% significant difference is identified.

With contrast to conventional approaches, marked by a modest accuracy of 59.47%, their limitations become apparent when faced with the intricacies of complex data, resulting in an increased susceptibility to inaccurate predictions (Pan et al., 2021). These models analyze key factors such as learning styles and engagement patterns, transforming course management strategies (Sadiq Hussain et al., 2020). Beyond assessment, predictive analytics enable personalized learning experiences, fostering an engaging virtual environment (Anderson et al., 2021). The insights guide proactive interventions, addressing challenges promptly and enhancing overall learning experiences (Willging and Johnson, 2009). These predictions also inform institutions for strategic decisions on resource allocation and curriculum design, marking a technological shift toward data-driven education (Liu et al., 2017). The amount of time students invest in various learning tasks can indicate their dedication and focus. Analyzing time allocation helps in predicting

Table 5.4 Statistical insights for independent samples, comparing the NDCG with the SSC.

	Algorithm	N	Mean	Std. deviation	Std. error mean
Time	NDCG	30	90.33	5.281	0.964
	SSC	30	59.47	9.043	1.651

In NDCG, the mean retention stands at 90.33%, surpassing the SSC's retention of 59.47%. The NDCG exhibits a standard deviation of 5.281, while the SSC has a standard deviation of 9.043. Additionally, the standard error mean for NDCG is 0.964, lower than the SSC's 1.651.

Source: Author

Table 5.5 T-test for statistical independent samples, comparing the NDCG with SCC at a 95% confidence interval.

Levene's test for equality of variances		T-test for equality of means								
									95% confidence interval of the difference	
		F	Sig.	t	df	Sig. (2-tailed)	Mean difference	Std. error difference	Lower	Upper
Retention	Equal variances assumed Equal variances not assumes	4.667	0.35	16.145	58	0.003	30.867	1.912	27.040	34.694
	Equal variances assumed Equal variances not assumed			16.145	46.718	0.003	30.867	1.912	27.020	34.714

The findings reveal that there is statistically significant difference between the NDCG and SSC, with a p-value of 0.003 ($p<0.05$).
Source: Author

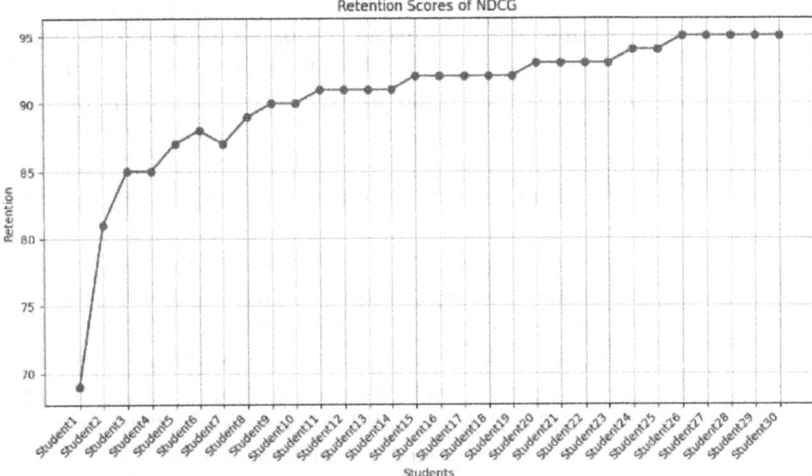

Figure 5.1 Line graph for retention rate for normalized discounted cumulative gain. x-axis: students, y-axis: retention rate
Source: Author

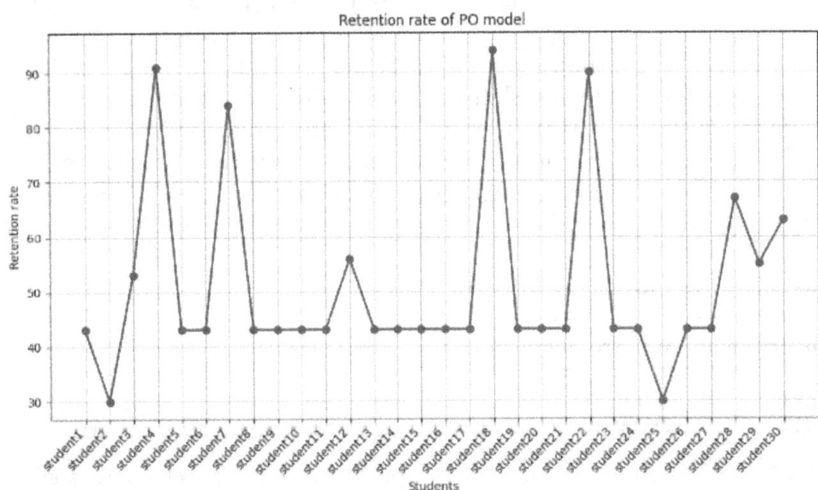

Figure 5.2 Line graph for retention rate for Silhouette score cluster. x-axis: students, y-axis: retention rate
Source: Author

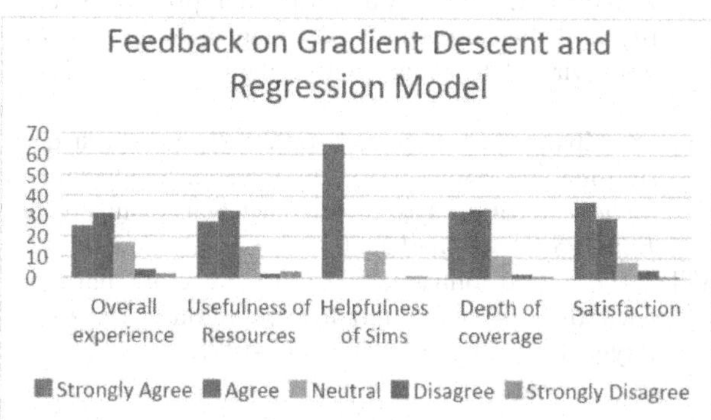

Figure 5.3 The mean retention comparison between novel normalized discounted cumulative gain technique and Silhouette score cluster model shows that NNDCG technique has a higher mean retention rate of 90.33%, compared to the 59.47% of SSC model. The standard deviation of NNDCG technique is also lower than that of the SSC model. On the x-axis: NDCG vs. SSC algorithm, and on the y-axis: mean retention. The error bar is represented by ± 2 SD.

Source: Author

achievements and adapting course structures accordingly (Li et al., 2022).

Factors affecting the effectiveness of our course management e-learning evaluation include the quality and diversity of the training dataset, influencing the predictive power. Limitations arise from the potential contextual specificity of findings and the need to adapt to future changes in the educational landscape. The future scope lies in refining models with real-time data, addressing adaptability challenges, and exploring innovative strategies for further improving retention rates in e-learning environments.

Conclusions

Normalized discounted cumulative gain technique emerges as superior solution for prediction of students' achievement factor with a retention rate of 90.33% outperforming traditional methods, notably those with a 59.47% retention rate.

References

Aljohani, N. R., Fayoumi, A., and Saeed-Ul, H. (2019). Predicting at-risk students using clickstream data in the virtual learning environment. *Sustain. Sci. Prac. Policy*, 11(24), 7238.

Anderson, J., Bushey, H., Devlin, M., and Gould, A. J. (2021). Promoting student engagement with data-driven practices. *Internat. Perspect. Support. Engag. Online Learn.*, 39, 87–103.

Aydog˘du, S¸. (2019). Predicting student final performance using artificial neural networks in online learning environments. *Edu. Inform. Technol.*, 25(3), 1913–1927.

Azzi, I., Jeghal, A., Radouane, A., Yahyaouy, A., and Tairi, H. (2019). A robust classification to predict learning styles in adaptive e-learning systems. *Edu. Inform. Technol.*, 25(1), 437–448.

Calli, L., Balcikanli, C., Calli, F., Cebeci, H. I., and Seymen, O. F. (2013). Identifying factors that contribute to the satisfaction of students in e-learning. *Tur. Online J. Distance Edu.*, 14(1), 85–101.

Castejon, J. L., Núñez, J. C., Gilar-Corbi, R., and Abellán, I. M. J. (2021). New challenges in the research of academic achievement: Measures, methods, and results. *Frontiers Media SA*. 12(2) 70–80

Coman, C., Ţîru, L. G., Meseşan-Schmitz, L., Stanciu, C., and Bularca, M. C. (2020). Online teaching and learning in higher education during the coronavirus pandemic: Students' perspective. *Sustain. Sci. Prac. Policy*, 12(24), 10367.

Gray, J. A. and DiLoreto, M. (2016). The effects of student engagement, student satisfaction, and perceived learning in online learning environments. *Internat. J. Edu. Leadership Prep.*, 11(1). 89–100 http://files.eric.ed.gov/fulltext/EJ1103654.pdf.

Hussain, S., Gaftandzhieva, S., Maniruzzaman, Md., Doneva, R., and Muhsin, Z. F. (2020). Regression analysis of student academic performance using deep learning. *Edu. Inform. Technol.*, 26(1), 783–798.

Hussain, S. and Muhammad, Q. K. (2021). Student-performulator: Predicting students' academic performance at secondary and intermediate level using machine learning. *Ann. Data Sci.*, 10(3), 637–655.

Kant, N., Prasad, K. D., and Kumari, A. (2021). Selecting an appropriate learning management system in open and distance learning: A strategic approach. *Asian Assoc. Open Univer. J.*, 16(1), 79–97.

Kotsiantis, S., Pierrakeas, C., and Pintelas, P. (2004) . Predicting students' performance in distance learning using machine learning techniques. *Appl. Artif. Intell. AAI.* https://doi.org/10.1080/08839510490442058. 21(7) 120–130

Lee, S. J., Lee, H., and Kim, T. T. (2018). A study on the instructor role in dealing with mixed contents: How it affects learner satisfaction and retention in e-learning. *Sustain. Sci. Prac. Policy*, 10(3), 850.

Liu, D. Y.-T., Kathryn, B.-A., Pardo, A., and Adam, J. B. (2017). Data-driven personalization of student learning support in higher education. *Learn. Analyt. Fundaments Appl. Trends*, 143–169.

Li, X., Zhang, Y., Cheng, H., Li, M., and Yin, B. (2022). Student achievement prediction using deep neural network from multi-source campus data. *Comp. Intell. Sys.*, 8(6), 5143–5156.

Lykourentzou, I., Giannoukos, I., Mpardis, G., Nikolopoulos, V., and Loumos, V. (2009). Early and dynamic student achievement prediction in e-learning courses using neural networks. *J. Am. Soc. Inform. Sci. Technol.*, 60(2), 372–380.

Pan, D., Wang, S., Jin, C., Yu, H., Hu, C., and Wang, C. (2021). Research on student achievement prediction based on BP neural network method. *Adv. Artif. Sys. Med. Edu. IV*, 293–302.

Rajabalee, Y. B. and Mohammad Issack, S. (2020). Learner satisfaction, engagement and performances in an online module: Implications for institutional e-learning policy. *Edu. Inform. Technol.*, 26(3), 2623–2656.

Reschly, A. L. (2020). Dropout prevention and student engagement. *Student Engag.*, 31–54.

Rozić, R., Ljubić, H., Grujić, T., and Skelin, A. K. (n.d.). Detection of at-risk students in virtual learning environment. https://ieeexplore.ieee.org/abstract/document/10271591/ 14(7) 50–67

Solé-Beteta, X., Navarro, J., Gajšek, B., Guadagni, A., and Zaballos, A. (2022). A data-driven approach to quantify and measure students' engagement in synchronous virtual learning environments. *Sensors*, 22(9), 3294.

Teng, Y., Zhang, J., and Sun, T. (2023). Data-driven decision-making model based on artificial intelligence in higher education system of colleges and Universities. *Expert Sys.*, 40(4), e12820.

Willging, P. A. and Johnson, S. D. (2009). Factors that influence students' decision to drop out of online courses. *J. Async. Learn. Netw.*, 13(3); 115–127.

6 Measuring influence of emotional empowerment on English language acquisition of freshman engineering students through FLE and FLCA scale

Jayanti Shinge[a], *Vishakha Mandrawadkar*[b], *Naveenkumar Aigol*[c] *and Khezia Olagundi*[d]

Department of Humanities and Social Sciences, KLE Technological University, Hubli, Karnataka, India

Abstract

The acquisition of English in students is significantly impacted by their emotional empowerment. Positive emotions like enjoyment and enthusiasm enhance engagement and motivation in language learning, while negative emotions such as anxiety can obstruct progress and overall acquisition. Emotional experiences are crucial in shaping behavior and motivation during language instruction, and the ability to regulate emotions positively affects language outcomes by aiding stress management, focus, and performance. Despite extensive research on emotions' role in language learning, gaps remain in understanding how they facilitate the acquisition process. This study explores the emotional needs of freshman engineering students in language learning using a quantitative approach, employing the foreign language enjoyment (FLE) and foreign language classroom anxiety (FLCA) scales on 120 students divided into two groups based on their prior exposure to English. Significant differences were found in the groups' perceptions and experiences, highlighting the influence of emotional empowerment on language acquisition.

Keywords: Acquisition, English, language, emotions, anxiety

Introduction

The acquisition of English involves not just cognitive skills but also emotional empowerment, which plays a crucial role in successful language learning. This paper explores the relationship between emotional well-being and language acquisition, building on foundational studies by Krashen (1982) and Gardner (1985) highlighted the importance of affective factors like motivation, anxiety, and self-confidence. Further research shows that traits such as self-efficacy, autonomy, and agency significantly boost learners' engagement and success in acquiring English (Dewaele, 2015; Dörnyei, 2019). Empowered learners are intrinsically motivated, resilient in the face of challenges, and proactive in their learning process, leading to higher proficiency levels (Bandura 1997; Clément 2004; Dörnyei and Ushioda, 2011; Langer, 2009). However, traditional assessments often overlook the emotional-cognitive interplay, prompting researchers to use diverse methods like self-report measures, observational techniques, and physiological tools to capture it in dynamic state (Cameron, 2011; Csizer and Dörnyei, 2005; Dewaele et al., 2011; Gardner, 2006).

By integrating these methods, researchers can develop a complete understanding of how emotional empowerment influences language learning, informing interventions and pedagogies that enhance learner empowerment and language acquisition.

Literature review

Recent research underscores the integral role of emotional well-being in English language acquisition, particularly how self-efficacy, autonomy, and agency enhance learners' engagement and success. Studies by Eslami and Dehghan (2019), Moghaddam et al. (2020), and Jabbari and Faraji (2022) explore the effects of motivational factors and emotion regulation on language learning using questionnaires in Iranian English as a Foreign Language EFL contexts. Observational methods by Walker and Jesson (2018), Jabbar and Bouton (2020), Schmidt and Alderson (2022) assess how affective factors impact classroom dynamics and learner engagement. Neuroscientific approaches by He et al. (2020), Zhao and Yang (2020) examine emotional impacts on language learning via techniques like neuroimaging and heart rate variability. Further, a meta-analysis by Jabbar and Bouton (2017) and subsequent studies highlight the positive correlation between self-efficacy and motivation, and its negative relationship with anxiety, emphasizing

[a]jayantishinge@gmail.com, jayanti_s@kletech.ac.in, [b]vishakha.mandrawadkar@kletech.ac.in, [c]naveenkumar.aigol@kletech.ac.in, [d]khezia.olagundi@kletech.ac.in

DOI: 10.1201/9781003587538-6

the crucial role of emotional empowerment in effective language learning. These findings collectively demonstrate that fostering emotional intelligence and supportive teacher–student relationships are essential for creating conducive learning environments and enhancing language proficiency.

Methodology

Participants

This study used a quantitative approach with a pre-, and post-test design to investigate the influence of emotional empowerment on the English language acquisition of fresh engineering students. Participants were recruited through purposive sampling from a first-year engineering program at KLE Technological University. The following two groups were formed:

Group A: Sixty students who studied English as their first language throughout their schooling.

Group B: Sixty students who studied English as their second language throughout their schooling.

Inclusion criteria for participants included:

- Currently enrolled as a fresh in the engineering program
- Willing to participate in the study and complete all required tests
- No diagnosed learning disabilities or cognitive impairments.

Instruments

Two validated scales (MacIntyre et al., 2016) were used to measure the study's key variables:

- **Foreign language enjoyment (FLE) scale:** Assesses students' feelings of enjoyment and engagement in language learning.
- **Foreign language classroom anxiety (FLCA) scale:** Measures students' anxiety levels related to language learning and classroom situations.

Procedure

The study followed these steps:

1. **Informed consent:** Participants received an information sheet about the study and gave their consent before participation.
2. **Pre-test:** Both groups completed the FLE and FLCA scales at the beginning of the semester.
3. **Intervention:** Both groups participated in their regular English language courses throughout the semester. However, Group B additionally received a series of interventions designed to foster emotional empowerment.
4. **Post-test:** Both groups completed the FLE and FLCA scales again at the end of the semester.

Interventions for emotional empowerment

The intervention framework includes self-efficacy building through goal setting, feedback, and celebrating achievements to enhance language skills. Learners are given autonomy to choose their tasks, materials, and strategies, and are involved in decision-making to foster independence. Emotion regulation is taught using mindfulness and relaxation techniques to manage emotions and build resilience. Positive teacher-student relationships are cultivated to provide emotional support, and a supportive learning environment is created to promote safety and inclusion. Additionally, social-emotional learning is integrated into the curriculum to develop crucial emotional and cognitive skills.

Data analysis

Descriptive analysis was used to deeply examine the dataset by employing frequency distribution and comparison techniques. This involved counting how often students appeared on scale items and comparing these counts across variables, groups, or time periods. This method highlighted changes and differences, helping

Table 6.1 Frequency distribution – Comparison before intervention given below (BI) and after intervention (AI) FLCAS (Group B) (MacIntyre et al., 2016).

FLCA scale (Group B)	BI frequency in %	AI frequency in %
1. Anxiety about FL class preparation	75	58
2. Comparison with other students	42	45
3. Physical manifestations of anxiety	60	63
4. Worry about making mistakes	43	45
5. Lack of confidence in speaking	38	55
6. Nervousness and confusion while speaking	47	49
7. Panic when speaking without preparation	57	61
8. Embarrassment in volunteering answers	45	60

Source: Author

to succinctly summarize complex data and pinpoint influential factors. Ultimately, descriptive analysis aided in decision-making and guided further research inquiries.

Comparison FLCAS – Group B

The FLCA scale data for Group B, shown in Table 6.1 and Figure 6.1, indicates a positive post-intervention trend, with notable decreases in anxiety-related behaviors like panic, embarrassment, and speaking worries. Confidence levels increased significantly, affirming the intervention's effectiveness. Although comparison with peers slightly rose, it did not lead to increased anxiety, while concerns about mistakes stayed steady. There were modest improvements in speaking fluency and composure, suggesting that further practice

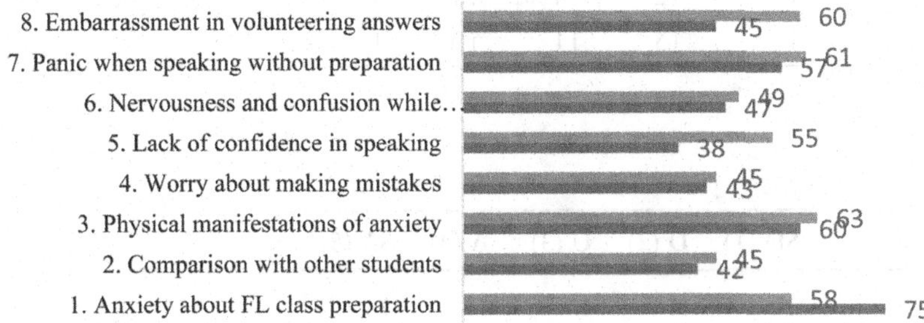

Figure 6.1 Bar chart for FLCAS analysis – Group B – Before and after intervention
Source: Author

Table 6.2 Frequency distribution – Comparison before intervention (BI) and after intervention (AI) FLE (Group B).

FLE scale (Group B)	BI frequency in %	AI Frequency in %
1. I can be more creative	52	58
2. I can laugh off my embarrassing mistakes in the FL	52	62
3. I do not get bored	63	67
4. I do enjoy it	52	58
5. I do feel as though I am a different person during the FL class	57	69
6. I did learn to express myself better in the FL	57	59
7. I am a worthy member of the FL class	53	57
8. I have learnt interesting things	60	66
9. In class, I do feel proud of my accomplishments	62	70
10. I feel, it is a positive environment	52	59
11. I feel it is cool to know a FL	47	51
12. It is fun	50	56
13. I think, making errors is part of the learning process	53	62
14. I feel, the peers are nice	52	63
15. I feel the teacher is encouraging	38	45
16. I feel the teacher is friendly	40	43
17. I think the teacher is supportive	43	45
18. There is a good atmosphere in the class	52	53
19. We form a tight group	47	49
20. We have common "legends", such as running jokes	40	48
21. We do laugh a lot	38	50

Source: Author

Table 6.3 FLCA scale (Group A).

FLCA scale (Group A)	**SD (1)**	**D (2)**	**U (3)**	**A (4)**	**SA (5)**
1. Anxiety about FL class preparation	9	9	9	17	16
2. Comparison with other students	9	8	9	15	19
3. Physical manifestations of anxiety	9	10	10	16	15
4. Worry about making mistakes	12	10	14	11	13
5. Lack of confidence in speaking	11	12	12	12	13
6.Nervousness and confusion while speaking	5	8	12	22	13
7. Panic when speaking without preparation	11	12	14	14	9
8. Embarrassment in volunteering answers	13	15	11	10	11

Source: Author

Table 6.4 FLE scale (Group A).

FLE scale (Group A)	SD (1)	D (2)	U (3)	A (4)	SA (5)
1. I can be more creative	10	9	15	13	13
2. I can laugh off my embarrassing mistakes in the FL	12	11	14	11	12
3. I do not get bored	21	13	8	9	9
4. I do enjoy it	14	18	12	9	7
5. I do feel If I am a different person during the FL class	16	17	10	8	9
6. I did learn to express myself better in the FL	11	17	11	12	9
7. I am a worthy member of the FL class	15	9	13	11	12
8. I have learnt interesting things	18	12	12	8	10
9. In class, I do feel proud of my accomplishments	14	15	12	8	11
10. I feel it is a positive environment	16	14	11	10	9
11. I feel it is cool to know a FL	12	19	12	9	8
12. It is fun	13	17	11	9	10
13. I think, making errors is part of the learning process	30	14	7	5	4
14. I feel the peers are nice	16	15	10	9	10
15. I feel the teacher is encouraging	11	16	14	9	10
16. I feel the teacher is friendly	20	19	8	9	4
17. I think the teacher is supportive	14	15	9	10	12
18. There is a good atmosphere in the class	16	17	10	9	8
19. We form a tight group	15	16	9	8	12
20. We have common "legends", such as running jokes	19	11	9	10	11
21. We do laugh a lot	12	16	10	9	13

Source: Author

might be needed for more significant gains. Overall, the intervention successfully reduced anxiety and enhanced confidence, but continued support is essential for addressing remaining challenges fully.

Comparison FLE – Group B

The FLE scale data from Table 6.2 shows that the intervention had a positive effect on Group B's experiences in their foreign language (FL) class. Post-intervention results show improvements in creativity, the ability to laugh off mistakes, enjoyment, self-confidence, and self-expression. Students felt prouder of their achievements and more engaged in learning interesting content. They also experienced a better classroom atmosphere, characterized by increased support from peers and teachers and a greater appreciation of making mistakes as part of learning. While there were gains in perceived teacher friendliness and supportiveness, these areas still have potential for further enhancement. Overall, the intervention significantly improved the positivity and engagement of the FL learning environment for Group B.

FLCAS – Group A

Group A (as detailed in Table 6.3) shows significant anxiety-related behaviors according to FLCA scale data. They experience nervousness, panic when speaking unprepared, and worry about making mistakes in FL classes. Many students in Group A also feel their peers are more proficient, adding to their anxiety. Despite varying attitudes towards mistakes and confidence levels, Group A generally struggles with lower confidence and higher anxiety compared to Group B. These variations in the FL learning environment likely stem from individual factors or teaching dynamics.

FLE – Group A

Group A (based on FLE scale data from Table 6.4) generally holds positive perceptions of their FL learning experience. They exhibit creativity, enjoy learning, and can laugh off mistakes, though engagement and attitudes towards mistakes vary. Despite some mixed feelings about teacher supportiveness, Group A shows more positive attitudes towards their FL learning experience compared to Group B, particularly in terms of enjoyment and classroom atmosphere. These differences may be influenced by individual preferences, teaching methods, or prior experiences, highlighting the diverse outlooks within FL learning groups.

Conclusions

The comparison between Group A and Group B, both exposed to varying levels of English, shows significant differences in their FLE and FLCA scores. Group B, which studies English as a secondary subject, consistently outperforms Group A, where English is the primary subject. They exhibit lower anxiety, higher confidence, and more positive perceptions of their learning experience, with lesser anxiety-related behaviors and higher enjoyment and creativity in the classroom. In contrast, Group A students show higher anxiety, varied engagement, and mixed perceptions, indicating a need for more support and tailored interventions to improve their learning outcomes. These results highlight the impact of previous language learning experiences on current FL learning success and the importance of addressing individual differences in educational strategies.

Future scope

Future research should track long-term changes in foreign language learning attitudes and behaviors, analyzing how interventions affect anxiety and confidence. Studies should also consider individual and cultural factors, use comparative and mixed-method methods, and explore the relationship between attitudes and academic performance. Collaboration between researchers and educators is crucial to apply these insights practically. Additionally, examining technology's role in learning and the impact of societal attitudes on motivation could guide educational policy.

References

Bandura, A. (1997). Self-efficacy: The exercise of control. Freeman.

Cameron, L. (2011). Researching language learning in the affective domain: New directions and new challenges. *Internat. J. Res. Method Edu.*, 34(3), 261–273.

Clément, R. (2004). Motivation for second language acquisition: A theoretical framework. *Multilin. Matt.*

Csizer, K. and Dörnyei, Z. (2005). Language anxiety and cognitive processes: A review of the literature. *Lang. Learn.*, 55(4), 553–604.

Dewaele, J. M. (2015). Emotion and motivation in language learning: The state of the art. *Multilin. Matt.*

Dewaele, J. M., Jarvis, S., and Dewaele, C. (2011). Foreign language anxiety research: Defining the construct and measuring its effects. Routledge.

Dörnyei, Z. (2019). The psychology of language learning. Oxford University Press.

Dörnyei, Z. and Ushioda, E. (2011). Motivation, language identity and the self. *Multilin. Matt.*

Gardner, R. C. (1985). Social psychology and second language learning: The role of affective variables. Edward Arnold.

Gardner, R. C. (2006). Motivational dispositions and second language acquisition. *Multilin. Matt.*

Krashen, S. D. (1982). 3rd Edition, Principles and practice in second language acquisition. Pergamon.

Langer, J. A. (2009). Mindfulness and second language acquisition. Routledge.

Eslami, Z. and Dehghan, M. (2019). The relationship between self-efficacy and motivation in second language learning. *Internat. J. English Lang. Lit.*, 8(4), 11–20. doi:10.14447/ijell.v8i4.7623.

Moghaddam, A. A., Gholipoor, E., and Sadeghi, Z. (2020). The impact of motivation, emotional intelligence, and self-efficacy on foreign language anxiety. *J. Edu. Appl. Psychol.*, 10(2), 301–312. doi:10.3109/20701792.2020.1710214.

Jabbari, S. and Faraji, H. (2022). Exploring the links between emotion regulation, motivation, and second language learning progress. *Internat. J. English Lang. Lit.*, 11(3), 1–11. doi:10.14447/ijell. v11i3.33401.

Walker, C. and Jesson, R. (2018). Investigating the role of affective factors in learner-teacher interactions in second language classrooms. *Lang. Awaren.*, 27(4), 366–385. doi:10.1080/09889950.2018.1490026.

Jabbar, Z. and Bouton, R. C. (2020). Observing motivation and emotion in language learning: A review of methodological approaches. *Second Lang. Learn. Teach.*, 17(2), 193–225. doi:10.1177/1753369319895230.

Schmidt, J. and Alderson, C. (2022). The relationship between teacher emotion and student engagement in second language learning. *Edu. Psychol. Prac.*, 38(1), 31–46. doi:10.1080/02667363.2021.1955142.

He, Y., Wu, J., and He, J. (2020). The effect of self-compassion on language anxiety and second language acquisition: A neurophysiological study. *Human Brain Mapp.*, 41(12), 3586–3596. doi:10.1002/hbm.25109.

Zhao, J. and Yang, Y. (2021). Investigating the relationship between emotion and second language learning using physiological measures: A review. *Front. Psychol.*, 12, 650850. doi:10.3389/fpsyg.2021.650850.

He, Y., Wu, J., and He, J. (2022). The effect of mindfulness training on emotion regulation and second language learning. *Front. Psychol.*, 13, 732245. doi:10.3389/fpsyg.2022.732245.

Jabbar, Z. and Bouton, R. C. (2017). Examining the relationship between self-efficacy, motivation, and foreign language anxiety: A meta-analysis. *Edu. Psychol. Prac.*, 33(4), 299–315. doi:10.1080/02667363.2016.1180982.

He, J., Wu, J., and Wu, Y. (2020). The impact of motivation and emotion regulation on second language proficiency. *Front. Psychol.*, 11, 582019. doi:10.3389/fpsyg.2020.582019.

Schmidt, J. and Alderson, C. (2022). The relationship between teacher emotional expression and student engagement in second language learning. *Edu. Psychol. Prac.*, 38(1), 31–46. doi:10.1080/02667363.2021.1955142.

Walker, C. and Jesson, R. (2018). Investigating the role of affective factors in learner-teacher interactions in second language classrooms. *Lang. Awaren.*, 27(4), 366–385. doi:10.1080/09889950.2018.1490026.

Jabbar, Z. and Bouton, R. C. (2019). Exploring the role of emotional intelligence in second language acquisition: A review of the literature. *Internat. J. Appl. Ling.*, 29(4), 655–674. doi:10.1177/0261970X18795044.

MacIntrye, P. D., Gregerson, T., and Mercer, S. (2016). Positive psychology in SLA. *Multilin. Matt.*

7 Achieving program outcomes in engineering education: Strategies for curriculum integration and assessment refinement

Arun Kumar A[1,a], *Rajakarunakaran S*[2] *and Ganesan L*[3]

[1]Department of Electrical and Electronics Engineering, Ramco Institute of Technology, Rajapalayam, Tamilnadu, India

[2]Department of Mechanical Engineering, Ramco Institute of Technology, Rajapalayam, Tamilnadu, India

[3]Department of Electronics and Communication Engineering, Ramco Institute of Technology, Rajapalayam, Tamilnadu, India

Abstract

This paper explores different strategies to enhance the attainment of program outcomes (POs) for engineering graduates, a critical aspect in aligning education with industry demands. The National Board of Accreditation mandates that engineering graduates must demonstrate proficiency in 12 POs, posing a challenge for academicians to effectively map course outcomes (COs) to these requirements. Traditional approaches in curriculum, often consists separate theory and practical courses which may not facilitate mapping of COs to many POs. To address this challenge, this paper presents the integration of theory and practical components within courses. By consolidating these elements, academicians can improve the alignment between COs and POs. Furthermore, this integration expands the scope for mapping COs to a broader range of POs. In addition to course integration, this paper highlights the design of transparent evaluation criteria or rubrics for capstone projects courses which enables thorough assessment of POs. Moreover, incorporating various assessment tools and assessing them appropriately contributes to a more holistic evaluation of students' knowledge, skills and attitudes. By adopting different direct assessment tools from the latest affiliated institute's curriculum this paper showcases practical implementations of several assessment tools and their impact on PO attainment. By adopting a comprehensive approach that integrates courses and refinement of assessment practices, academician's can effectively enhance the attainment of POs, equipping engineering graduates with the skills needed to thrive in today's dynamic professional landscape.

Keywords: Outcome-based education, program outcome, assessment and integration, curriculum integration

Introduction

In recent years, the landscape of engineering education has undergone intense transformations which are driven by rapid advancements in technology, globalization, and changing societal needs. The demand for highly competent engineering graduates equipped with the necessary knowledge, skills and attitude to tackle real-world challenges is vital. The expectations placed upon an engineering graduate have been evolved like demanding not only technical expertise but also interdisciplinary skills, critical thinking abilities and a global perspective (Crawley et al., 2007). In this context, the importance of aligning engineering education with industry demands and global standards is inevitable. The Washington Accord, an international agreement among engineering accrediting bodies, highlights the significance of ensuring that engineering graduates demonstrate proficiency in a defined set of program outcomes (POs). However, aligning educational programs with these POs poses a challenging challenge for academic institutions worldwide (Nawab et al., 2020).

Traditionally, engineering education has been structured around separate theory and practical courses where a limited integration between them for a specific course. This categorization often hinders the effective mapping of course outcomes (COs) to the diverse requirements outlined in the Washington Accord (Arunkumar et al., 2018, Kelley et al., 2016). Consequently, there is a pressing need for innovative strategies to bridge this gap and enhance the attainment of POs among engineering graduates.

Moreover, the increasing complexity of engineering projects and the interdisciplinary nature of modern challenges necessitate a more holistic approach to education. In order to get learning experiences that segregate theory from practice not only inhibits students' ability to transfer knowledge to real-world contexts but also limit their capacity to develop essential skills such as problem-solving, teamwork and communication. Recognizing these challenges, educators

[a]arunkumar@ritrjpm.ac.in

DOI: 10.1201/9781003587538-7

and stakeholders within the engineering community are called upon to innovate and restructure educational paradigms to better align with the multifaceted demands of modern engineering practices (Alias et al., 2014). By promoting integration, collaboration and alignment between educational objectives and industry needs, academic institutions can play a pivotal role in shaping the next generation of engineering professionals who are not only technically proficient but also adaptable, creative and ethically minded.

This paper explores various strategies aimed at aligning COs with POs to meet the evolving demands of the industry (Van den Beemt et al., 2020). One such strategy involves the integration of theory and practical components within courses, thereby facilitating a more seamless mapping between COs and POs. By consolidating these elements, academic institutions can enhance the relevance and applicability of their educational programs, better preparing students for the challenges they will face in their professional careers (Falloon et al., 2020; Margot et al., 2019).

Furthermore, the paper highlights the importance of transparent evaluation criteria and assessment practices in assessing students' attainment of POs. By designing clear rubrics for capstone projects and incorporating diverse assessment tools, academic institutions can ensure a more comprehensive evaluation of students' knowledge, skills, and attitudes (Muhsin et al., 2019). Practical implementations of these assessment tools, drawn from the latest curriculum frameworks, are showcased to illustrate their impact on PO attainment (Sharma et al., 2019).

Through a comprehensive approach that integrates curriculum design and assessment practices, this paper aim to provide academic institutions with actionable insights to enhance the attainment of POs among engineering graduates. By equipping students with the skills and competencies demanded by today's dynamic professional landscape, we can ensure the continued relevance and excellence of engineering education (Patel et al., 2019; Teixeira et al., 2020; Zeidmane et al., 2011).

A case study compares the effectiveness of conventional theory courses with integrated courses in an engineering education. Through quantitative analysis of student performance in terms of pass percentage and qualitative assessment of learning experiences have been presented. This case study evaluates the impact of course structure on knowledge acquisition and application. The results indicated that integrated courses, which combine theory and practical components, facilitate deeper understanding, better retention of concepts, and improved problem-solving skills compared to traditional theory-based courses (Miranda et al., 2021). The findings of the study proposed to promote integrated approaches in enhancing learning outcomes and preparing students for real-world engineering challenges.

A comparative analysis of conventional theory courses and integrated courses in Science, Technology, Engineering, and Mathematics (STEM) education has been done for a program (Becker et al., 2011). Utilizing both quantitative and qualitative methodologies, the study examines student performance, engagement levels, and perceived learning outcomes across different course structures. Findings reveal that integrated courses, which integrate theoretical concepts with practical applications, result in higher levels of student engagement, deeper understanding of course material and increased readiness for real-world problem-solving. The study highlights the benefits of integrating theory and practice in STEM education to enhance student learning experiences.

The impact of innovative assessment practices in engineering education has been analyses for a case study. It has been proposed that the emerging trends include authentic assessments, rubrics and competency-based evaluations. The incorporation of industry feedback is also crucial for ensuring assessment authenticity and relevance (Cruz et al., 2020).

A study compared different assessment methods for professional competencies in engineering education (Quelhas et al., 2019). Competency-based assessments have been proposed as a better assessment than traditional methods since it reflects industry expectations better. Industry professionals perceive graduates assessed with competency-based approaches make the students job-ready.

In this paper, contribution has been made to the existing literature by presenting a comparative analysis between conventional theory courses and integrated courses in engineering education. Specifically, the alignment of COs with POs and assess students' attainment of these POs for a sample course has been examined. Additionally, a novel review system has been proposed to evaluate students' attainment of POs for a project course by providing a comprehensive framework for assessing their knowledge, skills, and attitudes.

Program outcomes (POs)

The POs serve as a guiding framework for engineering education which covers the core competencies expected of graduates to grow well in professional locations. By defining these POs, a comprehensive analysis of curriculum alignment and assessment strategies aimed at enhancing students' attainment has been done. The list of POs is engineering knowledge (PO1), problem analysis (PO2), design/development of solutions (PO3), conduct investigations of complex problems (PO4), modern tool usage (PO5), engineer and society (PO6), environment and sustainability (PO7), ethics (PO8), individual and teamwork (PO9), communication (PO10), project management and finance (PO11) and lifelong learning (PO12).

Sample course – 1

In the third-semester curriculum for electrical and electronics engineering (EEE) students, "Electronic Devices and Circuits" stands as a core course has been selected for the analysis. With a class size of 60, this subject examines into the principles and applications of electronic devices and circuitry. This course is presented as a theory course initially later it is presented as an integrated course. Through a blend of theoretical lectures and practical laboratory sessions, students gain insight into semiconductor devices, transistor circuits, and amplifier configurations, equipping them with foundational skills for engineering practice. The sample course consists of five units. Each unit has been allotted with 9 lecture hours. The CO is framed for each unit. The details of the COs are presented in Table 7.1. The assessment plan comprises continuous internal assessment (60%) and an end-semester examination (40%). Within continuous internal assessment, direct assessment methods contribute 90%, while indirect assessment methods contribute 10%. Direct assessment includes internal assessment tests (60%) and assignments, case studies, or seminars (40%). Indirect assessment involves a course exit survey to gather feedback on students' learning experiences and perceptions of the course. This comprehensive assessment strategy as give in Table 7.3 ensures a balanced evaluation of students' knowledge, skills and understanding of electronic devices and circuits. As per the assessment plan the course attainment has been calculated after the end semester examinations. The attainment values for CO are 2.67 for theory course, 2.76 for laboratory course and 2.62 for integrated course. The theory course is indexed as C205, the laboratory course as C207 and the integrated course as C401.

From Table 7.2, each CO is assigned a code, such as C205.1, C205.2, etc., to uniquely identify the CO for each unit within the course curriculum. The numbers under each CO (2, 2, 2) represent the levels of attainment expected for that particular outcome. In this case, a level of 2 indicates a high level of achievement. The attainment calculation for each PO is provided, showing the level of achievement reached for that outcome. The notation "2/3" indicates that 2 out of 3 criteria for that PO have been met, with a corresponding score of 2.67, which is a high level of attainment. Similarly, "1/3" indicates that only 1 out of 3 criteria has been met, resulting in a lower score of 0.89. The overall attainment for each CO and PO is calculated based on the levels achieved for each outcome.

Table 7.1 List of COs – Electronic devices and circuit courses.

Courses	CO	CO statements
Theory course	CO1	Explain the structure and operation of basic electronic devices such as diodes
	CO2	Illustrate the characteristics of different electronic devices such as transistor and thyristor
	CO3	Choose and adapt the required components to construct an amplifier circuit
	CO4	Explore the working of multistage, differential and power amplifiers
	CO5	Perform design and analysis of feedback amplifiers and oscillators
Laboratory course	CO1	Explain the fundamentals of operation and characteristics of semiconductor devices
	CO2	Calculate the basic parameters of semiconductor devices and their limiting factors
	CO3	Demonstrate BJT amplifiers in various configurations including DC biasing techniques
	CO4	Sketch the frequency response characteristics of amplifiers
	CO5	Construct basic electronic circuits, particularly with application to diodes, field-effect transistors and bipolar junction transistors
Integrated course	CO1	Interpret the functioning of the PN junction devices
	CO2	Evaluate the functioning of transistor and thyristor for various applications
	CO3	Interpret the functioning of BJT and FET based amplifiers
	CO4	Evaluate the performance of multistage and differential amplifiers
	CO5	Evaluate the performance of feedback amplifiers and oscillators for the electronics applications

Source: Author

Table 7.2 Attainment of POs – Theory course.

CO/PO	1	2	3	4	5	6	7	8	9	10	11	12
Theory course												
C205.1	2	2	2	-	-	-	-	-	-	1	-	1
C205.2	2	2	2	-	-	-	-	-	-	1	-	1
C205.3	2	2	2	-	-	-	-	-	-	1	-	1
C205.4	2	2	2	-	-	-	-	-	-	1	-	1
C205.5	2	2	2	-	-	-	-	-	-	1	-	1
C205	2	2	2	-	-	-	-	-	-	1	-	1
PO attainment Calculation	2/3* 2.67	2/3* 2.67	2/3* 2.67	-	-	-	-	-	-	1/3* 2.67	-	1/3* 2.67
PO attainment	1.78	1.78	1.78	-	-	-	-	-	-	0.89	-	0.89
Laboratory course												
C207.1	2	-	-	-	1	-	-	1	2	2	-	-
C207.2	2	-	-	-	1	-	-	1	2	2	-	-
C207.3	2	-	-	-	1	-	-	1	2	2	-	-
C207.4	2	-	-	-	1	-	-	1	2	2	-	-
C207.5	2	-	-	-	1	-	-	1	2	2	-	-
C207	2	-	-	-	1	-	-	1	2	2	-	-
PO attainment calculation	2/3* 2.76	-	-	-	1/3* 2.76	-	-	1/3* 2.76	2/3* 2.76	2/3* 2.76	-	-
PO attainment	1.84	-	-	-	0.92	-	-	0.92	1.84	1.84	-	-
Integrated course												
C401.1	3	2	2	1	1	-	-	2	2	2	-	2
C401.2	3	3	3	1	2	-	-	2	2	2	-	2
C401.3	3	2	2	1	2	-	-	2	2	2	-	2
C401.4	3	2	2	2	1	-	-	2	3	2	-	2
C401.5	3	3	2	2	1	-	-	2	3	2	-	2
PO attainment calculation	3/3* 2.62	2.4/3* 2.62	2.2/3* 2.62	1.4/3* 2.62	1.4/3* 2.62	-	-	2/3* 2.62	2.4/3* 2.62	2/3* 2.62	-	2/3* 2.62
PO attainment	2.62	2.09	1.92	1.22	1.22	-	-	1.75	2.09	1.75	-	1.75

Source: Author

The sample laboratory course consists of 11 experiments. The last row corresponding to laboratory course in Table 7.2 shows the overall attainment for each CO and PO, which has been calculated based on the individual levels of attainment in the previous rows. The numbers provided in this column could represent an average or cumulative attainment score for each outcome. The sample integrated course consists of five units. Each unit has been allotted with 9 lecture hours and 3 practical hours. The CO is framed for each unit and its relevant experiment. As per the assessment plan shown in Table 7.3, the course attainment has been calculated after the end semester examinations.

Competency refers to the combination of specific knowledge, skills and abilities that a student is expected to obtain by the end of a course or program. It is a measure of the student's overall proficiency in a particular subject area. Competency-based COs has been framed for integrated course.

Table 7.4 presents the attainment levels of each PO for three different course formats: theory course, laboratory course, and integrated course. In the theory course, PO1, PO2, and PO3 have an attainment level of 1.78, while PO10 has an attainment level of 0.89, and PO12 has an attainment level of 0.89. The laboratory course shows varying attainment levels across different POs, with PO1 and PO9 at 1.84, PO5 and PO8 at 0.92, and PO10 and PO11 at 1.84. In contrast, the integrated course demonstrates higher overall attainment, particularly in PO1 at 2.62, PO2 at 2.09, PO3 at 1.92 and PO9 at 2.09. Overall, the integrated course exhibits the most balanced attainment across various POs compared to the theory and laboratory courses.

Table 7.3 Assessment plan.

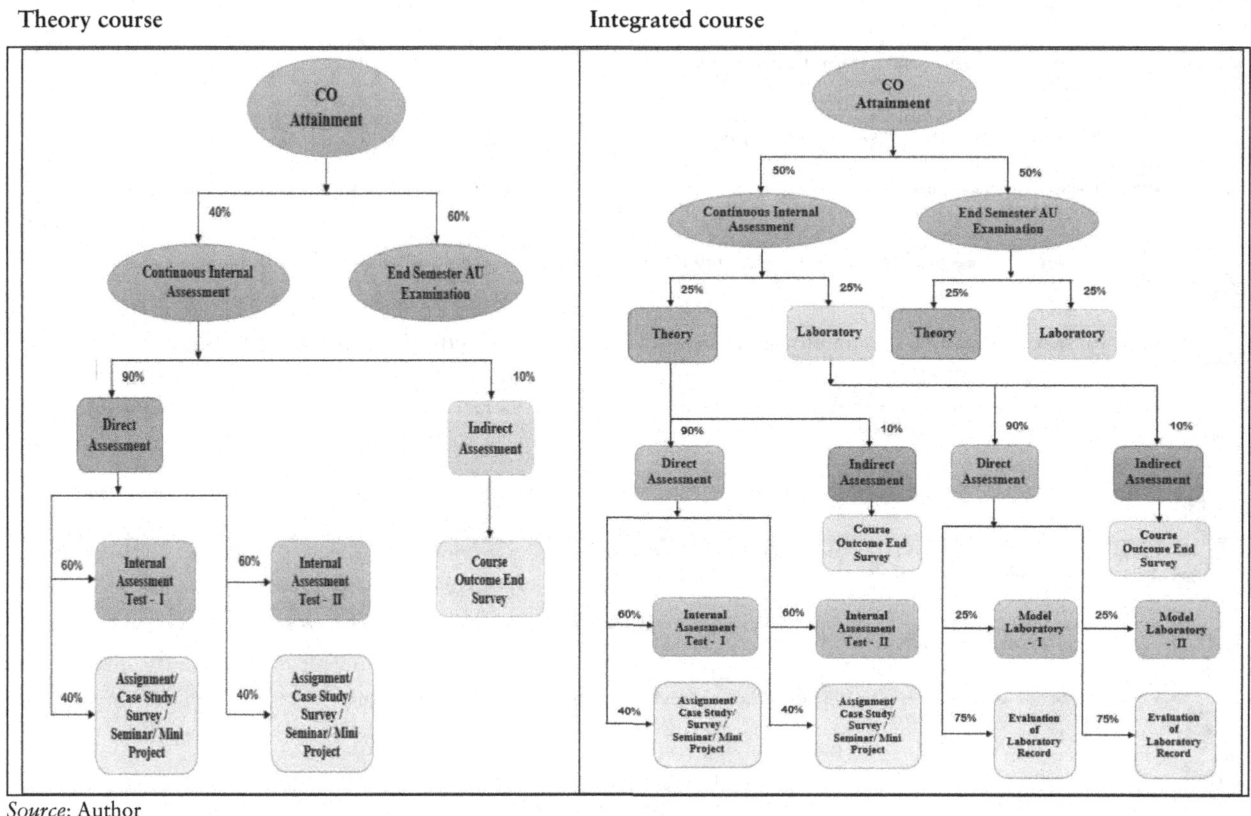

Source: Author

Table 7.4 Comparative analysis of PO attainment.

PO attainment	1	2	3	4	5	6	7	8	9	10	11	12
Theory course	1.78	1.78	1.78	-	-	-	-	-	-	0.89	-	0.89
Laboratory course	1.84	-	-	-	0.92	-	-	0.92	1.84	1.84	-	-
Integrated course	2.62	2.09	1.92	1.22	1.22	-	-	1.75	2.09	1.75	-	1.75

Source: Author

Sample course – 2

The project-based course is considered as a sample course – 2 which aimed at evaluating student performance in alignment with all 12 POs. The conventional methods of assessment often fall short in comprehensively measuring the attainment of program outcomes. Predominantly depending upon few criteria such as presentation skill and project completion, these assessments may neglect critical aspects which are essential for graduates to fully achieve in alignment with POs. Recognizing this deficiency, there arises there is a need to devise and implement a well-structured rubrics tailored to the specific goals and outcomes of the program. By integrating detailed rubrics on a 10-point scale into the assessment process, academicians can provide students with clearer guidelines and criteria, facilitating a more thorough evaluation of their performance. Consequently, the impact of implementing rubrics on the attainment of program outcomes by comparing the effectiveness of assessment before and after their introduction is presented in this section. The course outcomes for the project work are presented in Figure 7.1. In alignment with the curriculum the assessment process encompasses three distinct project reviews where strategically designed to evaluate student progress across the spectrum of POs from 1 to 12. These reviews serve as pivotal checkpoints throughout the project work. By allowing for targeted assessment of specific competencies and skill development the rubrics strengthened the evaluation process (Figure 7.2).

To ensure alignment with all POs, dedicated rubrics are meticulously crafted for each review, defining clear criteria and performance expectations from students.

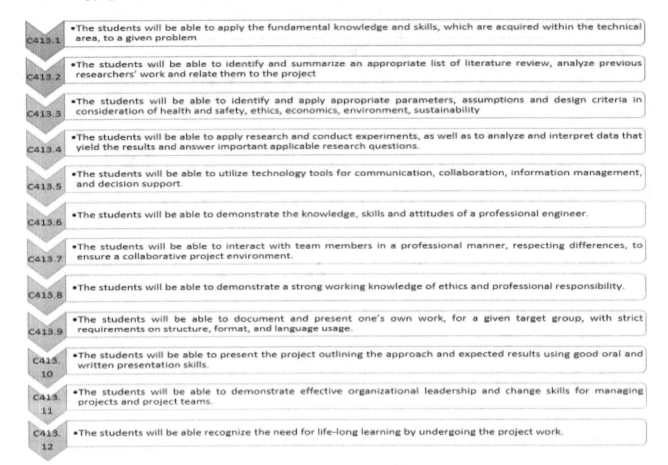

C413.1 • The students will be able to apply the fundamental knowledge and skills, which are acquired within the technical area, to a given problem

C413.2 • The students will be able to identify and summarize an appropriate list of literature review, analyze previous researchers' work and relate them to the project

C413.3 • The students will be able to identify and apply appropriate parameters, assumptions and design criteria in consideration of health and safety, ethics, economics, environment, sustainability

C413.4 • The students will be able to apply research and conduct experiments, as well as to analyze and interpret data that yield the results and answer important applicable research questions.

C413.5 • The students will be able to utilize technology tools for communication, collaboration, information management, and decision support.

C413.6 • The students will be able to demonstrate the knowledge, skills and attitudes of a professional engineer.

C413.7 • The students will be able to interact with team members in a professional manner, respecting differences, to ensure a collaborative project environment.

C413.8 • The students will be able to demonstrate a strong working knowledge of ethics and professional responsibility.

C413.9 • The students will be able to document and present one's own work, for a given target group, with strict requirements on structure, format, and language usage.

C413.10 • The students will be able to present the project outlining the approach and expected results using good oral and written presentation skills.

C413.11 • The students will be able to demonstrate effective organizational leadership and change skills for managing projects and project teams.

C413.12 • The students will be able recognize the need for life-long learning by undergoing the project work.

Figure 7.1 List of COs – Project work course
Source: Author

Course Code & Title : EE8811 Project Work (Before/After using rubrics); Course Index : C413												
CO/PO	1	2	3	4	5	6	7	8	9	10	11	12
C413.1	2/3	1/1										
C413.2	2/2	2/3										
C413.3	1/2	1/2	3/3				1/1	1/1			1/1	
C413.4	2/2	2/2	1/2	3/3								
C413.5	2/2	2/2	2/2	1/1	3/3				1/1	1/1		
C413.6	2/2	1/2	1/1	1/1	1/-	2/3						
C413.7	2/2	2/2	1/1	1/1		2/2	1/2		1/1	1/1		
C413.8	2/2	2/2	1/2	1/1		1/1		1/3				
C413.9	2/2	2/2	1/1	1/1	1/1			1/1	2/3	2/2	1/1	
C413.10	2/2	2/2	1/2	1/1	1/1			1/1	2/2	2/3	1/1	
C413.11	2/2	2/2	2/2	1/1	1/1		1/1	1/1	2/2	2/2	2/3	
C413.12	1/1	1/1	1/1	1/1	1/1		1/1		1/1	1/1	1/1	3/3
C413	1.83/ 2	1.5/2	1.4/ 1.7	1.22/ 1.2	1.33/ 1.4	1.67/ 2	1/ 1.25	1/ 1.4	1.5/ 1.6	1.5/ 1.6	1.2/ 1.4	3/3
PO Attainment	1.83/ 2.0	1.5/ 2.0	1.4/ 1.7	1.22/ 1.2	1.33/ 1.4	1.67/ 2.0	1/ 1.3	1/ 1.4	1.5/ 1.6	1.5/ 1.6	1.2/ 1.4	3/ 3

Figure 7.2 Attainment of PO
Source: Author

Table 7.5 Details of sample rubrics for PO1 and PO2.

Program outcome	Assessment criteria	Exceeds expectations (7–10)	Meets expectations (4–6)	Partially meets expectations (0–3)
Engineering knowledge	Application of engineering principles and concepts	Correctly applied and utilized the engineering principles, demonstrating an excellent grasp of the factors and constraints involved	Sufficiently applied and utilized the engineering principles, demonstrating the factors and constraints involved at accepted level	Insufficient use and application of engineering principles. Incomplete demonstrating of the factors and constraints
Problem analysis	Literature review	Clear and complete details, relevant, specific and thorough supporting evidence from literature	Sufficient details, adequate supporting evidence from literature	Sufficient details, minimum supporting evidence
	Identification and definition of problem	Describe (or sketch out) the problem and its major components to be examined. Complete definition and description of the project	Describe (or sketch out) the problem and its major components to be examined. Defines the problem inaccurately and/ or incompletely	Recognize that there is a problem or concept that needs to be solved. Defines the problem inaccurately and/or incompletely
	Identification of alternate solutions	Clear identification and evaluation of the alternatives	Accepted level of identification and evaluation of the alternatives	Few identification and evaluation of the alternatives
	Project objectives	Objectives clear, focused and innovative	Objectives present but not clear, focused or made explicit	Objectives not clearly stated or inappropriate
	Project specifications and constraints	Clearly list project specifications and offers realistic constraints	Identifies and list project specifications and offers constraints	Little understanding of the project specifications and its constraints
	Problem solving approach	Focuses the analysis on the desired result	Gathers information in an appropriate form	Information gathering is somewhat unorganized, but relevant
	Project work plan/ progress of work	Work plan clearly stated	Work plan exists but not clear	Work plan not clearly stated

Source: Author

A sample rubric for the assessment of PO1 and PO2 for a review is given in Table 7.5. It is also noted that, a similar rubric has also been designed for covering all the twelve POS. This rubric provides transparency for educators and students with assessment and feedback. Through this structured approach, the integration of tailored rubrics aims to enhance the precision and consistency of evaluation, ultimately nurturing a more robust and holistic assessment of student attainment of a program outcomes throughout the project course.

Before the implementation of rubrics, the assessment criteria primarily revolved around evaluating presentation skills, technical knowledge, project completion percentage, and the quality of publications derived from the projects. While these factors provided a baseline for assessment, they often fell short in capturing the holistic development of students' knowledge, skills and attitudes essential for POs. However, following the introduction of rubrics tailored to assess these multifaceted dimensions, significant improvements in PO attainment have been observed. This transformation emphasizes the effectiveness of incorporating rubrics, not only in elevating the consistency and correctness of evaluation but also in development of student performance and progression toward POs. This transparent evaluation of the project work will improve their practical knowledge, placement, publication, and other tangible outcomes.

Conclusions

This paper investigates into strategies aimed at enhancing the attainment of POs for engineering graduates since it is crucial for aligning education with industry needs. This paper proposed the change in type of course that is integrating theory and practical components within courses and refining assessment practices to better map COs to POs. The study conducts a comparative analysis between conventional theory courses and integrated courses, revealing the effectiveness in achieving higher overall attainment across various POs. Furthermore, a novel review system for assessing PO attainment in project course is introduced, which provided a transparent and effective assessment of students' knowledge, skills, and attitudes. By adopting these comprehensive strategies, academic institutions can better equip engineering graduates with the necessary competencies to excel in today's dynamic professional life.

References

Alias, M., Lashari, T. A., Akasah, Z. A., and Kesot, M. J. (2014). Translating theory into practice: integrating the affective and cognitive learning dimensions for effective instruction in engineering education. *Eur. J. Engg. Edu.*, 39(2), 212–232.

Arunkumar, S., Sasikala, S., and Kavitha, K. (2018). Towards enhancing engineering education through innovative practices in teaching learning. *Internat. J. Engg. Adv. Technol.*, 8(2), 153–159.

Becker, K. H. and Park, K. (2011). Integrative approaches among science, technology, engineering, and mathematics (STEM) subjects on students' learning: A meta-analysis. *J. STEM Edu. Innov. Res.*, 12(5), 23–37.

Crawley, E., Malmqvist, J., Ostlund, S., Brodeur, D., and Edstrom, K. (2007). Rethinking engineering education. *The CDIO App.*, 302(2), 60–62.

Cruz, M. L., Saunders-Smits, G. N., and Groen, P. (2020). Evaluation of competency methods in engineering education: A systematic review. *Eur. J. Engg. Edu.*, 45(5), 729–757.

Falloon, G. (2020). From digital literacy to digital competence: the teacher digital competency (TDC) framework. *Edu. Technol. Res. Dev.*, 68(5), 2449–2472.

Muhsin, M. A. A. and Ahmad, N. (2019). The emergence of Education 4.0 trends in teaching Arabic Islamic finance curriculum design: A case study. *Internat. J. Psychosoc. Rehab.*, 23(4), 1019–1029.

Kelley, T. R. and Knowles, J. G. (2016). A conceptual framework for integrated STEM education. *Internat. J. STEM Edu.*, 3, 1–11.

Margot, K. C. and Kettler, T. (2019). Teachers' perception of STEM integration and education: A systematic literature review. *Internat. J. STEM Edu.*, 6(1), 1–16.

Miranda, J., Navarrete, C., Noguez, J., Molina-Espinosa, J. M., Ramírez-Montoya, M. S., Navarro-Tuch, S. A., Bustamante-Bello, M. R., Rosas-Fernández, J. B., and Molina, A. (2021). The core components of education 4.0 in higher education: Three case studies in engineering education. *Comp. Elec. Engg.*, 93, 107278.

Nawab, M., Alblawi, A., Alsayyari, A., and Alharbi, S. (2020). Bridging the gaps in engineering curriculum through systems engineering approach. *2020 IEEE Global Engg. Edu. Conf. (EDUCON),* 1230–1236.

Patel, N. H., Franco, M. S., and Daniel, L. G. (2019). Student engagement and achievement: A comparison of STEM schools, STEM programs, and non-STEM settings. *Res. Schools*, 26(1), 1–11.

Quelhas, O. L. G., Lima, G. B. A., Ludolf, N. V. E., Meiriño, M. J., Abreu, C., Anholon, R., and Rodrigues, L. S. G. (2019). Engineering education and the development of competencies for sustainability. *Internat. J. Sustain. Higher Edu.*, 20(4), 614–629.

Sharma, R. (2019). Globalization issues and their impacts on engineering education. Edited by Mohsen, J. P., Mohamed, Y. I., Hamid, R. P. and Waldemar, K. *Global Adv. Engg. Edu.*, 93–100.

Teixeira, R. L. P., Silva, P. C. D., Shitsuka, R., de Araújo Brito, M. L., Kaizer, B. M. and e Silva, P. D. C. (2020). Project based learning in engineering education in close collaboration with industry. *2020 IEEE Global Engg. Edu. Conf. (EDUCON),* 1945–1953.

Van den Beemt, A., MacLeod, M., Van der Veen, J., Van de Ven, A., Van Baalen, S., Klaassen, R., and Boon, M. (2020). Interdisciplinary engineering education: A review of vision, teaching, and support. *J. Engg. Edu.*, 109(3), 508–555.

Zeidmane, A. and Cernajeva, S. (2011). Interdisciplinary approach in engineering education. *2011 IEEE Global Engg. Edu. Conf. (EDUCON),* 1096–1101.

8 Improving educational assessment: Evaluating student performance with gradient boosting regression and comparative analysis with linear regression to minimize mean squared error

P. Lakshmi Kiran Reddy[a] *and R. Jeena*[b]

[1]Department of CSE, Saveetha School of Engineering, Saveetha Institute of Medical and Technical Sciences, Chennai, India

Abstract

This study's objective is to evaluate students' performance using gradient boosting on performance and educational data, and to assess how well gradient boosting reduces the mean squared error (MSE) in comparison to linear regression. Two groups, each with twenty samples, were chosen for analysis in this study. G-power was used to determine the sample size, with a target pre-test score of 80% and a 95% confidence interval. This approach uses machine learning (ML) techniques, notably gradient boosting and linear regression that were trained on the collected data. The gradient boosting approach fared better than the linear regression model, according to the findings, which had an MSE of 2.46, with an MSE of 0.20. The statistical significance of gradient boosting and linear regression was evaluated using an independent sample T-test, which produced a p-value of $p=0.001$ ($p<0.05$). This implies that there is a statistically significant performance difference between the two models. This work highlights how well regression-based ML techniques—specifically, the gradient boosting and linear regression algorithms—predict student performance data. In terms of predicting accuracy, the gradient boosting algorithm outperforms the linear regression algorithm thanks to the incorporation of an autoregressive integrated moving average framework.

Keywords: Student performance, supervised learning, regression analysis, performance evaluation, education

Introduction

An essential component of contemporary education is the student performance evaluation, which acts as a tool for tracking and evaluating students' behavioral patterns, academic progress, and learning capacities. This procedure is very important in all kinds of educational settings, such as colleges, universities, and schools because it gives educators and other stakeholder's important information about students' educational paths and helps them make decisions that will help them grow. Student performance can be predicted using regression techniques like linear regression or more sophisticated approaches like gradient boosting, taking into account a variety of input parameters such academic records, attendance, extracurricular activities, and demographic data. Student performance evaluation has several useful applications that are implemented in various areas of the education sector.

Student performance data is used by educational institutions to monitor the progress of individual students, pinpoint areas that require development, and successfully modify their teaching methods to suit each student's individual needs. Performance data is used by government agencies and educators to evaluate the efficacy of programmes and policies in education, which leads to adjustments in the educational system and methods of allocating resources. The examination of student performance has given rise to numerous research projects. One noteworthy study (Alsubihat and Al-shanableh, 2023) predicts student performance using a blended heterogeneous model. In this research, several contemporary algorithms are compared and examined to predict students' academic achievement. All of the classifiers showed accuracy in its predictions through the analysis of the data. But out of all the classifiers tested, the CatBoost algorithm turned out to be the most accurate. Performance prediction using the gradient boosting approach is another important piece of work (Mahaboob Basha, et al., 2021). Two-fold is used in this work.

First, in an attempt to get over this restriction, it is investigated if framing the issue as a binary classification could lead to a more accurate prediction of low-performing pupils. Second, several human-interpretable features that quantify these factors are developed in order to obtain insight into the possible causes of subpar performance. Boosting algorithms have been used to assess student performance (Hamim

[a]pidaparthireddy1681.sse@saveetha.com. [b]jeenar.sse@saveetha.com

DOI: 10.1201/9781003587538-8

et al., 2022). The authors of this work give a comparative analysis of several boosting algorithms that have proven effective in a variety of fields. Furthermore, in order to enhance the preprocessing task and boost the models' performance, the authors integrated recursive feature elimination with Fisher score and information gain feature selection techniques. The results of the study demonstrate that while employing the information gain with recursive feature elimination strategy to predict students' performance in mathematics using a multi-label dataset, the light gradient boosting machine (LightGBM) algorithm outperformed other boosting algorithms.

Materials and methods

The SIMATS' Saveetha School of Engineering's Machine Learning Laboratory served as the study's location. Gradient boosting and linear regression models were used in this study. There were 20 samples in each group, for a total of 40 samples. Based on a 95% confidence interval and an 80% Gpower pre-test score, the sample size was determined. The compiler used to detect and diagnose gastrointestinal disorders was Python. The stats models Python library was used to carry out the statistical analysis. This dataset offers insightful information about secondary school student accomplishment at two Portuguese schools. It includes a wide range of attributes, such as social factors, student marks, demographic data, and elements pertaining to the institution.

With the use of questionnaires and school reports, the data was gathered to provide a thorough picture of student achievement. There are two separate datasets available, one concentrating on student performance in Portuguese (por) and the other on mathematics (mat). These datasets were utilized for regression and binary/five-level classification tasks (Cortez and Silva, 2008). The goal property G3, has a substantial relationship through qualities G2 and G1, or grades from the second and first periods, respectively. This is an important finding. All things considered, this dataset offers scholars and educators a wealth of data that they may use to better understand and raise student accomplishment in secondary education (Friedman, 2001).

Gradient boosting

In machine learning, gradient boosting is an effective ensemble learning method that may be applied to both regression and classification issues (Cortes and Vapnik, 1995). The working principle of this model is combining the predictions of several different models, typically decision trees, to create a model that is more precise and dependable.

Step 1: Load the data from the student evaluation. pd data = pd.read_csv("student-per.csv") is imported as pandas.

Step 2: Prepare the information Encode the category values on the label.

Step 3: Separate the dataset into sets for testing and training. train size is equal to int(len(data) * 0.8). data.iloc[:train_size], data.iloc[train_size:] = train_data, test_data

Step 4: Fit the gradient boosting model from the import of sklearn.ensemble in step four. Gradient BoostingGradientBoostingRegressor (n estimators = 100, learning rate = 0.1, max depth = 3, random state = 42) is the regression model. train_data["value"]), model.fit(train_data.drop(columns=["value"]), train_data)

Step 5: Use sklearn.metrics to validate the model. import the error_squared print ("Mean Squared Error:", mse) = mean_squared_error(test_data["value"], model.predict(test_data.drop(columns=["value"]))

Step 6: Project future values forecast_future = model.predict(data_future) print ("Forecasted values for subsequent steps:", future_forecast)

Linear regression

The statistical technique known as linear regression is used to model the relationship between a target or dependent variable and one or more predictors or independent variables (Morsy and Karypis, 2017). The independent and dependent variables are thought to be related linearly. The best-fitting straight line, or hyperplane in higher dimensions, that minimizes the difference between the model's projected values and the actual values is the target of linear regression.

Pseudo code

Step 1: Import pandas as pd data = pd.read_csv("student-per.csv") in order to load the student evaluation data.

Step 2: Prepare the data # Label encode the sklearn categorical values.preparing an import LabelEncoder data.select_dtypes(include=["object"]) categorical_columns = LabelEncoder() label_encoder.columns for col in categorical_columns: label_encoder.fit_transform(data[col]) = data[col]

Step 3: Using sklearn.model_selection import train_test_split, divide the dataset into training and testing datasets (step 3). train_data, test_data = train_test_split(data, test_size=0.2, random_state=42)

Step 4: Fit the linear regression model using the sklearn.linear_model import in step four.

Model linear regression = Model linear regression().fit(columns=["value"] in train_data.drop), train_data[Value])

Step 5: Use sklearn.metrics to validate the model. import the error_squared Model.predict(test_data.drop(columns=["value"])) = mean_squared_error(test_data["value"])) print(mse, "Mean Squared Error:")

Step 6: Project future values # Future forecasting is not a feature of linear regression by nature. Its main application is in prediction using current data.

Statistical analysis

This study's statistical analysis made use of the stats models Python library, a popular statistical computing tool. This library was used for a number of studies, such as calculating mean deviation and standard deviation (SD), establishing significant levels, running independent sample T-tests, and creating graphical representations. Specifically, the performance of the innovative gradient boosting and linear regression models were compared (Knowles, 2015). The study had 20 samples per group, and the primary aim was to identify disease-affected areas with higher accuracy rates (input parameters). The independent factors were connected to gastrointestinal illnesses. The analysis of these variables was done to see how well the linear regression and gradient boosting models performed.

Results and discussions

The project's conclusion shows that, in comparison to linear regression, gradient boosting performs better. The mean linear regression MSE is 2.46 and the mean gradient boosting MSE is 0.20 in the context of student performance analysis (Kotsiantis et al., 2004). These metrics show how well the algorithms predict student performance data, as well as how accurate those models are at each level (Iqbal et al., 2017).

The MSE comparison between the suggested and standard approaches is shown in Table 8.1. The group's mean and standard deviation are shown in Table 8.2, and the MSEs for the linear regression and gradient boosting were 0.05, 0.0177, and 0.52, 0.16, respectively. In contrast, the linear regression had an error of 0.167 and the gradient boosting had an error of 0.017.

The independent sample T-test results for DLinear regression and gradient boosting are displayed in Table 8.3. A statistically significant result of 0.001 ($p<0.05$) was obtained (Ashraf et al.).

Table 8.1 MSE comparison of gradient boosting (0.20) and linear regression (2.46).

S. No.	Test size	MSE rate	
		GB	LR
1	Test 1	0.2	2.9
2	Test 2	0.19	1.85
3	Test 3	0.32	2.38
4	Test 4	0.18	3.77
5	Test 5	0.24	2.18
6	Test 6	0.23	2.12
7	Test 7	0.13	2
8	Test 8	0.19	2.45
9	Test 9	0.24	2.26
10	Test 10	0.12	2.74
Average test results		0.20	2.46

Source: Author

Table 8.2 The mean MSE computation of gradient boosting and linear regression models.

Group		N	Mean	SD	SEM
Accuracy rate	GB	10	0.20	0.05	0.017
	LR	10	2.46	0.52	0.167

Source: Author

Table 8.3 Comparison between the gradient boosting and linear regression.

Group/accuracy	Levene's test for equality of variances		T-test for equality of means			
	F	Sig.	t	Df	Mean difference	Std. error difference
Equal variances assumed	162	1.92	12.7	15	2.26	0.177
Equal variances assumed	162	1.92	12.7	15	2.26	0.177

Source: Author

Table 8.4 Group statistics.

Group N		Mean	Std. deviation	Std. error Mean
Accuracy	GB	81.449	3.23219	1.02211
	LR	79.877	2.70546	0.85554

Source: Author

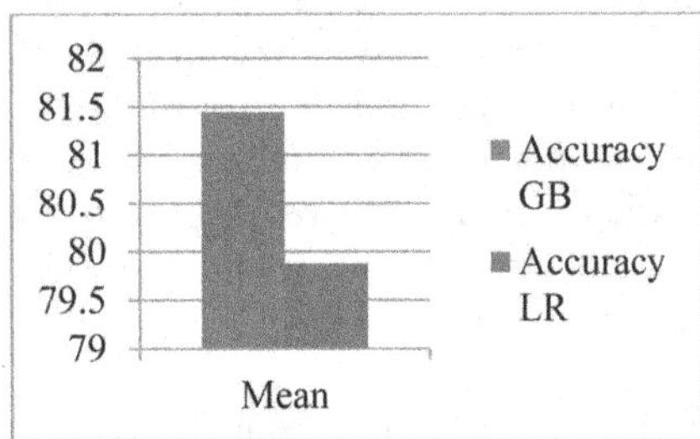

Figure 8.1 Mean MSE comparison: Gradient boosting versus linear regression model
Source: Author

The MSE and mean MSE calculations for the suggested over-selected input and the traditional technique are shown in Figure 8.1. Compared to the current method, which yielded a mean MSE of 2.46, the proposed method achieved a mean MSE of 0.20.

In forecasting and analyzing time-dependent data, the project emphasizes the value of using student performance evaluation analysis approaches, particularly gradient boosting (autoregressive integrated moving average) and linear regression (seasonal gradient boosting) models (Ke et al., 2017).

The mean linear regression MSE in this study is 2.46, while the mean gradient boosting MSE (mean squared error) is reported as 0.20.

The MSE of gradient boosting model is marginally lower than that of linear regression models, which are distinguished by their capacity to capture linear trends and temporal dependencies. Table 8.4 shows the mean, standard deviation and standard error mean of GB and LR.

In contrast, a more sophisticated method of forecasting student performance evaluations is offered by linear regression models, which include seasonal components to account for sporadic swings in the data. Linear regression methods improve predicted accuracy by directly modeling seasonal fluctuations, but gradient boosting models are more straightforward and easily interpreted when attempting to identify underlying trends in student performance evaluation data.

The choice between the gradient boosting and linear regression models depends on the specific features of the data from the student performance evaluation and the analysis's objectives.

In order to further increase forecasting accuracy and robustness, future research projects may investigate hybrid techniques that combine the advantages of linear regression and gradient boosting models (Lal and Kumar, 2017). Furthermore, both gradient boosting and linear regression models' performance can be improved in a variety of student performance evaluation predicting tasks by advances in feature engineering and model optimization techniques (McFarland et al., 2017). In summary, this research advances the analysis approaches for student performance evaluations and sheds light on how well gradient boosting and linear regression models predict temporal patterns in data (Motazm Khorshid et al., 2015) x.

Conclusions

With the use of gradient boosting (autoregressive integrated moving average) and linear regression (seasonal gradient boosting) models, this study presents a thorough framework for evaluating student performance. The models' predictive accuracy in accurately projecting student performance evaluation data is indicated by their respective mean gradient boosting MSE of 0.20 and mean linear regression MSE of 2.46. The efficiency of both gradient boosting and linear regression models in identifying temporal correlations and seasonal fluctuations in the data is demonstrated by their comparison (Jeena and Logesh, 2022). Gradient boosting models are a good choice for capturing linear trends and non-seasonal patterns because of their simplicity and interpretability. On the other hand, by adding seasonal components, linear regression models increase the power of gradient boosting and allow for more accurate modeling of periodic variations in the data from student performance assessments. Both the gradient boosting and linear regression models exhibit strong performance, highlighting their usefulness for a range of forecasting tasks related to student performance evaluations. The predicted accuracy and resilience of these models could be improved by more research into model parameters and optimization strategies, guaranteeing its use in practical contexts. Furthermore, hybrid methods that combine the benefits of linear regression and gradient boosting models can provide better predicting performance for intricate datasets used in student performance assessments. Overall, this study advances the methods for evaluating student performance and shows that linear regression and gradient boosting work well for predicting and analyzing temporal trends in data. Subsequent investigations can concentrate on enhancing these models and investigating innovative methods to tackle particular difficulties in applications related to forecasting student performance assessments.

References

Alsubihat, D. and Al-shanableh, N. (2023). Predicting student's performance using combined heterogeneous classification models. *Internat. J. Engg. Res. Appl.*, 13. 206–218. 10.9790/9622-1304206218.

Mahaboob Basha, Sk., Preethi, K. S. S. S., Pooja, M., and Meghana, G. (2021). Classifying students performance using gradient boosting algorithm technique. *UGC Care Group I Listed J.*, 11(01), 364–373.

Hamim, T., Benabbou, F., and Sael, N. (2022). Student profile modeling using boosting algorithms. Internat. *J. Web-Based Learn. Teach. Technol.*, 17, 1–13. 10.4018/IJWLTT.20220901.oa4.

Cortez, P. and Silva, A. (2008). Using data mining to predict secondary school student performance. In Brito, A. and Teixeira, J. (eds.). *Proc. 5th FUt. BUsin. TEChnol. Conf. (FUBUTEC 2008)*, 5–12. ISBN 978-9077381-39-7.

Friedman, J. H. (2001). Greedy function approximation: A gradient boosting machine. *Ann. Stat.*, 1189–1232.

Cortes, C. and Vapnik, V. (1995). Support-vector networks. *Mach. Learn.*, 20(3), 273–297.

Morsy, S. and Karypis, G. (2017). Cumulative knowledge-based regression models for next-term grade prediction. *Proc. 2017 SIAM Internat. Conf. Data Min.*, 552–560.

Knowles, J. E. (2015). Of needles and haystacks: Building an accurate statewide dropout early warning system in Wisconsin. *J. Edu. Data Min.*, 7(3), 18–67.

Kotsiantis, S., Pierrakeas, C., and Pintelas, P. (2004). Predicting students' performance in distance learning using machine learning techniques. *Appl. Artif. Intell.*, 18(5), 411–426.

Iqbal, Z., Qadir, J., Mian, A. N., and Kamiran, F. (2017). Machine learning based student grade prediction: A case study. arXiv preprint arXiv:1708.08744, 1–22.

Jeena, R. and Logesh, R. (2022). Optimum selection of virtual machine using improved particle swarm optimization in cloud environment. *Internat. J. Comp. Netw. Appl. (IJCNA)*, 9(1), 125–134. DOI: 10.22247/ijcna/2022/211631, 125–134.

McFarland, J., Hussar, B., de Brey, C., Snyder, T., Wang, X., Wilkinson-Flicker, S., Gebrekristos, S., Zhang, J., Rathbun, A., Barmer, A., et al. (2017). Undergraduate retention and graduation rates. *Condition Edu. 2017*. 144.

Ashraf, A., Anwer, S., and Khan, M. G. A comparative study of predicting student's performance by the use of data mining techniques. *Am. Sci. Res. J. Engg. Technol. Sci.*, 2018, 122–136.

Ke, G., Meng, Q., Finley, T., Wang, T., Chen, W., Ma, W., Ye, Q., and Liu, T. Y. (2017). Lightgbm: A highly efficient gradient boosting decision tree. , 31st Conference on Neural Information Processing Systems (NIPS 2017), Long Beach, CA, USA., 1–9.

Lal, A. and Kumar, C. R. S. (2017). Hybrid classifier for increasing accuracy of fitness data set. *IEEE 2017 2nd Internat. Conf. Converg. Technol.* pp. 929

Motazm Khorshid, M. H., Tarek, H. M., Abou-El-Enien, Ghada, M. A., and Soliman. (2015). Hybrid classification algorithms for terrorism prediction in the Middle East and North Africa. *Internat. J. Emerg. Trends Technol. Comp. Sci.* 23–29.

9 Improving accuracy in assessing student adaptability in online education: A comparative study of random forest and decision tree models

G. NithinKumar[a] *and Mahaveerkannan R.*[b]

Department of Computer Science and Engineering, Saveetha School of Engineering, Saveetha Institute of Medical and Technical Sciences, Saveetha University, Chennai-602105, Tamil Nadu, India

Abstract

Aim: The objective of this study is to assess the performance and accuracy of prediction-oriented modeling tools, notably decision tree (DT) and random forest (RF), in evaluating students' adaptability. Materials and methods: The experiment entailed the cooperative effort of two parties to carry out the plan. The main algorithms utilized were DT and RF. However, only 1000 out of the total 1206 samples were employed in this investigation. The key parameters consisted of a pre-test G-power of 80%, a confidence interval (CI) of 95%, and a sample size of 0.05%. Result: The shift to online instruction has necessitated those pupils to adjust. The recent improvement to RF resulted in an accuracy of 87.87%, whereas the current DT technique produced a slightly lower accuracy of 87.20%. The acquired data has substantial consequences for statistical analysis. Significantly, the p-value (0.799) is well below the threshold of 0.05. Conclusion: Educational institutions are struggling to adjust to the requirements of the "New Economy." The RF algorithm surpasses the DT technique in accurately forecasting students' adaptability.

Keywords: Random forest, decision tree, student, prediction, online education, class, instruction, educational, learning

Introduction

In order to enhance their readiness for employment, students in vocational high schools are encouraged to cultivate valuable skills and adopt a mindset focused on personal and professional development. It is expected that students will secure employment that is directly related to their chosen career path upon graduation. The source of this information is the UNESCO International Centre for Technical and Vocational Education and Training in the year 2021. The number of students who successfully finish vocational high school programs in each skill competency directly impacts the imbalance between supply and demand in the labor market, as well as the level of competitiveness in that market. The high unemployment rate among those with a Vocational High School certificate can be attributed to this particular factor. Students encounter significant levels of stress, ambiguity, and uncertainty when they begin to explore and select a career path. Simultaneously fulfilling the required courses to initiate one's professional journey, workplace flexibility involves being adaptable to manage alterations, novel obligations, and unforeseen circumstances (Mahboob et al., 2016). The occupation of any individual is of utmost importance in their lives. Healy suggests that an individual's career can be divided into three distinct stages: preoccupation, occupational, and post occupation. This perspective aligns with Super's concept that an individual's professional trajectory adheres to a predictable pattern spanning from birth to death (Schiffer, 2016). Various reasons, including migration, technological advancements, international competition, changing markets, environmental issues, and transnational politics, contribute to the increasing complexity of the ever-changing nature of professional obstacles (Tarkar, 2020).

The prevalence of pneumonia has contributed to the growing popularity of online schooling. Nevertheless, students are not finding pleasure in engaging in online learning activities due to the substantial disparity from conventional classroom instruction. Multiple research (Hofer et al., 2021) has discovered that students encounter difficulties in adapting to online education. Google Scholar retrieved 100 publications on this topic, whilst IEEE Explore generated 62 papers. The implementation of lockdown measures in response to the virus, along with the subsequent social isolation, has caused numerous institutions to transition their courses to an online format. Even prestigious universities have started providing online courses (Mahboob et al., 2016). Nevertheless, these hurriedly executed online classes are failing to produce the intended outcomes. Studies, such as the one conducted (Sahlaoui

[a]192011172.sse@saveetha.com, [b]mahaveerakannanr.sse@saveetha.com

DOI: 10.1201/9781003587538-9

et al., 2024), demonstrate that students' academic achievement is negatively impacted when they engage in online learning as opposed to attending a conventional classroom environment. In addition, online learning has a substantial influence on students' comprehension of the subject matter and their capacity to independently manage their learning process (Vijayalakshmi and Venkatachalapathy, 2019). Students from low-income families may face significant challenges throughout the shift to online learning due to their limited access to the essential resources (Nnadi et al., 2024). The current wave of globalization is partially propelled by previous transformations, including recent technological and economic progressions such as the internet, industrial automation, and widespread utilization of digital technology (Dass et al., 2021). The current process of adjusting to previous changes is currently being impacted by emerging disruptive technologies such as genetic breakthroughs, the Internet of Things (IoT), artificial intelligence (AI), advanced robotics, materials science, 3D printing, and additional advancements in IoT (López Zambrano et al., 2021).

In light of the extensive transmission of the COVID-19 virus, certain educational institutions have initiated the provision of courses exclusively through online platforms (Sharrock and Parkerson, 2020). The advent of online teaching presents educators, education officials, and educational institutions with fresh prospects, alongside pupils, according to the International Association for the Evaluation of Educational Achievement and UNESCO (2022). Scientists determined that decision tables were the most effective when evaluating forecast accuracy, response speed, and confidence. Prior research has mostly concentrated on DTs that incorporate rules (Orabona, 2019). According to the Tarkar (2020), this study suggests that despite the numerous benefits of online learning, high school students are not performing well due to the current atmosphere of high academic stress, and would benefit more from a traditional classroom environment.

Materials and methods

The Saveetha Institute of Medical and Technical Sciences' programming division was in charge of conducting the experiment. There exist two overarching categories of individuals. Afterwards, a RF algorithm was used, and then a DT algorithm was applied. The sample size was obtained using the clinical software. We created a sample size calculator using data from previous studies. The calculator has a power of 80% and a confidence level of 95%, with a threshold set at 0.05.

Other available choices consist of a 512 MB RAM Intel Core i5 CPU, Windows 11 Professional operating system, Jupyter Notebooks, and Google Colab. In order to assess accuracy, the data obtained from Kaggle was analyzed using IBM SPSS. After training with 1000 data points, the performance accuracy of both algorithms was assessed using an independent T-test in SPSS.

Random forest

Within the field of machine learning, RF is a technique to supervised learning that has received a considerable degree of respect within the field. It is commonly utilized in tasks like as categorization and prediction, to name just two examples of the same sort of activity. The application of ensemble learning makes it possible to improve the performance of models and to find solutions to problems that are difficult to solve. This technique makes use of a wide variety of distinct classifiers across the board. The RF classifier is a type of classifier that is comprised of a large number of DTs that have been trained on different subsets of the dataset during the training process. For the purpose of improving the accuracy of the dataset, the predictions that are generated by these DTs are then utilized. In contrast to relying on a single DT, RF takes into consideration all of the predictions and selects the one that happens the most frequently in order to decide the final result. This is done in order to ensure that the best possible outcome is achieved.

Flowchart

Ionospheric spread-F (SF) is a frequently observed event in the F-layer, characterized by disruptions in electron density and known for its frequent incidence. Aberrations in the structure of the ionosphere impede the propagation of electromagnetic waves. Automated identification of ionospheric spread-F and statistical analysis of its development are extremely useful for studying the physical processes of ionospheric inhomogeneity and forecasting ionospheric anomalies. Both of these traits hold great significance, and when taken together, they are exceedingly crucial. This article explores and formulates three different methods for automatically detecting spread-F, utilizing several methods of machine learning, including DT, RF, and CNN, among others (Figures 9.1 and 9.2).

Random forest algorithm

Step 1: Upload the info that you prepared (labels y, characteristics x).

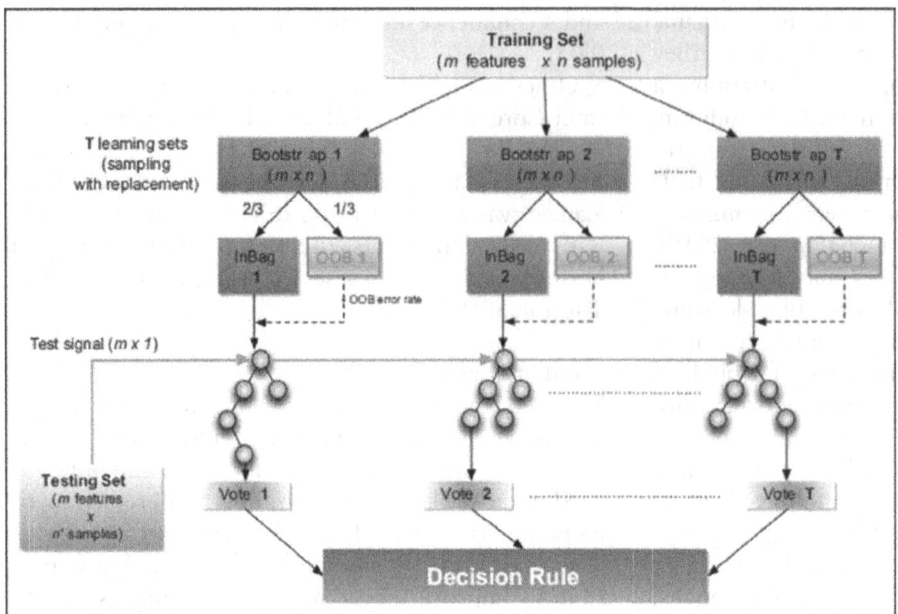

Figure 9.1 Ionospheric spread-F (SF)
Source: Author

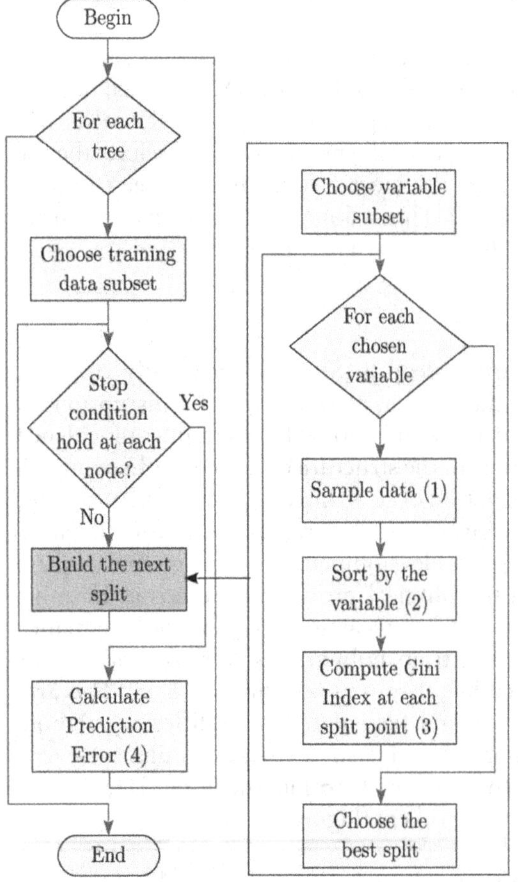

Figure 9.2 RF algorithm to evaluate the model's performance
Source: Qiu, B., & Fan, W. (2021)

Step 2: Extract the historical data (y_train, y_test, X_train, and X_deployment are your options) into testing and training sets.

Step 3: Create a RF classifier and configure the features of the system, which include the maximum depth, the number of light gradient boostings (n_estimators), and other settings.

Step 4: Train the RF classifier by making use of the training set of data (X_train and Y_train).

Step 5: Based on the experiment's findings, make forecasts by using g(x)=f0(x)+f1(x)+f2(x).

Step 6: Evaluate the effectiveness of the prediction model by making use of the appropriate metrics, which include recall, accuracy, and precision.

Step 7: If you want to evaluate the model's performance, print or save the evaluation results.

Decision tree

Decision trees, a type of supervised learning algorithm, are mostly employed for classification purposes. Tree-structured classifiers are composed of nodes that store attributes of the dataset, branches that reflect decision rules, and leaf nodes that include outcomes. There are two distinct types of nodes that are used in the construction of DTs. These are known as decision nodes and leaf nodes. Decision nodes differ from leaf nodes in that they have the ability to make decisions and can branch out in multiple directions.

Decision tree algorithm

Step 1: The file should be exported with the characteristics X and the labels Y.

Step 2: Every single one of the training and testing sets is made up of the following components: X_train, X_test, y_train, and y_test, which should be separated from the dataset.

Step 3: Construct a DT classifier and configure its parameters, including the maximum depth that can be achieved and the impurity measure (which may be the Entropy, sometimes known as the Gini index, is a measure of, for example).

Step 4: It is crucial to make use of the training data, which consists of X_train and y_train, in order to effectively finish the training of the DT classifier. By doing so, you will be able to successfully complete the training.

Step 5: Conclusions should be drawn based on the results of the tests conducted by E(X)=new Product.

Step 6: Due to the fact that this is an essential phase, the performance of the model needs to be reviewed by employing an appropriate metric, such as recall, accuracy, or precision.

Step 7: Make a note of or save the findings of evaluation is out in order to ascertain whether or not the model is successful in its intended purpose.

Statistical analysis

The statistical research that was carried out with IBM SPSS Version 26 employed an independent sample T-test in order to study the mean accuracies of the algorithms that were applied. This was done in order to determine whether or not the algorithms involved were accurate. This was done in order to ascertain whether or not the algorithms offered accurate results. It was necessary to carry out this procedure in order to determine whether or not the algorithms provided reliable findings.

Results

Table 9.1 presents a comparison of the accuracy of RF and DT in machine learning datasets. The table may be found under the following heading. The results of independent sample T-tests using DT and RF are displayed. The table provides a comparison between RF and DT in terms of significance, mean, and standard error. Both visualizations were created using SPSS. Based on the data, it has been observed that there is no statistical significance ($p<0.05$, $p=0.799$), contradicting the findings of the study. An study of the results obtained by the DT and RF classifiers was carried out with the assistance of a statistical program known as SPSS. This program was applied in order to determine the mean, standard deviation, and standard error. The findings of the research indicate that the RF model obtains an accuracy rate of 87.87%, while the DT model achieves an accuracy rate of 87.20%. Both models are considered to be very accurate. Both models are considered to be predictive models, according to the widespread consensus.

In Figure 9.1, you will see a representation of the data analysis that was carried out with the programming language SPSS. With a confidence interval (CI) of 95% and a margin of error of ±2 standard deviations, the RF model achieves an accuracy of 87.87%, whilst the DT approach that was devised obtains an accuracy of 87.20%. Both of these percentages are based on the results of the analysis. Using its predictive skills, the RF model obtains an accuracy of 87.87%, which is significantly higher than the other models.

Discussion

Decision tree outperformed RF by 87.87% in the machine learning study. With a statistical significance of ($p=0.799$) ($p<0.05$), the RF was able to attain an accuracy of 87.20% when comparing the two groups. Online education doesn't seem to be very beneficial since pupils have a hard time adjusting to it. Many electrical gadgets lack adaptability. Students act more lethargic and chronically late when they are no longer guided by an instructor. Online education is challenging for young people and teenagers.

Table 9.1 The unique RF and DT algorithm accuracy values.

S. No.	Random forest	Decision tree
1	88.04	87.27
2	88.01	87.25
3	87.76	87.29
4	88.03	87.12
5	87.69	87.18
6	88.05	87.14
7	87.78	87.11
8	87.75	87.29
9	87.96	87.44
10	87.63	87.27
Accuracy	87.87	87.20

Source: Author

Table 9.2 New RF and DT Algorithms T-tests conducted on groups of independent samples.

	A test that Levene developed to determine whether or not the variances are equal		**A T-test to determine whether or not the means are comparable**					**A confidence interval for the difference that is established at a level of reliability of 95%**	
	F	**Sig.**	**t**	**df**	**The signature with 2-tails**	**The distinction between the mean values**	**A standardized error difference analysis**	**Bring down**	**At the top of the**
The presumption that all variances are constant	0.067	0.799	10.88	18	.000	0.675	0.621	0.545	0.806
The presumption that all variances are constant			10.88	17.98	0.000	0.675	0.621	0.545	0.806

Source: Author

Table 9.3 Statistics on RF and DT classifier accuracy.

	Iterative procedure	N	Mean	Standard deviation (SD)	Normalized error mean
Preciseness	RF	10	87.87	0.136	0.043
	DT	10	87.20	0.141	0.044

Source: Author

It is beneficial to one's future preparation and professional advancement to have the option to change one's career trajectory. Previous research has shown that being able to switch occupations whenever necessary has several benefits (Sahlaoui et al., 2024). Feeling more in charge of one's life and content in general, setting more attainable and consistent professional objectives, being receptive to new experiences, being more dedicated to one's work, and, in the long run, being more entrepreneurial are all benefits. The ability to adapt one's work to new situations is crucial in today's competitive job market (Vijayalakshmi and Venkatachalapathy, 2019). One of the biggest obstacles that professionals face is the perpetual flux and uncertainty that characterizes today's workplace (Tables 9.2 and 9.3).

Table shows sig, mean, and standard error differences between Novel RF and DT methods ($p=0.799$, $p<0.05$) using SPSS.

The dataset's SPSS mean, standard deviation, and standard error for RF and DT were calculated using both algorithm values. Above, RF has 87.87% accuracy and DT 87.20%.

Online students have a harder time adjusting to the different learning tools, and online teachers struggle to connect with students in a way that guarantees success. Paper is superior than digital displays for learning, according to one research. A student's ability to self-regulate is crucial. Learning is a challenge for the majority of kids. Online practical courses are hindered for students due to instructor teaching issues (Orabona, 2019). Characteristics' effects on the predictions of classification and regression models were described using a general technique. The method builds on a general strategy for identifying which features in prediction models are situationally important. As a solution to the issues with generic techniques that fail to account for feature interactions, this method aims to deliberately disrupt all feature subsets (Sharrock and Parkerson, 2020). We used the average situational significance of a feature's value to build a model representation that demonstrates the impact of attributes on the model's forecasts. Earlier methods for explaining additive models were either model-specific or more generic, this technique expands upon the latter (Street, 2019) (Figure 9.3).

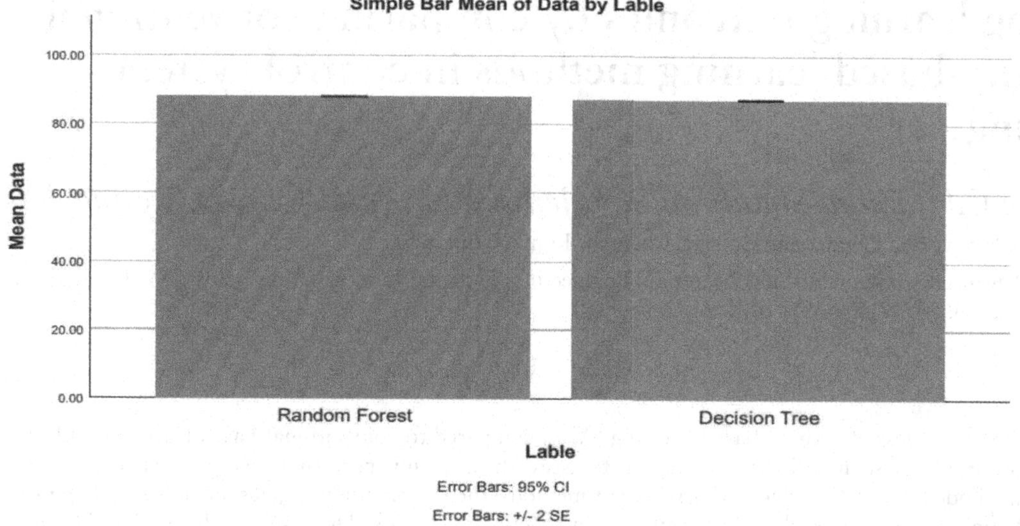

Figure 9.3 SPSS statistical study to determine data correctness yields 87.87% for the suggested technique and 87.20% for DT with an error bar of 95% CI and ±2 SD.
Source: Author

Conclusions

For the most part, students study on the tiny displays of mobile phone terminals. The drawback of mobile phone terminals makes it difficult for students to adjust to online education, leading to less than ideal "online courses" for students. In a vote, 87.87% went for novel RF and 87.20% for DT. Novel RF performs better than DT in identifying diets, according to this research. A significant disparity is seen, as shown by a p-value of 0.799 ($p<0.05$).

References

Tarkar, P. (2020). Impact of COVID-19 pandemic on education system. *Internat. J. Adv. Sci. Technol.*, 29(9), 3812–3814.

Hofer, S. I., Nistor, N., and Scheibenzuber, C. (2021). Online teaching and learning in higher education: Lessons learned in crisis situations. *Comp. Human Behav.*, 121, 106789.

Mahboob, T., Irfan, S., and Karamat, A. (2016). A machine learning approach for student assessment in E-learning using Quinlan's C4. 5, Naive Bayes and Random Forest algorithms. *2016 19th Internat Multi-topic Conf. (INMIC)*, 1–8.

Sahlaoui, H., El Arbi Abdellaoui, A., Agoujil, S., and Nayyar, A. (2024). An empirical assessment of smote variants techniques and interpretation methods in improving the accuracy and the interpretability of student performance models. *Edu. Inform. Technol.*, 29(5), 5447–5483.

Vijayalakshmi, V. and Venkatachalapathy, K. (2019). Comparison of predicting student's performance using machine learning algorithms. *Internat. J. Intell. Sys. Appl.*, 11(12), 34.

Nnadi, L. C., Watanobe, Y., Mostafizer Rahman, Md., and John-Otumu, A. M. (2024). Prediction of students' adaptability using explainable AI in educational machine learning models. *Appl. Sci.*, 14(12), 5141.

Dass, S., Gary, K, and Cunningham, J. (2021). Predicting student dropout in self-paced MOOC course using random forest model. *Information*, 12(11), 476.

Qiu, B., & Fan, W. (2021). Machine learning based short-term travel time prediction: Numerical results and comparative analyses. *Sustainability*, 13(13), 7454.

López Zambrano, J., Torralbo, J. A. L., and Morales, C. R. (2021). Early prediction of student learning performance through data mining: A systematic review. *Psicothema*. Vol. 33 (nº 3). 456–465, https://doi.org/10.7334/psicothema2021.62

Sharrock, E. and Parkerson, C. (2020). Investing in the birth-to-three workforce: A new vision to strengthen the foundation for all learning. *Faculty and Staff Papers and Presentations*. Bank Street College of Education. pg. No 1–79 (book) https://educate.bankstreet.edu/cgi/viewcontent.cgi?article=1000&context=bsec

Orabona, F., & Pál, D. (2018). Scale-free online learning. Theoretical Computer Science, 716, 50–69.

Orabona, F. (2019).

10 Evaluating learning outcomes by comparing conventional and activity-based learning methods in control system engineering

Sheeba Joice C.[1,a], *Hitha Shanthini S.*[1,b], *Jenisha C.*[1,c] *and John D. Rodney*[2,d]

[1]Department of ECE, Saveetha Engineering College, Chennai, Tamil Nadu, India

[2]Department of Advanced Components and Materials Engineering, Sunchon National University, 255, Jungang-ro, Suncheon-si Jellanamdo, 57922, Republic of Korea

Abstract

This study evaluates the effectiveness of activity-based learning (ABL) compared to conventional-based learning (CBL) in the control systems engineering course. Its primary objective is to assess the academic performance of learners exposed to these distinct teaching methodologies. Using rigorous assessment methods, the research investigates participants' comprehension, retention, and application of control systems concepts in practical scenarios. The results indicate that ABL enhances the development of critical practical skills essential for success in control systems engineering, alongside improving theoretical understanding. In particular, 93.45% of learners in the ABL group have successfully passed, whereas 81.23% of learners in the CBL group have achieved passing grades. These findings underscore the necessity for a shift in teaching paradigms and offer valuable insights for optimizing teaching methods to achieve learning objectives in this specialized domain.

Keywords: Activity-based learning (ABL), conventional-based learning (CBL), parameters, assessment

Introduction

Post-pandemic, there's decreased interest in conventional-based learning (CBL), necessitating the adoption of new instructional methods, especially in engineering education. This study underscores the importance of pedagogical enhancements, digital tools, and activity-based learning (ABL) to create a learner-centered environment (Cropley and Sitnikova, 2005). Engineers play a vital role in addressing emerging social challenges, but improvements in engineering education are essential for their preparedness (Hamdan and Saripudin, 2023).

In today's complex educational landscape, integrating teaching and learning is crucial, necessitating a curriculum overhaul to equip engineering graduates with the critical thinking skills essential for understanding technology's societal impacts (Roque et al., 2023). Modern learners find traditional teaching methods less effective, emphasizing the need for educators to adopt diverse approaches tailored to learner preferences (Dixit et al., 2024). While technology enables access to vast resources, educators remain indispensable for facilitating meaningful connections, knowledge sharing, and guidance, with ABL standing out as an effective strategy to meet evolving learner expectations (Selvi and Joice, 2022).

Various pedagogical approaches have been explored to enrich learning environments in different courses. Pair programming (Phil et al., 2014) improves outcomes. Cooperative learning, mastery learning (Alabbadi and Qureshi, 2016), and project-based learning (Mason et al., 2017), engage learners effectively. The active learning and teaching model (Ramachandran, 2017), and laboratory course (Lokare et al., 2020) challenges addressed. Active and blended learning (Abirami et al., 2021) was introduced for undergraduates to enhance performance. Teachers using a learner-centered approach foster lifelong learning habits and assess development, while activity-based learning enhances critical thinking, problem-solving, and real-world skills, enriching the overall learning experience (Mahadev and Kamerikar, 2020). The research shows that active learning and new teaching methods work well for learners, helping them learn better in different subjects. In the implementation section, we explain how we applied ABL to second-year ECE learners, including detailed explanations and assessment parameters. The observation and analysis section presents findings, ultimately concluding that ABL is the most effective approach.

[a]sheebajoice@saveetha.ac.in, [b]hithashanthini@saveetha.ac.in, [c]jenishac@saveetha.ac.in, [d]rodney95@scnu.ac.kr

DOI: 10.1201/9781003587538-10

Implementation

The second-year ECE control system engineering learners were taken as test cases for the implementation of ABL. The learners were taken from the two batches. The batch 1 learners were taken from the third semester of the 2023–2024 academic year. The batch 2 learners were taken from the fourth semester of the 2023–2024 academic year. The batch 1 followed the CBL in control system engineering. The batch 2 followed the ABL in control system engineering. Each batch consisted of 70 learners with four categories. The learners were categorized by conducting a common internal assessment (CIA). Figure 10.1 shows the four categories of learners.

- **Winners:** Are those who ask questions, know their goals, and can study by themselves
- **Active runners:** Are those who are dedicated, but find difficulty in studying
- **Survivors:** Are those who are very poor in studies
- **Outsiders:** Are those who are disconnected from the class

The batch 2 learners in the ABL program were assigned a control system-based exercise to boost their active engagement rather than simply delivering material to them passively. Their task required them to find a practical use of a control system in real-time, take a picture of this application with the learner and their location, and then analyze the specific control system being used. The task was submitted as assignment 1, along with a block diagram and explanation, which let the learners explore topics, make connections, and practically apply their knowledge. Furthermore, participants were obligated to create a presentation displaying their discoveries, film this presentation as a video, and submit it to a YouTube channel for assessment, which constituted assignment 2. By showcasing their work, they not only enhanced their comprehension but also increased their levels of self-assurance.

To guarantee equitable evaluation and congruence with program objectives, meticulously crafted parameters were utilized to analyze both assignments 1 and 2, as seen in Figures 10.2 and 10.3, respectively. Table 10.1 presents a summary of assessments I and II of learners' activity, as determined by these parameters (Soniya Agrawal and Chandasree Das, 2018).

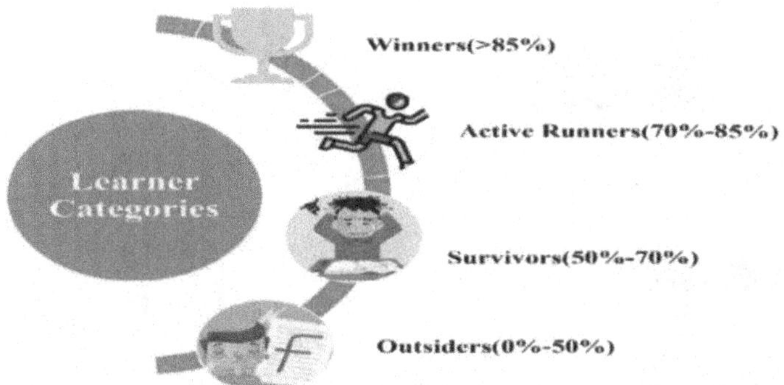

Figure 10.1 Four categories of learners
Source: Author

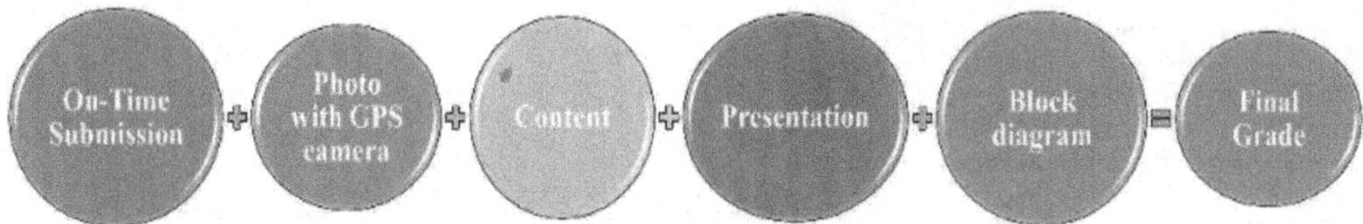

Figure 10.2 Parameters for evaluating the ABL for assessment I
Source: Author

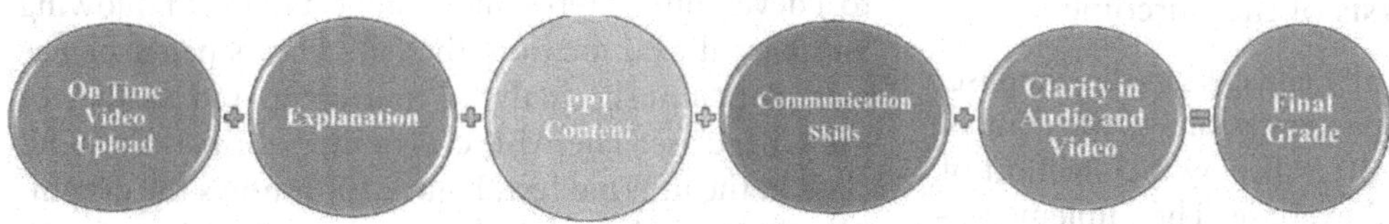

Figure 10.3 Parameters for evaluating the ABL for assessment II
Source: Author

Table 10.1 The summary of the assessment I and II based on parameters

Learner Category	Evaluation Parameters									
	Assessment I					Assessment II				
	On Time Submission	Photo with GPS Map Camera	Content	Presentation	Block Diagram	On-Time Video Upload	Explanation	PPT Content	Communication Skills	Clarity in Audio and Video
Outsiders 0% - 50%	1	0	1	0	0	0	0	1	0	2
Survivors 50% - 70%	5	1	0	0	2	5	1	0	0	1
Active Runners 70% - 85%	18	36	22	10	12	19	26	12	15	14
Winners >85%	46	33	47	60	56	46	43	57	55	53

Source: Author

Figure 10.4 A few samples from the learner's activity I
Source: Author

Figure 10.5 A sample from the learner's activity II
Source: Author

Observations and analysis of the outcome

Every learner enthusiastically engaged in the exercise, exploring the practical implementations of the control system. Figures 10.4 and 30.5 display a collection of samples from the learners' exercise. The implementation of the ABL technique resulted in reduced distractions and increased focus among the learners, leading to a developing interest in the subject matter. Following an internal examination, the ABL class's performance was meticulously analyzed and compared to that of the CBL class. The ABL class outperformed the CBL class in the internal test. Figure 10.6 shows the graphical representation of the learners' performance growth after the ABL method.

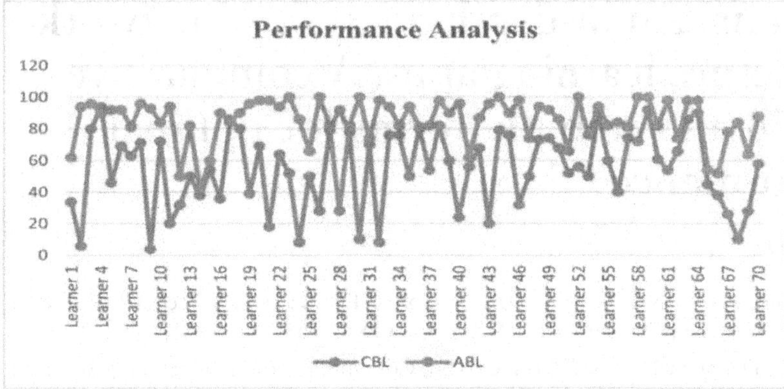

Figure 10.6 Performance comparison graph of CBL and ABL in the internal test
Source: Author

Conclusions

Ultimately, the study examined the effectiveness of traditional and activity-based teaching approaches in the field of control system engineering. After conducting a thorough analysis of performance measures, it became clear that the ABL approach offered significant benefits compared to the CBL method. ABL not only facilitated increased learner involvement and hands-on learning, but it also led to higher academic achievement, as demonstrated by the internal assessment outcomes. The outsiders and survivors from the learners' category were decreased. Thus, the results emphasize the efficacy of interactive, experiential teaching methods in engineering education, especially in difficult topics like control systems. Adopting ABL techniques has the potential to improve learner performance and foster a more profound comprehension of engineering principles in the academic field.

References

Abirami, A. M., Pudumalar, S., and Thiruchadai Pandeeswari, S. (2021). Active learning strategies and blended learning approach for teaching under graduate software engineering course. *J. Engg. Edu. Transform.*, 35(1), 42–51.

Alabbadi, A. A. and Qureshi, R. (2016). The proposed methods to improve teaching of software engineering. *Internat. J. Modern Edu. Comp. Sci.*, 8(7), 13–21. https://doi.org/10.5815/ijmecs.2016.07.02.

Cropley, D. and Sitnikova, E. (2005). Teaching and learning in engineering education: Constructive alignment. *4th ASEE/AaeE Global Colloquium on Engineering Education*,1293–1302 .

Dixit, A. C., C. A. B., A. S. M., and N. P. K. (2024) . Innovative pedagogical approaches for diverse learning styles and student-centric learning. *J. Engg. Edu. Transform*,178-188,Issue 37.

Hamdan, A. H. and Saripudin, M. (2023). Designs for research, teaching, and learning: A framework for future education. *Innov. Edu. Teach. Internat.*, 60(5), 795–796. https://doi.org/10.1080/14703297.2023.2241319.

Lokare, V. T., Jhetam, I., Kiwelekar, A. W., and Netak, L. D. (2020). A blended approach for teaching laboratory course: Internet technology. *J. Engg. Edu. Transform.*, 34(1), 41. https://doi.org/10.16920/jeet/2020/v34i1/151189.

Mahadev, S. P. and Kamerikar, U. A. (2020). Learning by doing through project based active learning technique. *J. Engg. Edu. Transform.*, 33(Special issue), eISSN 2394–1707.

Mason, R. T., Masters, W., and Stark, A. (2017). Teaching agile development with DevOps in a software engineering and database technologies practicum. *Proc. 3rd Internat. Conf. Higher Edu. Adv*,1353–1362. https://doi.org/10.4995/HEAD17.2017.5607.

Phil, M., Maguire, R., Hyland, P., and Marshall, P. (2014). Enhancing collaborative learning using pair programming: Who benefits? *All Ireland J. Teach. Learn. Higher Edu. (AISHE-J)*, 6(2), 1411–1425.

Ramachandran, M. (2017). Technology enhanced active learning in software engineering. *Proc. 9th Internat. Conf. Comp. Support. Edu.*, 242–248. https://doi.org/10.5220/0006257602420248.

Roque, I. M.-S., Aza-Blanc, G., Hernández-Arriaza, M., and Prieto-Navarro, L. (2023). Learning approaches and high-impact educational practices at university: A proposal for a reduced scale of the student process questionnaire. *Innov. Edu. Teach. Internat.*, 1–15. https://doi.org/10.1080/14703297.2023.2251447.

Selvi, M. and Joice, C. S. (2022). Performance analysis of conventional and innovative teaching learning methodologies in engineering. *J. Engg. Edu. Transform.*, 36(S1), 110–114. https://doi.org/10.16920/jeet/2022/v36is1/22183.

Soniya, A. and Das, C. (2018). A comparative study on parameters and its impact on program outcomes for the project work of under graduate program. *J. Engg. Edu. Transform.*, 31(Special Issue) eISSN 2394–1707.

11 Navigating data streams: An in-depth analysis of network efficiency in data structure learning apps. A comparative study between android and web platforms for optimal educational app performance

T. Neshanth[1,a] *and P. Dass*[2,b]

[1]Research Scholar, Department of Information Technology, Saveetha School of Engineering, Saveetha Institute of Medical and Technical Sciences, Saveetha University, Chennai-602105, Tamil Nadu, India

[2]Corresponding author: Research Guide, Department of Electronics and Communication Engineering, Saveetha School of Engineering, Saveetha Institute of Medical and Technical Sciences, Saveetha University, Chennai-602105, Tamil Nadu, India

Abstract

Aim: The project's goal is to examine and compare data stream management in data structure learning apps on android and web platforms, with a particular emphasis on network efficiency. The goal is to determine the ideal ways to improve the performance of educational apps for data structure learning **Materials and methods:** Based on the article, a G power of 80% was used to assess data consumption's in both android and web apps, using 12 sample sizes each. **Result:** Data consumption was investigated, and the web development data utilization is 19.3333 MB, whereas the android application data utilization is 22.2500 MB. The statistical difference between android and web applications is considerable, with p=0.452 (p<0.05). **Conclusion:** Android application has data consumption of 22.2500 MB and web development data consumption has 19.3333 MB.

Keywords: Android app, data usage, data structure, consumption, room database, research, performance

Introduction

The application uses real-time data to track user interactions and progress within the quiz platform. This comprises information about activities done, stages progressed through, difficulty levels achieved, and any interactive feedback supplied during gameplay. Real-time analytics can also track user engagement indicators including time spent on tasks, success rates, and general performance trends. This information is required for dynamically modifying the learning experience, delivering individualized feedback, and increasing user engagement and understanding of data structures.

Exploring the exciting industry of data structure app development. With over 72,50,000 publications available on Google Scholar, the amount of research on this topic has surged. Notably, IEEE and Springer contribute significantly, with 220,349 and 77,524 papers, respectively. The article (De Floriani et al., 2005) which has the 103 citations, This paper surveys data structures for encoding level of detail (LOD) models, categorized by dimensionality: point-, triangle-, and tetrahedron-based. It reviews single-level structures, general ones for irregular meshes, and specialized structures for semi-regular data. The article by Siddiqa et al., (2017) has 144 citations. This paper explores big data storage technologies, addressing inefficiencies in current solutions and industry standards like Amazon, Google, and Apache. It presents a comprehensive investigation and taxonomy to assist in selecting suitable storage mechanisms, analyzing performance using Brewer's CAP theorem, and discussing future research challenges. The aim is to expedite the deployment of reliable and scalable storage systems for various real-world applications. The article by Bendre and Thool (2016) has 93 citations. This paper examines the background and future prospects of big data, addressing its massive volume and diverse sources. It covers system architecture, analytics phases, and introduces an open-source framework to tackle big data challenges. Various applications of big data are discussed, along with illustrative examples. The article by Pourabbas and Shoshani (2015) has 7 citations, where the paper introduces the composite data model (CDM), merging object classes, multidimensional objects, and hierarchical structures into a unified schema. The composite query language (CQL) facilitates expressive queries across these

[a]192021009.sse@saveetha.com, [b]dass@saveetha.com

DOI: 10.1201/9781003587538-11

models, leveraging the explicit semantics of each. The tool generates key frames of major data structure transformations for review and analysis, bridging the gap between abstract representation and practical learning.

In order to provide insights that might guide decision-making and optimization efforts to enhance the performance of both platforms in the local artisan marketplace environment, the study attempts to identify disparities in data utilization between web and android applications.

Materials and methods

The software bundle known as XAMPP, which stands for cross-platform (X), Apache (A), MySQL (M), PHP (P), and Perl (P), is a powerful tool designed to make setting up a local web server environment easier for users of Windows, Linux, and macOS. It meets a wide range of development requirements by combining the web server Apache, the database management system MySQL, the PHP scripting language for server-side scripting, and the Perl general-purpose programming language. Developers can now create and test dynamic websites and web applications offline before putting them live on servers thanks to this flexible technology. Moreover, XAMPP includes phpMyAdmin, an online MySQL database management tool, which enhances its value for database-related applications. Its extensive documentation and support resources help customers with installation and troubleshooting, and it makes managing several websites and databases at once easier. By incorporating XAMPP into their work, web developers may create and test websites and apps in a dependable local environment. Developers can use XAMPP, which includes Apache, MySQL, PHP, and Perl, to create dynamic websites with the use of relational databases and server-side scripting. While MySQL facilitates efficient data storage and retrieval, PHP makes it simple to integrate database queries into web applications. The process of incorporating Room Database into an android studio application is developing entity classes marked with {@Entity} to symbolize database tables and adding dependencies for Room. Developers specify methods for database operations by creating data access objects (DAOs) annotated with {@Dao}. Next, they extend "RoomDatabase" and give it DAOs to create an instance of the Room Database. Ultimately, developers acquire a Room Database instance and execute database operations via DAO methods. A reactive user experience can be achieved by using RxJava or LiveData to monitor changes to database tables and update the user interface (UI) appropriately.

Data structure android application

Group 1 uses the android studio environment to develop android applications with an emphasis on data consumption. The profiling tools in android studio, such as android profiler, allow developers to keep an eye on and assess the performance of their apps. Developers use techniques to get the timestamps at the beginning and end of a certain process or activity. The amount of data used for that specific job is represented by the interval between these timestamps. The Logcat feature in android studio then allows developers to log or print the data utilization so they may examine and monitor the application's performance in real time. N=12 is the sample size.

Data structure web development

A similar methodology is used in group 2 to evaluate data use for web applications developed in PHP with XAMPP SQL. Developers can use the develop option to log the start and end times of a Python request-handling process, which they can then use to calculate data consumption. Unlike android, web development often uses the error.log file from the server to document data utilization. Developers are able to monitor and assess server design and online application performance by logging data utilization statistics for each request in the error.log file.

Statistical analysis

A potent program for statistical analysis and data management is IBM SPSS. Thanks to its intuitive interface, users can work with large datasets, carry out in-depth statistical analysis, and create informative visualizations. Analyze obtained data on a regular basis to determine typical data consumption for both Android and web platforms. This may entail aggregating data consumption indicators over certain time periods (e.g., daily, weekly) and generating averages.

Discussion

Both the web and android versions of the data structure app's data usage were investigated. Android users used 15.9231 MB of data on average, whereas web users used 20.8333 MB on average. The calculated threshold for statistical significance exceeded the usual level of 0.05, at 0.000. This indicates that the mean reaction times between the android and web platforms do not differ significantly in a non-statistical way.

The first step in creating a web application using Visual studio code and XAMPP is to set up the platform and launch the Apache and MySQL services. To update the user interface and observe database changes, use Room's LiveData or RxJava integration. Using android studio's features debug and improve the application. Lastly, release the software through alternative distribution channels or the Google Play Store. With this method, it is possible to create effective android apps that integrate Java with Room Database.

Users will find the learning experience more pleasurable and productive as a result of the gamified approach, which promotes ongoing engagement and motivation. Result obtains a data consumption of 35 MB from the study of an additional application, which is greater utilization than the data structure's 29 MB.

Summary statistics (Table 11.1)

Table 11.1 N = 12. Furthermore, after 12 iterations, the data used for web and android development is determined. Android development has required a longer reaction time than web development.

S. No.	Android data utilization	Web development data utilization
1	25	23
2	18	17
3	20	22
4	24	18
5	15	20
6	21	19
7	26	21
8	23	20
9	20	16
10	18	23
11	17	18
12	22	21

Source: Author

Post-crisis estimations

Summary statistics (Table 11.2, Figures 11.1–11.3)

Table 11.2 Web development uses 19.3333 GB of data on average, while android development uses 22.2500 GB

	Algorithm	N	Mean	Std. deviation	Std. error mean
Data utilization	Android	12	22.2500	10.95549	3.16258
	Web development	12	19.3333	7.34022	2.11894

Source: Author

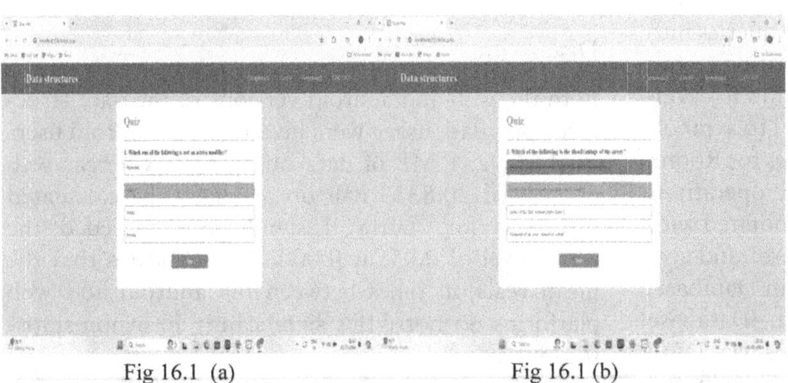

Figure 11.1 (a) This image will show which is the correct answer as in green color and the chosen wrong answer will be turned to red. (b). The page in the image serves as a quiz platform for data structures, displaying questions along with four answer options. Upon selecting an option, it reveals whether the choice is correct or incorrect

Source: Author

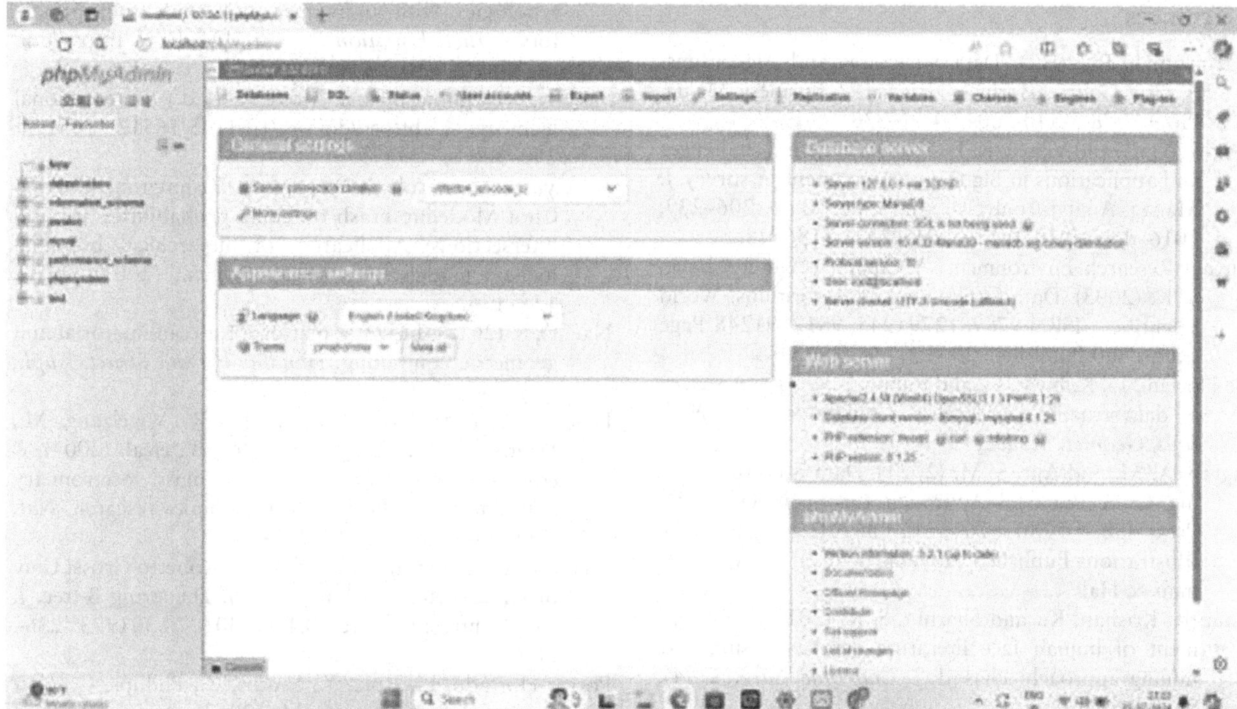

Figure 11.2 Datum is kept for data structure applications in the database inside the relevant table. This table provides an orderly framework for effective information handling in the web application
Source: Author

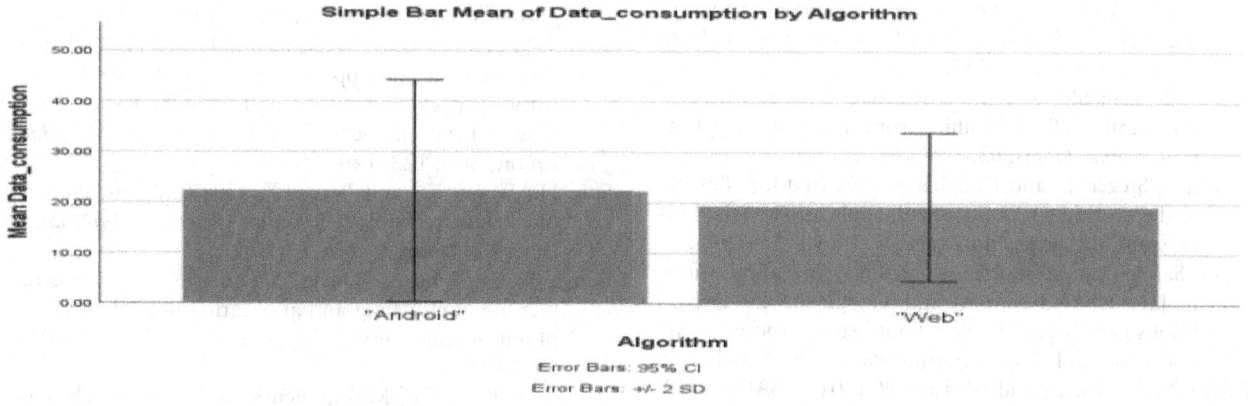

Figure 11.3 The bar chart displays the average data utilization for the data structure application on both the android and web development platforms
Source: Author

Conclusions

This project has tested the data utilization of android and web development the data used for web and android development is determined. Android development has required a longer reaction time than web development. The gamified method encourages continuous engagement and motivation, making the learning process more enjoyable and effective for users. The analysis of an extra application yielded a result of 35 MB of data consumption, which is more than the 29 MB of data structure utilization. After conducting an extensive examination, it was found that the web development platform exhibited a consumption of 20.8333 MB, surpassing that of android (15.9231 MB). Android application has data consumption of 22.2500 MB and web development data consumption has 19.3333 MB. The results show that the online application operates better than android with a statistically significant difference ($p=0.452$, $p<0.05$).

References

Adamson, I. T. (2012). Data Structures and Algorithms: A First Course. Springer Science & Business Media. ISBN : 978-3-540-76047-4 , 1–336.

Bendre, M. R. and Vijaya, R. T. (2016). Analytics, challenges and applications in Big Data environment: A survey. *J.* Manag. Analyt.Analytics, vol. 3, no. 3, pp. 206–239, 2016. doi: 10.1080/23270012.2016.1186578.

Virtual Research Environment: A Complete Guide.Chang, S. K. (2003). Data Structures and Algorithms. World Scientific. ISBN:9789812791245, 9812791248 Page count: 360 Published: 2003

De Floriani, L., Kobbelt, L., and Puppo, E. (2005). A survey on data structures for level-of-detail models. *Adv. Multires. Geomet. Model.*, 49–74.

Jagtap, A. M. and Ajit, S. M. (2021). Data Structures Using C: A Practical Approach for Beginners. CRC Press ISBN 9780367616311, 362 Pages 193 B/W Illustrations,Published November 9, 2021 by Chapman & Hall

Kaur, P., Krishan, K., and Sharma, S. K. (2018). Development of human face literature database using text mining approach: Phase I. *J. Craniofac. Surg.*, 29(4), 966–969.

Lin, J. and Zhang, H. (2020). Data structure visualization on the web. *2020 IEEE Internat. Conf. Big Data (Big Data).* https://doi.org/10.1109/bigdata50022.2020.9378249, 3272–3279.

Pourabbas, E. and Shoshani, A. (2015). The composite data model: A unified approach for combining and querying multiple data models. *IEEE Trans. Knowl. Data Engg.*, 27(5), 1424–1437.

Martin Kronbichler Katharina Korman Generic interface for parallel cell-based finite element operator application. *Comp. Fluids*, 63, 135–147.

Benek, J., Steger, J., and Dougherty, F. C. (n.d.). A flexible grid embedding technique with application to the Euler equations. https://doi.org/10.2514/6.1983–1944.

Sartaj Sahni Data structures, algorithms, and applications in Java (second edition). n.d. Google Books. https://books.google.com/books/about/Data_Structures_Algorithms_And_Applicati.html?id=aew7q_cMJRIC.

Färber F., Jürgen, P., and Christof, B. (2012). SAP HANA database. ACM SIGMOD Record. https://doi.org/10.1145/2094114.2094126, 45–51.

Finkel, R. A. and Bentley, J. L. (1974). Quad trees a data structure for retrieval on composite keys. *Acta Inform.*, 4(1), 1–9.

Graz, F. A. T., Schiebbstattgasse, and Austria. (1991). Voronoi diagrams—A survey of a fundamental geometric data structure. *ACM Comput. Sur. (CSUR).* https://doi.org/10.1145/116873.116880, 345–405.

Hanan, S. (1984). The quadtree and related hierarchical data structures. *ACM Comput. Sur. (CSUR).* https:/ https://doi.org/10.1145/356924.356930, 188–260.

Henry, K. L. and Muthén, B. (2010). Multilevel latent class analysis: An application of adolescent smoking typologies with individual and contextual predictors. *Struct. Equation Model. Multidis. J.* https://doi.org/10.1080/10705511003659342, 193–215.

Herlihy, M., Eliot, J., and Moss, B. (n.d.). Transactional memory. https://doi.org/10.1145/165123.165164, 289–300.

Do-Gyeong Kim, Yuhwa Lee, Simon Washington, Keechoo Choi Modeling crash outcome probabilities at rural intersections: Application of hierarchical binomial logistic models. *Acc. Anal. Preven.*, (2007). 39(1), 125–134.

Naeher, S. (2018). LEDA, a platform for combinatorial and geometric computing. *Handbook Data Struct. Appl.*, 653–666.

Pedrioli, P. G. A., Eng, J. K., Hubley, R., Vogelzang, M., Deutsch, E. W., Raught, B., Pratt, B., et al. (2004). A common open representation of mass spectrometry data and its application to proteomics research. *Nat. Biotechnol.*, 22(11), 1459–1466.

Pisa Italy, Paolo Ferragina Univ di, and Roberto Grossi Univ di Firenze Florence Italy. (1999). The string B-tree. *J. ACM.* https://doi.org/10.1145/301970.301973 236–280.

Porras-Hurtado, L., Ruiz, Y., Santos, C., Phillips, C., Carracedo, A., and Lareu, M. (2013). An overview of structure: Applications, parameter settings, and supporting software. *Front. Genet.*, 4, 48396.

Rajeev, R., Raman, V., and Satti, S. R. (2007). Succinct indexable dictionaries with applications to encoding Kary trees, prefix sums and multisets. *ACM Trans. Algorith.* https://doi.org/10.1145/1290672.1290680.

Sosa, J. M., Huber, D. E., Welk, B., and Fraser, H. L. (2014). Development and application of MIPARTM: A novel software package for two- and three-dimensional microstructural characterization. *Integr. Mat. Manufac. Innov.*, 3(1), 123–140.

Sun, H., Wang, X., and Wang, X. (2018). Application of blockchain technology in online education. *Internat. J. Emerg. Technol. Learn.*, 13(10), 252.

Jianya Gong, Penggen Cheng, Yandong Wang, Three-dimensional modeling and application in geological exploration engineering. *Comp. Geosci.*, 30(4), (2004). 391–404.

Jon Louis, B. (1975). Multidimensional binary search trees used for associative searching. *Comm. ACM.* https://doi.org/10.1145/361002.361007, 509–517.

Alexander, Y., Wang, X., Zeldovich, N., Frans Kaashoek, M. (n.d.). Improving application security with data flow assertions. Massachusetts Institute of Technology – Computer Science and Artificial Intelligence Laboratory SOSP'09, October 11–14, 2009, Big Sky, Montana, USA. Copyright 2009 ACM 978-1-60558-752-3/09/10. https://doi.org/10.1145/1629575-1629604.

Wickham, H. (2011). The split-apply-combine strategy for data analysis. *J. Statist. Softw.*, 40, 1–29.

12 Differentiated instruction and assessment for English language learners from theory to practice

K. Sasidhar[a] *and G. Aruna*[b]

Department of Foreign Languages and Linguistics, Sri Venkateswara University, Tirupathi, Andhra Pradesh, India

Abstract

Differentiated instruction and assessment (DIA) is a method that recognizes and addresses the diverse abilities of English language learners (ELLs) within the ESL (English as a second language) classroom at the tertiary level. It includes tailoring instruction to meet the individual learning abilities, styles, language proficiency levels, and cultural backgrounds of each student. It is a flexible and responsive teaching approach that prioritizes the individual growth and needs of each English language learner. It acknowledges that students enter the classroom with varying levels of language proficiency and background knowledge, and it aims to establish a conducive academic setting where all learners can thrive. DIA method motivates the student to set his/her goals in acquiring skills in English language. In this regard, the teacher has to provide authentic materials and conduct organized assessments to manage both slow and fast learners in an ESL classroom. In this regard, the study examines how well students do after using the DIA method to teach English and provides evidence that the approach is both practical and effective in fostering critical thinking abilities and communicative competencies among the undergraduate students studying at the engineering colleges.

Keywords: Differentiated instruction, differentiated assessment, mixed ability classroom, teaching and learning process, materials

Introduction

Language competencies have become a basic requirement for employees of multinational firms in the manufacturing, distribution, and service sectors of the economy. As a result, learning written and spoken English has become a fundamental need. According to this viewpoint, it is imperative that English language learners (ELLs) must acquire critical thinking abilities in addition to communication competencies through a comprehensive approach. However, during the journey of building essential skills, the surrounding environment presents challenges for students seeking to acquire language abilities. As a result, the students struggle to acquire higher order and lower order thinking skills which are often evaluated to progress in their academic and professional endeavors. Their inadequate proficiency in English is due to deficiency of relevant materials, lack of exposure and guidance.

Significance of the study

The purpose of the study is to address the problems of the learners and teachers in the ESL (English as a second language) setting. Along with communicative competencies, undergraduate engineering students should master critical thinking skills as they are tested at various instances. After completion of engineering, the majority of the students' goal is to procure a job either in a government sector or in a multinational company in their chosen field. As a part of the employment recruitment process, the students have to take two steps. First, the aspirant has to pass the written exam where his critical thinking levels are tested and if the student could clear the written exam, he has to face oral interview in which his communicative competencies are evaluated. Given that critical thinking skills are evaluated primarily through English language, such as in reading comprehension, sentence completion or analytical writing, students need to excel in these areas. In this context, students have been encountering obstacles in achieving their goals.

Imparting critical thinking skills alongside communicative competencies in a heterogeneous classroom is recognized as a challenging task for ESL instructors (Mohd Said, 2019). In a mixed ability classroom, the learners have diverse language proficiency levels. Further, they differ in their past learning experiences, and they vary in their learning styles. Besides these challenges, assessing and grading these learners is a major problem for the teachers. Thus, the teachers are always in search of a suitable teaching learning method to impart the necessary skills for the ELLs especially at the tertiary level.

[a]sasidher336@gmail.com, [b]aruna.s@gmail.com

DOI: 10.1201/9781003587538-12

Literature review

In the pursuit of imparting English language skills to ELLs, ESL researchers have identified two predominant approaches: teacher-centric methods and learner-based strategies (Khasawneh, 2022). Teacher-centric methods involve instructional techniques where the teacher takes a dominant responsibility in delivering language content, often through lectures, guided activities, and direct language instruction (Eve, 2022). Conversely, learner-based strategies place a higher priority on students' responsibility and active participation in the learning process (Carmel, 2020). This approach encourages ELLs to engage with the language through interactive activities, group discussions, and self-directed learning. Both methodologies offer distinct advantages, and the effectiveness of each may vary based on the individual abilities and preferences of ELLs (Anburaj, 2014).

As mentioned by Jack C. Richards and Theodore S. Rodgers, "Among all the English language teaching methods Communicative Language Teaching (CLT) method is widely being practiced by ESL teachers" (Richards and Rodgers, 2014). Though the method is learner centric, and activity based, it is found successful to enhance communicative competencies. The activities of CLT, including role plays, describing objects, places, giving directions, etc. focus on improving communicative competencies of the ELLs compared to critical thinking skills which are equally important for the learners (Dos Santos, 2020). Further, these teaching methods are appropriate only if the classroom size is less. In Indian classrooms, the students have varied abilities and mostly the classroom size will be more than 60 students. The learner-centric methods emphasize creation of an active and immersive learning atmosphere (Radhika, 2020). Peer activities and language games are integral component of the learner-centric methods. These activities serve the purpose of preparing students mentally and emotionally for the learning process ahead. However, at a later stage, the teachers found that learner centric could not address the problems of either large classroom or disruption in a mixed ability classroom. In order to address the requirements of varied learners in the general classroom setting, the concept of differentiated instruction was established (Chapman and King, 2003, 2005; O'Meara, 2010; Tomlinson, 1999, 2003). Differentiated instruction and assessment (DIA) is an adapted educational approach specifically designed to foster critical thinking skills and enhance communication competencies within a diverse tertiary-level classroom setting.

Differentiated instruction and assessment method

The DIA method is implemented through differentiated instruction, varied practice and learning materials, differentiated assessments and feedback. This approach ensures that each student receives instruction that is accessible and meaningful, promoting engagement and understanding. Moreover, providing varied practice activities and learning materials further supports this approach by offering students multiple opportunities to interact with each other in their learning process. Additionally, employing differentiated assessments allows teachers to accurately evaluate student progress. Furthermore, offering targeted feedback that is specific to each student's performance and learning goals enhances the effectiveness of the DIA method, guiding students towards continuous improvement and personalized growth trajectories.

Implementing DIA in ESL classroom

The implementation of the DIA method requires a systematic approach. The initial step involves conducting a pre-test on English language to assess learner competencies. This pre-test works as a significant tool for both teachers and students. For teachers, it provides insight into each learner's abilities beforehand, enabling them to tailor instruction effectively. Meanwhile, learners benefit by gaining a clear understanding of their current performance level and identifying specific areas requiring improvement to attain their learning objectives. Ultimately, the pre-test empowers teachers and learners alike to navigate the educational journey with targeted strategies and personalized goals, ensuring a more focused and productive learning experience. Based on their language proficiency level in the pre-test, ESL teachers could classify the students into slow and fast learners. The pre-test also serves as a valuable tool for teachers to strategically form collaborative student groups. This approach enhances collaborative learning experiences where students can benefit from peer interactions and mutual support. As a part of DIA, the ESL teachers have to conduct formative assessments to know the learner's performance at regular intervals of time. These assessments play a major role in enabling teachers to dynamically regroup students based on their performance levels. Additionally, formative assessments empower teachers to provide constructive feedback to learners and adapt instruction to cater to the needs of both slower and faster learners effectively.

To improve communication skills and cultivate critical thinking abilities, teachers should encourage the use of differentiated materials that cater to the varying pace of learning among students. This approach ensures that both slower-paced and faster-paced learners have opportunities to utilize materials and activities that are appropriately challenging and supportive of their individual learning abilities. By providing differentiated materials, instructors can effectively engage all participants in developing essential skills, fostering a learning environment that accommodates diverse learning abilities. During feedback sessions, ESL teachers play a critical role in addressing learners' difficulties and providing appropriate solutions to support their language development. These feedback sessions serve as valuable opportunities for teachers to assess individual progress, identify specific areas of challenge, and offer tailored guidance for improvement.

Impact of DIA – Research evidence

The study was conducted involving students from five engineering colleges in Visakhapatnam. Pre- and post-tests were administered, and the results were analyzed to assess the impact of a DIA method. Following a teacher development program focused on DIA, ESL teachers were instructed to implement in enhancing critical thinking and communicative skills within mixed-ability classrooms. Teachers were advised to utilize differentiated instructional materials for content delivery and were supplied with tailored practice and assessment resources to accommodate both slower and faster learners effectively.

Activities and tasks to foster critical thinking and communicative competencies

Activities like reading comprehension and sentence completion were conducted for the students to enhance their critical thinking skills. Further activities include debate, analytical and argumentative writing to develop their critical thinking skills and communicative competencies. Reading comprehension exercises provided students with opportunities to engage deeply with texts, extracting meaning, and synthesizing information. By engaging with complex passages, students learnt to analyze content, identify main ideas, and to draw logical conclusions. Sentence completion tasks further refined critical thinking by requiring students to understand context and apply language skills effectively. Moreover, activities like debates played a pivotal role in fostering critical thinking and communication. Through debates, students learnt to formulate persuasive arguments, substantiate their viewpoints with evidence, and counter opposing perspectives. This not only sharpens their critical thinking abilities but also enhances their capacity to articulate and defend ideas effectively. Analytical, and argumentative writing tasks encouraged students to explore intricate subjects, examine issues from multiple perspectives, and craft cohesive arguments.

Methodology

A study was conducted on undergraduate students studying engineering in Computer Science from 3rd June, 2023 to 30th September, 2023. The total number of students includes 290 students from five engineering colleges as illustrated in Table 12.1. A pre-test is conducted on the students and based on their performance students were divided into slow and fast learners. The ELLs were grouped with the combination of slow and fast learners to encourage peer interactions. After the implementation of DIA method on the participants, a post-test is conducted to observe the performance of the students. The marks obtained in the pre- and post- test were analyzed to evaluate the impact of DIA method. In the study, dependent variables are the marks obtained by the students both in pre- and post-test and independent variables are the teaching method and teacher's instruction. Learners

Table 12.1 Number of Research Participants from Each College.

S. No.	College	No of students
1	College A	57
2	College B	59
3	College C	56
4	College D	60
5	College E	58
	Total	290

Source: Author

and teachers are the active participants where teachers cascade the instruction and learners receive to improve their performance levels.

The students who score ≤60 marks are considered as slow learners and the students who obtain ≥61 marks are considered as fast learners. Based on the marks obtained by the marks in the study, statistical analysis was made to observe the efficacy of the DIA method. Pre- and post-test were conducted for 100 marks on reading comprehension, sentence completion, debate, analytical and argumentative writing.

Statistical analysis

To evaluate the impact of DIA method, the Chi-square test is performed college wise.

The formula to calculate the Chi-square statistic for a contingency table is:

$$\chi^2 = \Sigma\,(O - E)^2/E,$$

where χ^2 is the chi-square statistic, O is the observed frequency in each cell, E is the expected frequency in each cell. The sum is taken over all cells in the contingency table.

College A

	Pre-test	Post-test
Slow learners	39	23
Fast learners	18	34

Python code to calculate Chi-square statistic

```
from scipy. stats import chi2_contingency.
# Contingency table data
observed = [[39, 23], [18, 34]] # Rows: Pre-Test, Post-Test; Columns: Slow Learners, Fast Learners of College A
# Calculate chi-square statistic, p-value, degrees of freedom, and expected frequencies
chi2_stat, p_val, dof, expected = chi2_contingency(observed)
print("Chi-square statistic:", chi2_stat)
print("p-value:", p_val)
print("Degrees of freedom:", dof)
print("Expected frequencies:\n", expected)
```

Output

```
Chi-square statistic: 7.955955334987593
p-value: 0.004792940297127079
Degrees of freedom: 1
Expected frequencies:
[[31. 31.]
[26. 26.]]
```

Null hypothesis (H0): The proportion of slow and fast learners in the pre-test group is the same as in the post-test group, with no impact of the DIA method on the participants.

Alternate hypothesis (H1): The proportion of slow and fast learners in the pre-test group differs from the proportion in the post-test group, with the impact of the DIA method on the participants.

Level of significance (LOS): At 5% level of significance, the Chi-square table value with 1 degree of freedom is 3.8415.

Inference: For the proposed problem, the obtained Chi-square value is 7.9556, which is greater than the Chi-square table value, i.e. 3.8415, and it rejects the null hypothesis.

From the value obtained, it is observed that the DIA method has an impact.

College B

	Pre-test	Post-test
Slow learners	39	22
Fast learners	20	37

Output (Python code)

```
Chi-square statistic: 8.68794938165085
p-value: 0.0032032080825925065
Degrees of freedom: 1
Expected frequencies:
[[30.5 30.5]
[28.5 28.5]]
```

Null hypothesis (H0): The proportion of slow and fast learners in the pre-test group is the same as in the post-test group, with no impact of the DIA method on the participants.

Alternate hypothesis (H1): The proportion of slow and fast learners in the pre-test group differs from the proportion in the post-test group, with the impact of the DIA method on the participants.

Level of significance (LOS): AT 5% level of significance, the Chi-square table value with 1 degree of freedom is 3.8415.

Inference: For the proposed problem, the obtained Chi-square value is 8.6880, which is greater than the Chi-square table value, i.e. 3.8415, and it rejects the null hypothesis.

From the value obtained, it is observed that the DIA method has an impact.

College C

	Pre-test	Post-test
Slow learners	37	22
Fast learners	19	34

Output (Python code):
Chi-square statistic: 7.020147105852255
p-value: 0.008059760893656585
Degrees of freedom: 1
Expected frequencies:
[[29.5 29.5]
[26.5 26.5]]

Null hypothesis (H0): The proportion of slow and fast learners in the pre-test group is the same as in the post-test group, with no impact of the DIA method on the participants.

Alternate hypothesis (H1): The proportion of slow and fast learners in the pre-test group differs from the proportion in the post-test group, with the impact of the DIA method on the participants.

Level of significance (LOS): AT 5% level of significance, the Chi-square table value with 1 degree of freedom is 3.8415.

Inference: For the proposed problem, the obtained Chi-square value is 7.0201, which is greater than the Chi-square table value, i.e. 3.8415, and it rejects the null hypothesis.

From the value obtained, it is observed that the DIA method has an impact.

College D

	Pre-test	Post-test
Slow learners	42	27
Fast learners	18	33

Output (Python code):
Chi-square statistic: 6.683716965046889
p-value: 0.009729752787309698
Degrees of freedom: 1
Expected frequencies:
[[34.5 34.5]
[25.5 25.5]]

Null hypothesis (H0): The proportion of slow and fast learners in the pre-test group is the same as in the post-test group, with no impact of the DIA method on the participants.

Alternate hypothesis (H1): The proportion of slow and fast learners in the pre-test group differs from the proportion in the post-test group, with the impact of the DIA method on the participants.

Level of significance (LOS): AT 5% level of significance, the Chi-square table value with 1 degree of freedom is 3.8415.

Inference: For the proposed problem, the obtained Chi-square value is 6.6837, which is greater than the Chi-square table value, i.e. 3.8415, and it rejects the null hypothesis.

From the value obtained, it is observed that the DIA method has an impact.

College E

	Pre-test	Post-test
Slow learners	39	21
Fast learners	19	37

Output (Python code):
Chi-square statistic: 9.977380952380951
p-value: 0.00158474938011549
Degrees of freedom: 1
Expected frequencies:
[[30. 30.]
[28. 28.]]

Null hypothesis (H0): The proportion of slow and fast learners in the pre-test group is the same as in the post-test group, with no impact of the DIA method on the participants.

Alternate hypothesis (H1): The proportion of slow and fast learners in the pre-test group differs from the proportion in the post-test group, with the impact of the DIA method on the participants.

Level of significance (LOS): AT 5% level of significance, the Chi-square table value with 1 degree of freedom is 3.8415.

Inference: For the proposed problem, the obtained Chi-square value is 9.9774, which is greater than the

Chi-square table value, i.e. 3.8415, and it rejects the null hypothesis.

From the value obtained, it is observed that the DIA method has an impact.

Conclusions

In the study conducted on the five colleges, the effectiveness of the DIA method was assessed through the performance of ELLs in the post-test evaluation. By comparing scores from both the pre- and post-test assessments, the study showed substantial improvement among the students. Specifically, the analysis revealed a decrease in the total number of slow learners and an increase in fast learners following the implementation of the DIA method. This outcome highlights the suitability of the DIA approach in nurturing critical thinking skills and communicative competencies among learners within mixed-ability tertiary classrooms. The shift in learner profiles suggests that DIA effectively supports diverse learners, enabling improved engagement and learning outcomes in environments where abilities vary widely. Consequently, the findings underscore the potential of the DIA method to enhance pedagogical practices in tertiary education, particularly for promoting inclusive and effective teaching strategies in mixed-ability settings.

References

Ahn, S.-Y. and Kang, H.-S. (2017). South Korean university students' perceptions of different English varieties and their contribution to the learning of English as a foreign language. *J. Multilin. Multicul. Dev.*, 38(8), 712–725.

Allwright, R. L. (2003). Exploratory practice: Rethinking practitioner research in language teaching. *Lang. Teach. Res.*, 7, 113–141.

Alrabai, F. (2016). Factors underlying low achievement of Saudi EFL learners. *Internat. J. English Linguist.*, 6(3), 21–37.

Anburaj, G., Christopher, G., and Ming, N. (2014). Innovative methods of teaching English language. *IOSR J. Human. Soc. Sci.*, 19(8), 62–65.

Block, D. (2010). Globalization and language teaching. The handbook of language and globalization. 287–304.

Brown, D. (2000). Principles of language learning and teaching (4th ed.). New York, USA: Pearson Education. 139–141.

Carmel, M. S., et al. (2020). Learner-centric approach to learning English. New Approaches & Methods in ELT. 1, 117.

Chapman, C. and King, R. (2005). 11 practical ways to guide teachers toward differentiation (and an evaluation tool). *J. Staff Dev.*, 26(4), 20–25.

Dendrinos, B. (1992) . The EFL textbook and ideology. Athens: Grivas. 46.

Dos Santos, L. M. (2020). The discussion of communicative language teaching approach in language classrooms. *J. Edu. e-Learn. Res.*, 7(2), 104–109.

Eve, P. H. D. (2022) . Teaching LSRW skills through the test, teach, test (TTT) method. *Contemp. ELT Strateg. Engg. Pedag. Theory Prac.*. 159–161

Harmer, J. and Ana, A. (2008). Your Turn. Textbook 2. Wien: Langenscheidt. 12.

Khasawneh, M. (2022). The relationship of curriculum, teaching methods, assessment methods, and school and home environment with learning difficulties in English language from the students' perspectives. *J. Innov. Edu. Cult. Res.*, 3(1), 41–48.

Larsen-Freeman, D. (1986). Techniques and principles in language teaching. Oxford: Oxford University Press. 203–205.

Lee, J. S. and Lee, K. (2019). Perceptions of English as an international language by Korean English- major and non-English-major students. *J. Multilin. Multicul. Dev.*, 40(1), 76–89.

Nunan, D. (1989). Designing tasks for the communicative classroom. New York: Cambridge University Press. 96–104.

O'Meara, J. (2010). Beyond differentiated instruction. Thousand Oaks, CA: Corwin Press. 103–119.

Radhika, Ch. (2020) . Paradigm shift from teacher centric to learner centric classroom. 109–112.

Richards, J. and Theodore, R. (2007). Approaches and methods in language teaching. Cambridge: Cambridge University Press. 153.

Thornbury, S. (2000). How to teach grammar. Harlow: Longman Pearson Education. 129.

Tomlinson, C. A. (1999). The differentiated classroom: Responding to the needs of all students. Alexandria, VA: Association for Supervision and Curriculum Development. 102–122.

Thornbury, S. (2000),Tomlinson, C. A. (1999) and Tomlinson, C. A. and Allan, S. D. (2000). Leadership for differentiating schools & classrooms. 36.

13 Upskilling strategies for HR in the age of AI: Empowering organizational transformation

Vanitha A.[1,a] *and Latha K.*[2,b]

[1]Associate Professor, Department of Management studies, Saveetha Engineering College, Chennai, Tamil Nadu, India

[2]Associate Professor, Department of Management Studies, SRM Valliammai Engineering College, Chennai, Tamil Nadu, India

Abstract

Organizations must change to keep up with the rapidly developing nature of work in an era defined by fast technological innovation. This study suggests that the secret to realizing organizational potential lies in the strategic integration of artificial intelligence (AI) into human resource management (HRM) procedures, especially through all-inclusive upskilling initiatives. By clarifying the mutual benefits of AI and upskilling, the research explores how these components might operate together to promote an environment of ongoing education, adaptability, and creativity in the workplace. The study takes a multifaceted approach, using knowledge from case studies, scholarly literature, and industrial practices. The study aims to offer practical suggestions for HRM practitioners by thoroughly examining AI-driven upskilling. It highlights the crucial role that strategic planning plays in assisting HRM practitioners in managing the opportunities and difficulties that arise from integrating AI into the workplace. In the end, this study hopes to add to the conversation around HRM in the AI era by providing a road map for businesses looking to use upskilling's transformative power for long-term organizational growth.

Keywords: Strategic HRM, artificial intelligence, continuous learning, organizational transformation, upskilling initiatives, workplace innovation

Introduction

Modern human resource management (HRM) strategies and advances in artificial intelligence (AI) technology have become pivotal drivers of organizational success in the rapidly evolving business landscape. A new era of HRM practices has been ushered in with the integration of AI, which is characterized by enhanced efficiency, data-driven decision-making, and the potential for profound organizational change. As organizations leverage AI not only to optimize existing human resource (HR) processes but also to shape their workforce proactively in anticipation of future challenges and opportunities, the concept of strategic HRM gains increasing significance.

AI has revolutionized many aspects of HRM, from talent acquisition to performance management to learning and development. Recruiting platforms powered by artificial intelligence, for example, analyze vast datasets, streamline candidate selection processes, and identify top talent with unprecedented precision. Similarly, AI-driven performance management systems provide real-time feedback and actionable insights, enhancing employee productivity, engagement, and performance. As a result of these technological advancements, AI can reshape traditional HRM practices and facilitate organizational agility and competitiveness.

The integration of AI in HRM, however, presents challenges and complexities. The integration of AI technologies into HR processes has raised concerns about data privacy, algorithmic bias, and workforce displacement. In sensitive areas such as performance evaluation and talent management, using AI in decision-making requires careful scrutiny and responsible implementation. Also, AI-driven automation raises questions about the future of work and the need to address potential skill gaps and ensure workforce readiness.

Upskilling and reskilling initiatives emerge as strategic imperatives for organizations to adapt and transform in the age of AI. To prepare their workforce for AI-augmented environments, organizations must invest in continuous learning and development programs as AI reshapes job roles and skill requirements. In order to foster a culture of innovation, adaptability, and lifelong learning within the organization, upskilling must not only focus on technical training but also include critical thinking, problem-solving, and interpersonal skills.

With these considerations the evolving role of strategic HRM in the age of AI, highlighting the

[a]vanithaa@saveetha.ac.in, avanitha.inc@gmail.com, [b]lathak.mba@srmvalliammai.ac.in, lrk.latha@gmail.com

DOI: 10.1201/9781003587538-13

importance of upskilling initiatives for enabling organizational transformation. The purpose of this research is to reveal key insights, challenges, and best practices associated with the strategic integration of AI into HRM as well as the imperative for organizational agility, competitiveness, and sustainability by examining relevant research and theoretical frameworks.

Review of literature

Organizational performance is enhanced and organizational change is facilitated by aligning HR practices with organizational goals. As HR processes are streamlines, decision-making is enhanced, and workforce management is optimized using AI, strategic HRM takes on increased significance in the age of AI. According to Barney and Wright (2017), HRM plays an essential role in enabling organizational agility and adaptability in the face of AI-driven disruption.

AI is integrated into HRM through a variety of applications, including talent acquisition, performance management, learning and development, and employee engagement. AI-based recruitment platforms, such as those utilizing natural language processing algorithms and machine learning algorithms, facilitate efficient candidate screening and selection (Rudnitsky et al., 2020). Performance management systems powered by AI also provide real-time feedback and data-driven insights to enhance employee productivity and development (Parry and Tyson, 2018). In enhancing HRM practices, AI has the potential to be transformative.

Adapting organizational roles and skills to AI and ensuring workforce readiness require upskilling initiatives. According to Cappelli and Tavis (2018), upskilling plays an important role in preparing employees for AI-driven changes, emphasizing the importance of continuous learning and development. AI-augmented workplaces increasingly value soft skills such as creativity, adaptability, and emotional intelligence, which are cultivated through effective upskilling programs (Marr, 2018).

HRM integration of AI presents challenges relating to data privacy, algorithmic bias, and workforce displacement, despite its potential benefits. Using AI in decision-making processes, particularly in areas such as talent management and performance evaluation, requires careful consideration (Davenport et al., 2019). HR leaders who are navigating AI-driven transformation should also ensure equitable access to upskilling opportunities and minimize job displacement risks.

Study objectives and research questions

Objectives

- To explore the strategic integration of AI into HRM procedures through upskilling initiatives for organizational transformation and growth.
- To offer practical suggestions for HRM practitioners by thoroughly examining AI-driven upskilling and its role in fostering ongoing education, adaptability, and creativity in the workplace.
- Highlighting the crucial role of strategic planning in assisting HRM practitioners in managing the challenges and opportunities of AI integration, providing a roadmap for businesses.

Research questions

- How can AI and upskilling initiatives work together to promote continuous education, adaptability, and creativity in the workplace?
- What are the practical implications of AI-driven upskilling for HRM practitioners?
- What role does strategic planning play in assisting HRM practitioners in managing the challenges and opportunities of AI integration?

Importance of the study

This study is crucial as it emphasizes the strategic integration of AI into HRM procedures, particularly through upskilling initiatives, to enable organizational transformation. By clarifying the mutual benefits of AI and upskilling, the research highlights how these components can work together to foster continuous education, adaptability, and creativity in the workplace, ultimately unlocking organizational potential. The study's multifaceted approach, drawing from case studies, scholarly literature, and industrial practices, offers practical insights for HRM practitioners by thoroughly examining AI-driven upskilling. It underscores the critical role of strategic planning in helping HRM practitioners navigate the opportunities and challenges arising from AI integration, providing a roadmap for businesses to leverage upskilling for long-term organizational growth. The research sheds light on the imperative for organizational agility, competitiveness, and sustainability through upskilling initiatives in the age of AI, aligning HR practices with organizational goals for enhanced performance and change facilitation.

Section for data gathering and analysis

Research design: The methodology involves a multifaceted approach utilizing case studies, scholarly literature, and industrial practices to offer practical suggestions for HRM practitioners in the context of AI-driven upskilling.

Data collection methods: Data is gathered from various sources such as case studies, scholarly literature, and industrial practices to provide a comprehensive understanding of AI-driven upskilling initiatives in HRM.

Analytical techniques: The analysis includes examining the integration of AI into HRM procedures through upskilling initiatives to promote ongoing education, adaptability, and creativity in the workplace.

Conceptual model for aligning upskilling strategies with organizational goals

Assess organizational needs: Conduct a comprehensive analysis of current skill gaps and future requirements to tailor upskilling programs effectively (Figure 13.1)

Develop customized training programs: Design upskilling initiatives that specifically target identified skill deficiencies, ensuring alignment with strategic objectives

Leadership support and communication: Secure leadership buy-in for upskilling initiatives and communicate the connection between enhanced skills and organizational success to drive employee engagement.

Continuous evaluation and adaptation: Regularly evaluate the impact of upskilling efforts on organizational goals and adjust strategies to meet evolving business needs.

Integration of soft skills development: Include training on critical thinking, problem-solving, and interpersonal skills alongside technical upskilling to promote holistic employee development and organizational growth.

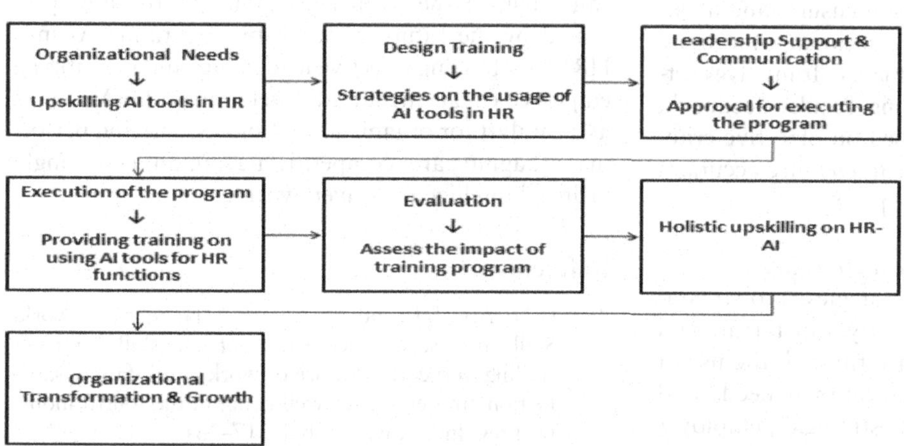

Figure 13.1 Self Proposed Conceptual model
Source: Author

Key dimensions of AI in HRM

AI integration in HRM presents pivotal dimensions for organizational transformation. With AI, HR processes can be optimized and decision-making can be enhanced, from talent acquisition to performance management. A culture of innovation and strategic alignment with organizational objectives are necessary for successful AI integration. As AI-driven environments become more prevalent, upskilling initiatives play a critical role in enabling a successful transformation. Organizations can remain competitive in the evolving work landscape by investing in both technical and soft skill development.

AI integration in HRM

Through AI integration in HRM, processes such as talent acquisition and performance management are revolutionized by leveraging algorithms to screen candidates efficiently and provide real-time feedback. As technology advances, it improves decision-making, enhances workforce management, and ensures organizational competitiveness.

Upskilling initiatives

The upskilling initiative addresses the growing skill requirements stemming from AI and technological advancements. Continual learning programs, reskilling efforts, and development opportunities aim to enhance employees' competencies, foster adaptability, and promote lifelong learning within the organization.

Organizational transformations

The transformation of an organization involves making strategic changes to its structures, processes, and cultures as a result of internal and external influences. For companies to remain competitive in dynamic markets, they need to embrace innovation, foster agility, and drive sustainable growth, leveraging technologies like AI to facilitate change.

Ethical considerations in AI-powered HRM
The ethical considerations in AI-powered HRM include fairness, transparency, and accountability. Data privacy, algorithmic bias, and the ethical use of AI in sensitive HR functions, including talent management, performance evaluation, and performance management, are addressed.

Future of work and workforce adaptation
Organizations need to adapt their workforce strategies as AI and automation shape the future of work. By upskilling employees, encouraging flexibility, and reimagining job roles, organizations can harness the potential of technology while ensuring workforce resilience and readiness.

Leadership and change management
Managing organizational change in the age of AI requires leadership and change management. To achieve strategic goals and sustain competitive advantage, effective leaders inspire vision, drive innovation, and cultivate a culture of change. They guide employees through transitions, overcome resistance, and foster engagement.

Employee engagement and satisfaction
Organizational success and performance depend on employee engagement and satisfaction. Employees' emotional attachment to their work affects productivity, retention, and organizational culture. Work-life balance initiatives and recognition programs promote engagement and enhance job satisfaction.

Performance management effectiveness
Effective performance management ensures alignment of individual goals with organizational objectives, driving productivity and performance. It involves setting clear expectations, providing regular feedback, and evaluating performance based on objective criteria, leveraging AI-driven systems to enhance accuracy, fairness, and transparency in the process.

Talent acquisition and recruitment strategies
The goal of talent acquisition strategies is to streamline recruitment processes, identify top talent, and enhance the candidate experience through the use of AI technologies. To meet organizational needs and objectives, they include sourcing strategies, employer branding initiatives, and innovative recruitment methods.

Learning and development programs
The purpose of learning and development programs is to enable employees to acquire new skills, enhance performance, and adapt to changing job demands. A culture of continuous learning and growth is fostered within the organization through training initiatives, professional development opportunities, and mentoring programs.

Key findings on HR practice in the age of AI

- Strategic integration of AI into HRM through upskilling initiatives enhances organizational agility and adaptability
- AI-driven upskilling promotes ongoing education, adaptability, and creativity in the workplace, fostering a culture of innovation
- Ethical considerations in AI-powered HRM include fairness, transparency, and accountability, emphasizing the need for responsible AI implementation
- AI integration in HRM optimizes decision-making processes, talent management, and performance evaluation, requiring a strategic alignment with organizational objectives
- Continuous learning and development programs are essential for preparing the workforce for AI-augmented environments, emphasizing the importance of upskilling initiatives

Conclusions

A holistic approach to organizational transformation is imperative in the age of AI due to the multifaceted nature of strategic HRM. Organizations must invest in upskilling initiatives, prioritize ethical considerations, and foster a culture of innovation and adaptability as AI continues to reshape HR practices. Organizations can navigate the challenges and opportunities presented by the future of work by integrating AI into HRM, embracing effective leadership, and prioritizing employee engagement and development. HRM serves as a catalyst for organizational success, driving performance, agility, and competitiveness in an increasingly artificial intelligence-driven world.

References

Abe, E. N., Abe, I. I., and Adisa, O. (2021). Future of work: Skill obsolescence, acquisition of new skills, and upskilling in the 4IR. Future of work, work-family satisfaction, and employee well-being in the fourth industrial revolution. IGI Global, 217–231.

Barney, J. B. and Wright, P. M. (2017). On becoming a strategic partner: The role of human resources in gaining competitive advantage. *Human Res. Manag.*, 56(1), 173–188.

Cappelli, P. and Tavis, A. (2018). The corporate implications of longer lives. *Harvard Busin. Rev.*, 96(5), 84–91.

Davenport, T. H., Guha, A., and Grewal, D. (2019). How AI will change the way we make decisions. *Harvard Busin. Rev.*, 97(1), 68–76.

Gomathi, S., Rajeswari, A., and Kadry, S. (2023). Emerging HR practices—digital upskilling: A strategic way of talent management and engagement. Disruptive artificial intelligence and sustainable human resource management. River Publishers, 51–63.

Kasih, E., Qalbia, F., and Novrizal, N. (2022). Empowering talent in the age of artificial intelligence: Innovations in human resource management. *Internat. Conf. Edu. Soc. Sci. Technol.*, 1(2), 287–295.

Kumar, W. and Shabir, G. (2024) AI Empowerment: Revolutionizing job performance through artificial intelligence integration, 1–10.

Li, L. (2022). Reskilling and upskilling the future-ready workforce for industry 4.0 and beyond. *Inform. Sys. Front.*, 1–16.

Marr, B. (2018). How artificial intelligence is revolutionizing business in 2019. Forbes. Retrieved from https://www.geeksforgeeks.org/how-artificial-intelligence-is-revolutionizing-the-business-landscape/.

Ogedengbe, D. E., James, O. O., Afolabi, J. O. A., Olatoye, F. O., and Eboigbe, E. O. (2023). Human resources in the era of the fourth industrial revolution (4ir): Strategies and innovations in the global South. *Engg. Sci. Technol. J.*, 4(5), 308–322.

Pandey, A., Balusamy, B., and Chilamkurti, N. (2023). *Disruptive artificial intelligence and sustainable human resource management: Impacts and innovations-The future of HR*. CRC Press, 26–38.

Parry, E. and Tyson, S. (2018). Desired goals and actual outcomes of e-HRM. *Human Res. Manag. J.*, 28(1), 27–48.

Pradhan, I. P. and Saxena, P. (2023). Reskilling workforce for the artificial intelligence age: Challenges and the way forward. The adoption and effect of artificial intelligence on human resources management, Part B. Emerald Publishing Limited, 181–197.

Rudnitsky, A., Bendersky, C., and Cortes, C. (2020). Learning representations for HR analytics, 16–25.

Sofia, M., Fraboni, F., De Angelis, M., Puzzo, G., Giusino, D., and Pietrantoni, L. (2023). The impact of artificial intelligence on workers' skills: Upskilling and reskilling in organisations. *Inform. Sci. Internat. J. Emerg. Transdis.*, 26, 39–68.

14 Unleashing innovative engagement in programming education

Sneha Pokharkar[a], *Shridhar Khandekar*[b], *Dipti Sakhare*[c], *Vrushali Waghmare*[d] *and Mahesh Goudar*[e]

Department of E and TC Engineering MIT Academy of Engineering, Pune, Maharashtra, India

Abstract

In education, fostering community and student success is a priority, with educators increasingly incorporating games into the curriculum to enhance learning outcomes. Learning programming poses challenges, including mastering syntax, logic, and problem-solving skills, navigating evolving languages and tools, and fostering persistence and a growth mindset. Engaging students through online programming contests promotes creativity and broadens perspectives, encouraging them to think innovatively and explore alternative viewpoints.

Keywords: Component, programming, quizzes, out-of-box thinking, contest, problem solving, competitive, innovation, class engagement

Introduction

In engineering and programming education, the use of teaching tools is essential for creating dynamic learning environments and fostering critical skills development. From traditional lectures to technology-driven approaches like simulation tools and virtual reality, these tools facilitate hands-on learning, collaboration, and creativity, bridging the gap between theory and practice. Similarly, in programming education, tools like integrated development environments (IDEs), online coding platforms, visual programming tools, and simulation/emulation tools enhance students' problem-solving skills and software development proficiency, preparing them for success in their respective fields (Digital Professional Development space, n.d.; Mucundanyi and Woodley, 2021; Salleh et al., 2011; Poloju and Naidu, 2020; Ghavifekr and Rosdy, 2015).

Challenges students face while learning programming

Programming can present students with a number of difficulties. The following are some typical obstacles encountered by students while learning programming:

- Programming requires a comprehension of abstract and complex concepts, such as algorithms, data structures, and syntax rules. These concepts may initially be difficult for students to grasp because they require a shift in thinking and problem-solving strategies.
- Errors in syntax and debugging: Writing code requires adhering to specific syntax rules, and even minor syntax errors can lead to frustrating bugs.
- Debugging and troubleshooting code can be time-consuming and requires a methodical approach to identify and correct errors.
- To be successful at programming, one must be able to think logically and deconstruct difficult problems into simpler ones.
- The creation of efficient algorithms and problem-solving strategies can be difficult for beginners (Jayashree and Kale, 2018).
- Time management and perseverance: Programming requires time and perseverance to learn. Students may struggle to effectively manage their time to regularly practice coding and complete programming assignments or projects.
- Learning to code is best accomplished through practice and experimentation. Some students may have trouble transferring their theoretical understanding to more concrete, real-world contexts (Ever and Hirsh, 2013).
- Programming is an ever-evolving field, with new languages, frameworks, and tools appearing on a regular basis. Keeping up with the latest technology developments, best practices, and trends can be challenging for students (Robles, 2013).
- Programming necessitates experimentation, and errors are a normal part of the learning process. Some students may grapple with a fear of failure,

[a]sneha.pokharkar@mitaoe.ac.in, [b]sakhandekar@mitaoe.ac.in, [c]dysakhare@mitaoe.ac,in, [d]vrushali.waghmare@mitaoe.ac.in, [e]mdgouadar@mitaoe.ac.in

DOI: 10.1201/9781003587538-14

which can impede their coding development and willingness to take risks.

- Some students may experience imposter syndrome, in which they question their abilities and believe they do not belong in the field of programming. This can affect their motivation and confidence to pursue programming.

Approaches of teaching programming

The structured programming approach facilitates the construction of programs that are accurate, simple to comprehend, and straightforward to modify and keep up-to-date.

A problem solving technique involves writing computer programs that consists of the following four stages: understanding, designing, writing, and reviewing.

Students in an IBL classroom are actively encouraged to pose and investigate questions, conduct research on various topics, and take an active role in the process of their own education under this method of teaching. It inspires analytical thinking, the solving of problems, and a more in-depth comprehension of the material being studied.

The cooperative learning strategy places an emphasis on the students working together to complete group projects. Students learn from and collaborate with one another through the process of working on projects, conversations, and activities in groups. In the classroom, it helps to cultivate a sense of community, as well as teamwork and communication skills.

The term "experiential learning" refers to a teaching method that places an emphasis on learning by direct participation in real-world activities. It incentivizes students to actively participate in activities, experiments, and field trips that are directly related to the topic at hand. It fosters real-world application, encourages introspection, and increases overall comprehension.

Differentiated teaching strategies accommodate for the fact that each student has unique needs and learning preferences. It entails changing teaching tactics, content, and evaluations to meet the strengths, interests, and readiness levels of individual students. It encourages individualized education as well as teaching that are centered on the learner.

Project-based learning (PBL) is a mode of instruction that requires students to carry out extensive, multi-faceted projects in order to investigate and display their level of comprehension regarding a specific subject matter or issue. Thinking critically, developing research abilities, being creative, and learning to educate oneself independently are all fostered by this activity.

Innovative attempt is to gain knowledge through exposure to actual world examples, hands-on coding experience, and concept recycling.

Methodology

In order to increase class engagement and encourage active participation in our programming course, it was decided to implement live quizzes. The objective of these quizzes is to create an interactive and dynamic learning environment in which students can implement their programming skills in the real world. Incorporation of live exams into our curriculum will stimulate critical thinking, promote healthy competition, and provide students with immediate feedback, thereby fostering a deeper understanding of the subject matter. Students will be more engaged, more motivated, and better equipped to comprehend the complexities of programming concepts through this approach.

To achieve all above objectives an online platform named QUIZIZZ having following features was selected:

- (Quiz + lesson) editor: Lessons and quizzes at an instructor's speed, where the pace is set by the instructor and the class works through each question together. Lessons and quizzes that students may do at their own pace, with real-time results and a global leaderboard (Quizizz, n.d.).
- (Test + instruction) editor: You can transfer questions from other quizzes and lectures into any of six different question kinds (Quizizz, n.d.).
- Reports: Obtain in-depth information on your class and individual students for each test you administer. Distribute to guardians so they can keep tabs on their children's development.
- Alter the difficulty and pace of your quizzes to suit your needs with the available customization options.
- Share your quiz with other teachers and/or enlist them as collaborators to improve it (Quizizz, n.d.).

Figures 14.1 and 14.2 shows live dashboard of quizizz platform. Figure 14.3 shows result of individual student consisting name, accuracy, points and score whereas Figure 14.4 shows quiz summary. In Figure 14.5, participants' data generated after the quiz is shown. After conducting quiz for all the concepts, it is observed that quiz conduction should be used as a supplement to overall learning process, they

Figure 14.1 Live dashboard of quizizz-1
Source: "Quizizz,"quizizz.com

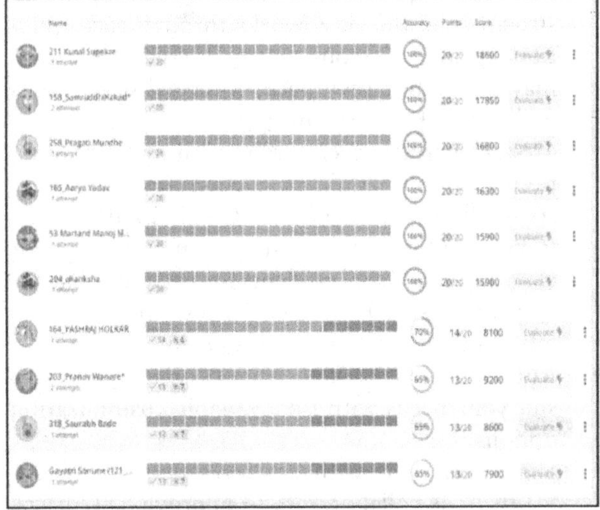

Figure 14.3 Result of individual student
Source: "Quizizz,"quizizz.com

are most effective when combined with other learning resources such as textbooks, tutorials, and hands-on coding practice. It provides a way to assess and reinforce the understanding, however they should not be the sole method of learning programming.

To enhance the students' confidence collaboration with others can be better solution. This will be achieved by joining coding communities, participating in forums, or contributing to open-source projects. As an enhancement after the quizzes, programming contest on CODECHEF platform was scheduled. Different problem statements along with the details of output format and subtask were provided and students will write the code, run and submit. The score will be based on submission. Figure 14.6 shows the sample problem statement, output format and subtasks in problem statement. Students will write code in the section given, run and submit the code. Marks will be given based on

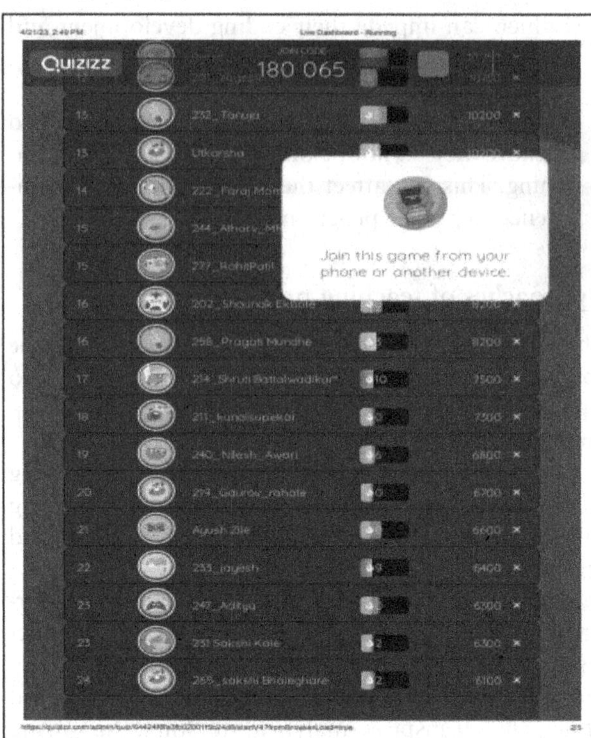

Figure 14.2 Live dashboard of quizizz-2
Source: "Quizizz,"quizizz.com

Name	Value
Game Started On	Tue 18 Apr 2023,01:26 PM
Game Type	Live Quiz
Participants	63
Total Attempts	63
Class Accuracy	72%
Game Ends On	Tue 18 Apr 2023,01:45 PM

QUIZIZZ

Figure 14.4 Result of a class
Source: "Quizizz,"quizizz.com

Rank	First Name	Last Name	Attempt #	Accuracy	Score	Correct	Incorrect	Unattempted	Total Time Taken	Started At	Info
1	139_Prathmesh		40	100 %	36150	39	0	0	10:13	Tue 18 Apr 2023,01:28 PM	ne Mobile on Android
2	167_Lavlesh		40	97 %	34350	39	1	0	08:16	Tue 18 Apr 2023,01:28 PM	ne Mobile on Android
3	127_Akshar		40	100 %	32400	40	0	0	11:34	Tue 18 Apr 2023,01:28 PM	obile Safari on iOS
4	126	Mihir Fulari	40	95 %	31400	38	2	0	09:57	Tue 18 Apr 2023,01:27 PM	ne Mobile on Android
5	158_Samruddhi		40	87 %	31150	35	5	0	08:35	Tue 18 Apr 2023,01:27 PM	obile Safari on iOS
6	146_Anushka		40	97 %	30300	39	0	1	09:16	Tue 18 Apr 2023,01:27 PM	ne Mobile on Android
7	173_Shravani		40	90 %	30000	36	4	0	08:03	Tue 18 Apr 2023,01:28 PM	ne Mobile on Android
8	119_Tanmay	Patle	40	84 %	29400	33	6	0	11:51	Tue 18 Apr 2023,01:27 PM	ne Mobile on Android
9	178_Vitthal	Patil	40	82 %	29200	33	7	0	07:53	Tue 18 Apr 2023,01:26 PM	ne Mobile on Android
10	130_Omkar		40	90 %	28500	36	4	0	09:03	Tue 18 Apr 2023,01:27 PM	ne Mobile on Android

Figure 14.5 Participants data generated after the quiz
Source: "Quizizz,"quizizz.com

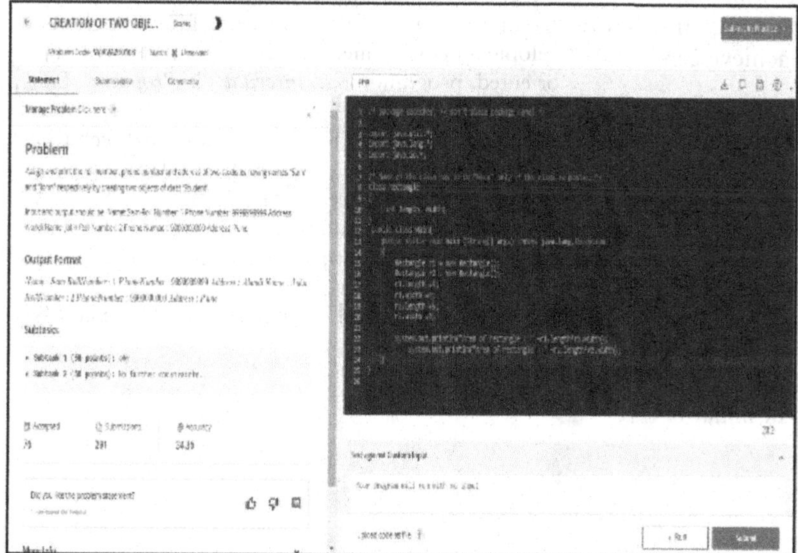

Figure 14.6 Sample problem statement in contest
Source: "Quizizz,"quizizz.com

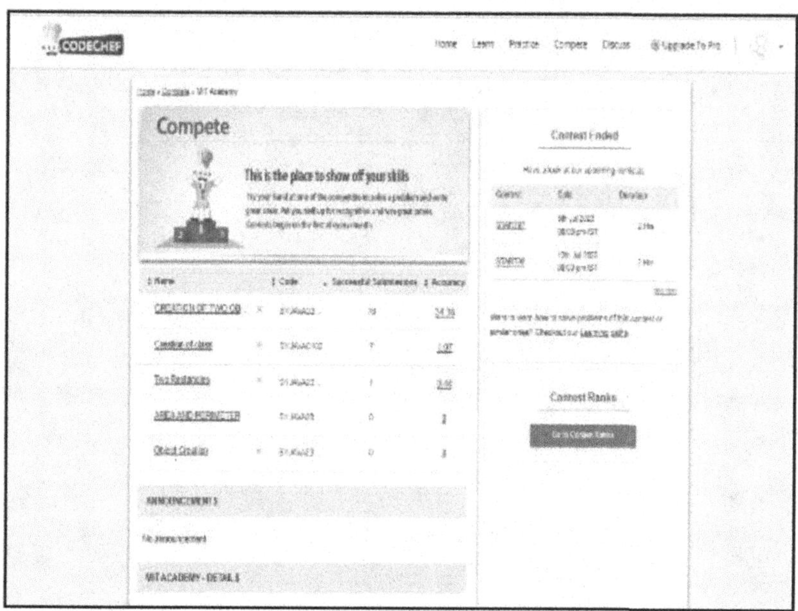

Figure 14.7 Designed contest with 5 problem statement
Source: "Quizizz,"quizizz.com

the submission. Figure 14.7 shows the designed contest with 5 problem statements (Codechef, n.d.).

Conclusions

Interactive quizzes reinforce learning, while programming contests provide motivation, problem-solving opportunities, and valuable competitive experience, enhancing student engagement in programming courses. These contests encourage students to set goals, challenge themselves, and develop problem-solving, time management, collaboration, and networking skills. Through experiential learning, recognition, and rewards, students gain confidence to clear programming tests, participate in hackathons, and achieve certifications in programming skills.

References

Digital professional development space. (n.d.). dpd-mlii.mfu.ac.th. https://dpd-mlii.mfu.ac.th/index.php/ddt-landing-page/. (accessed July. 3, 2023).

Mucundanyi, G. and Woodley, X. (2021). Exploring free digital tools in education. *Internat. J. Edu. Dev. Inform. Comm. Technol.*, 17(2), 96–103.

Salleh, S., Shukur, Z., and Judi, H. (2013). Analysis of research in programming teaching tools: An initial review. *Proc. Soc. Behav. Sci.*, 103, 127–135. 10.1016/j.sbspro.2013.10.317.

Poloju, Kiran Kumar, and Vikas Rao Naidu. "Impact of E-tools in Teaching and Learning for Undergraduate Students." In *Innovations in Electronics and Communication Engineering: Proceedings of the 8th ICIECE 2019*, pp. 783–790. Springer Singapore, 2020.

Ghavifekr, S. and Rosdy, W. A. W. (2015). Teaching and learning with technology: Effectiveness of ICT integration in schools. *Internat. J. Res. Edu. Sci. (IJRES)*, 1(2), 175–191.

Jayashree, A. P. and Kale, S. P. (2018). Impact of ICT tools in logic development of computer programming skills. *J. Engg. Edu. Transform.* 33(1), 7–15.

Ever, M. and Hirsh, L. (2013). Interpreter and applied development environment for learning concepts of object oriented programming. *Internat. J. Program. Lang. Appl. (IJPLA)*, 3(3), 17–27.

Robles, A. C. M. (2013). The use of educational web tools: An innovative technique in teacher education courses. *Internat. J. Modern Edu. Comp. Sci.*, 5, 34–40. 10.5815/ijmecs.2013.02.05.

"Quizizz," quizizz.com. https://quizizz.com/admin/reports. (accessedFeb7, 2023).

"Codechef",codechef.com. https://www.codechef.com/SYJAVAMIT.

15 Effective assessment of teaching and learning methods in management studies

P. Periasamy[1,a], *R. Ravimohan*[1,b], *Ramasundaram G.*[2,c], *Deepak R.*[1,d], *Sangeetha P.*[1,e] *and Srinidhi S.*[3,f]

[1]Associate Professor, Saveetha Engineering College, Chennai, Tamil Nadu, India

[2]Professor, Saveetha Engineering College, Chennai, Tamil Nadu, India

[3]I year M.B.A. Student, Saveetha Engineering College, Chennai, Tamil Nadu, India

Abstract

This article provides a systematic, cross-sectional examination of the antecedents of the effectiveness of different teaching methods in management research. Teaching methods are broad methods used to teach students how to achieve academic success, helping students understand course content and learn how to apply the content to specific situations, Therefore, this paper aims to conduct an empirical study examining the effects of policy on the academic performance of the students in the MBA program, evaluation of interesting skills adopted by teachers, and support provided by the teaching and learning process methods which are the best teaching methods used for effective learning and are topics related to the studies and methods offered in the MBA program. This article uses qualitative methods to organize and synthesize findings from studies in educational psychology, educational technology, higher education, and management. This particular research was initiated to map the relationship through student evidence to better understand and deepen what the department is trying to achieve and what it has achieved. Primary data has been collected from MBA students using a structured questionnaire. A total of 162 samples were collected from 240 students. Chi-square test, T-test, ANOVA and regression were used to analyze primary data. SPSS version 2023 was used for analysis. This paper represents research findings that demonstrate a positive relationship between outcomes and teachings – learning methods, curriculum and assessment. Additionally, future learning methods are needed to support and improve teaching and learning management.

Keywords: Challenges and solutions, MBA curriculum, skills assessment, teaching methods

Introduction

New teaching methods are sought in business education, especially in the Master of Business Administration (MBA) program. As the environment becomes more dynamic, educational institutions will change how they teach to ensure that students have theoretical knowledge and the practical skills needed to manage today's business operations. This chapter focuses on research on teaching methods that promote engagement, active learning and retention among MBA students. Developing the critical management skills needed in the 21st-century world, considering the non-business role of business graduates, teachers and professional education (Farashahi and Tajeddin, 2018). While existing literature supports the practice of blended learning in higher education including business schools, empirical evidence that it is effective in MBA context is scarce. Therefore, this study aims to conduct an empirical study examining the effects of policy on the academic performance of students in the MBA program. Conducted to contribute to the ongoing debate about new teaching methods in business schools, our research aims to demonstrate, through research and empirical studies, the effectiveness of these methods in improving deep learning, professional development, and student success in MBA programs (Malik et al, 2024). Although educational institutions use different methods and curriculum to implement effective teaching methods and research shows some benefits of some teaching methods, educational institutions are still searching for the best way to achieve these goals. Research rarely compares different learning strategies based on different learning goals (Zonen and Kalyan, 2012). As of now, the survey compares the best admission process among leading students of Saveetha College of Engineering.

Literature review

According to Bloom's (1956) educational theory, the entire learning process consists of office stages: understanding, applying, analyzing, processing and evaluating. In addition to the intellectual level, which

[a]periasamyp@saveetha.ac.in, [b]ravimohanr@saveetha.ac.in, [c]ramasundaramg@saveetha.ac.in, [d]deepakrajagopalan@saveetha.ac.in, [e]sangeethap@saveetha.ac.in, [f]ssrinidhi917@gmail.com

DOI: 10.1201/9781003587538-15

focuses on the acquisition of basic knowledge, the four continuous levels focus on practical activities and knowledge development. Teaching methods such as lectures, video presentations, guest lectures, structured learning, group discussions and experiential learning according to (Armstrong, 2010) are often used in higher education. This research examines teaching and learning methods that aim to increase the effectiveness of non-traditional activities. The proliferation of digital media has increased student confusion during learning and made it difficult for teachers to stay engaged with the students. Therefore, many secondary schools have turned to work-based learning targeting business school students to solve this problem (Sreemahadevan, 2024). This approach can be achieved by encouraging participation and collaboration which improves critical thinking, problem solving skills and interpersonal skills valued in the business world (Zamir et al., 2021). Zamir examined teaching methods for developing critical thinking in higher education and showed that: University professors mainly rely on the method based on lectures and reports. Research shows that students prefer distance learning, self-improvement and higher cognitive skills, teamwork and self-confidence throughout the course (Sankar and Raju, 2011). The foreign program is highly effective in terms of positive relationship, trust building, strategic communication, motivation, cultural adaptation and performance (Arachchige et al., 2020). Studying abroad through experiential learning improves individual and team behavior. In addition to the urban environment, this off-the-job program uses experiential learning to achieve multiple outcomes (Thisara, 2020). The MBA curriculum should be tailored to industry needs and include practical knowledge, experience and research to bridge the gap between academic learning and professional practice. The aim of management education should prioritize the development of general strength and management skills, enhance and enhance learning, develop creative thinking, provide adequate learning resources, and develop problem-solving skills.

The variables involved in the study are depicted in Table 15.1.

Research objectives

To examine the relationship between teaching and learning methods and learning outputs in management department of Saveetha Engineering College.

1. To examine the relationship between the curriculum and learning outputs in management department of Saveetha Engineering College.
2. To examine the relationship between assessment methods and learning outputs in management

Table 15.1 Variables and items included in the study

Learning methods	Curriculum	Skill assessment	Challenges would be confidently faced (outcomes)
Lectures	Build deep general and management skills	Newspaper cutting and interpretation	Time management
Video presentations	Makes learning more and faster	Quiz	Understanding the concepts
Guest lectures (from experts outside)	Generates creative thinking	Case studies	Verbal and non-verbal communication
Action learning (Like OBT)	Provides more resources to learn	PowerPoint presentation on the given topic	Written communication
Group discussions	Improves problem solving skills	Video presentation on the given topic	Ability to work in team
Internship-based learning		Field study	Ability to plan your own work
Project-based learning		Writing assignment on a given topic	Networking skills
		Debates	Analytical skills
			Group discussion in a forum
			Confidence in tackling problems
			Placement
			Career demands

National Institute for Learning Outcomes Assessment. (2011). *NILOA Transparency Framework*.
Source: Author

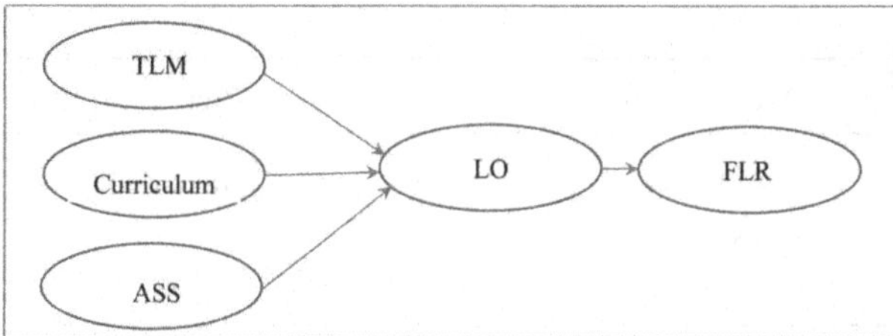

Figure 15.1 Conceptual framework

TLM – Teaching learning methods; ASS – Assessment; LO – Learning outcome/output; FLR – Future learning recommendations

Source: Author

department of Saveetha Engineering College (Figure 15.1).

Hypothesis

"H_0: There is no significant difference between teaching and learning methods and learning outcomes".

"H_0: There is no significant difference between the present curriculum and learning outcomes".

"H_0: There is no significant difference between skill assessments and learning outcomes".

"H_0: There is no significant difference between gender and learning outcomes".

"H_0: There is no significant difference between the undergrad degree studied by learners and learning outcomes".

Methodology

This section introduces the materials used in the study and describes the aspects of data collection that helped support the hypothesis. About 162 samples were collected from 240 students using a questionnaire using Google form. Chi-square test, ANOVA, T-test and regression were conducted to analyze the primary data. Exploratory factor analysis is used to identify underlying patterns in the data of individual variables and factors.

Results and discussion

The undergraduate degree of MBAs who participated in the study belonging to Commerce was 58%, Arts at 13%, Science at 14.2% and Engineering at 14.8%, respectively (Table 15.2). The regression analysis confirms that there is no substantial difference between the under-graduates and teaching learning methods, curriculum, assessment methods, and outputs.

The Durbin Watson statistic for autocorrelation in the regression model's output shows an acceptable range between 1.50 and 2.50, with a value of 2.34 for teaching and learning methods, 2.26 for curriculum, 2.24 for assessment, and 2.22 for learning output towards under-graduation, where successive error differences are small.

Table 15.2 Demographic characteristics

General information	No. of students	Percentage
Gender distribution		
Male	80	49.4
Female	82	50.6
Under graduation		
Commerce	94	58
Arts	21	13
Science	23	14.2
Engineering	24	14.8

Source: Author

The Durbin Watson statistic, with a value of 1.80 for teaching and learning methods, 1.81 for curriculum, 1.80 for assessment, and 1.77 for learning output, towards gender difference, shows there is an autocorrelation in the results. The output in learning has obtained very little error with autocorrelation in teaching learning methods, assessment, and curriculum with Durbin Watson values of 2.20. Cronbach's alpha value of 0.867 shows good reliability in the data collected, with a high level of internal consistency for our scale.

The ANOVA table (Table 15.3) exhibits that the null hypothesis is true, which means that there is no significant difference between the under-graduation degree of the learners and the challenging outcomes, skill assessments, curriculum and teaching learning methods.

The ANOVA table (Table 15.4) displays that the null hypothesis is true and that there is no significant difference between the gender of the learners and the challenging outcomes, skill assessments, curriculum and teaching learning methods.

The R value of 0.665 in the regression model shows there is a better fit in the regression model. An ANOVA value of 0.000 establishes a positive correlation between the variables: teaching and learning methods, curriculum, assessment methods, and outputs. This is one of the rare studies involving own learners narrowed down to first-year learners at the end of two semesters, and henceforth, the results will be useful in formatting the curriculum further along with introducing new teaching and learning methods KMO (Kaiser–Meyer–Olkin test) test values of 0.85,

Table 15.3 ANOVA – UG and learning output

Model		Sum of squares	df	Mean square	F	Sig.
1	Regression	15.245	4	3.811	3.077	0.018[b]
	Residual	194.489	157	1.239		
	Total	209.735	161			

[a]Dependent variable: UG (under graduation); [b]Predictors: (constant), OP, ASS, CU, TLM (outcome, assessment, curriculum, teaching learning methods)
Source: Author

Table 15.4 ANOVA – Gender and learning output

Model		Sum of squares	df	Mean square	F	Sig.
1	Regression	0.914	4	0.228	0.906	0.462[b]
	Residual	39.580	157	0.252		
	Total	40.494	161			

[a]Dependent variable: MF (male/female). [b]Predictors: (constant), OP, ASS, CU, TLM (outcome, assessment, curriculum, teaching learning methods)
Source: Author

Table 15.5 Model summary

Model	R	R square	Adjusted R square	Std. error of the estimate	Durbin-Watson
1	0.665[a]	0.443	0.432	0.40194	2.209

[a]Predictors: (constant), ASS, CU, TLM (outcome, assessment, curriculum, teaching learning methods). [b]Dependent variable: OP (outcome)
Source: Author

0.80, 0.85 and 0.86 for teaching and learning methods, curriculum, assessments and outputs, respectively indicate a high level of sampling accuracy, suggesting that the dataset is highly suitable for factor analysis. The value of the commonalities indicates that the analysis of the variance in the variable is elucidated by the factors mined in the analysis. These values explain that there is a correlation among the variables (Table 15.5).

Conclusions

Business management is considered one of the fattest growing fields of social sciences. The process of global acceleration due to the great expansion of communication systems and the continuous development of various technologies has largely determined the level of management in business. This is reflected in persistent and serious problems between teachers and business schools, especially in the field of business. Critics have identified significant gaps in the current curriculum that provide new opportunities for future learning with MBA student. The study has established that there is a positive and significant association amid teaching learning methods, curriculum and assessment techniques and outcomes in the research methods evaluated.

Suggestions

Each teaching method can be evaluated separately before and after its implementation; Therefore, data must be analyzed and interpreted before and after the study. Simulation, collaborative learning, games, and problem-based learning can be extended to analyze quality education.

Limitations

This study is exclusive to the students of Saveetha College of Engineering and Business Management. This study does not examine the relationship between the researcher's evidence and other members of the university and other colleges or universities.

References

Farashahi, M., and Tajeddin, M. (2018). Effectiveness of teaching methods in business education: A comparison study on the learning outcomes of lectures, case studies, and simulations. The International Journal of Management Education, 16(1): 131–142.

Zamir, Dr., Zhang, Y., Sarwar, U., Maqbool, S., Fazal, K., and Zafeer, H. M. I. (2021). Teaching methodologies used for learning critical thinking in higher education: Pakistani teachers' perceptions. International Transaction Journal of Engineering, Management, and Applied Sciences and Technologies, 12: 1–10.

Mallik, Dr., and Aithal, S. (2024). Exploring the impact of emerging educational technology in MBA programs: Enhancing brand equity through virtual reality. International Journal of Management, Technology, and Social Sciences, 9: 216–238.

Arachchige, U., and Sathsara, K. L. (2020). The impact of outbound training (OBT). International Journal of Scientific and Technology Research, 3.

Thisara, L. S. (2020). The impact of outbound training (OBT). International Journal of Scientific and Technology Research, 9: 1-4.

Annandale, M., Belkasim, S., Bunt, B., Chahine, I., Beer, J., Jacobson, T., Jagals, D., Kamanga, E., Mackey, T., Olivier, J., Reyneke, M., and van Tonder, D. (2021). Learning through assessment: An approach towards self-directed learning.

Sankar, C. S., and Raju, P. K. (2011). Use of the Presage-Pedagogy-Process-Product Model to Assess the Effectiveness of Case Study Methodology in Achieving Learning Outcomes. Journal of STEM Education: Innovations and Research, 12(7).

Mallik, Dr., and Aithal, S. (2024). Exploring the impact of emerging educational technology in MBA programs: Enhancing brand equity through virtual reality. International Journal of Management, Technology, and Social Sciences, 9: 216–238.

Kolb, A. Y., and Kolb, D. A. (2014). Learning styles and learning spaces: Enhancing experiential learning in higher education. Academy of Management Learning and Education, 4(2): 193–212.

Male, S. A., and King, R. (2019). Enhancing learning outcomes from industry engagement in Australian engineering education. Journal of Teaching and Learning for Graduate Employability, 10(1): 101–117.

Tembrevilla, G., Phillion, A., and Zeadin, M. (2023). Experiential learning in engineering education: A systematic literature review. Journal of Engineering Education, 112(2), 195–218.

Verstraete, D., Wong, K. C., Lehmkuehler, K., Netzel, T., and Hendrick, P. (2012). A global experiential design studio for engineering students. In Proceedings of the ASME 2012 International Mechanical Engineering Conference and Exposition IMECE 2012: 1–8.

Aguanta, C. B., Augusto, M. A. T., Bajenting, J. V., Buayaban, K. C., Cruz, E. J. P., Fantonial, N. F., Kwan, J. A. M., Legaspino, J., Acut, D. P., and Picardal, M. T. (2024). Factors affecting students' concept retention in learning science online using instructional videos. Journal of Education and Learning (Edu Learn), 18(2), 499–511.

Ahmad, N. A., and Lidadun, B. P. (2017). Enhancing oral presentation skills through video presentations International Journal of Social Sciences, 3(2), 385–397.

Avendo, D., Djunaidi, and Marleni. (2023). The influence of the buzz group discussion learning method on student learning outcomes in the English subject SDN 69 Palembang. International Journal of Learning, Teaching, and Educational Research, 2(3), 231–240.

https://www.jagannath.org/blog/innovative-teaching-methods-in-management-education

https://www.glowbl.com/blog/en/training-the-different-teaching-methods/

https://www.linkedin.com/pulse/bridging-worlds-crafting-mba-curriculum-meets-needs-vamsidhar-cxkyc/

National Institute for Learning Outcomes Assessment. NILOA Transparency Framework. National Institute for Learning Outcomes Assessment; 2011. Accessed July 31, 2024. https://www.learningoutcomesassessment.org/ourwork/transparency-framework/

16 Technological impact and English proficiency level of engineering students

Jainab Zareena[a]
Department of Management Studies, Saveetha Engineering College, Chennai, Tamilnadu, India

Abstract

The digital technology in learning has advanced to a greater extent. Such tools are applied in the classrooms to facilitate teaching–learning process. The present study reviews those studies that highlight the role of technology, and the learning outcomes in the current scenario. The data was obtained from two hundred students studying first year engineering program in different Universities/Colleges at Chennai, the capital of Tamil Nadu, South India. Inferences are drawn through the statistical analysis of data. The respondents were divided into two groups, A and B. Comparatively, group A students (mean = 42.92, $p<0.01$; mean = 42.97, $p<0.01$) are found to be good than the group B students (mean = 14.98, $p<0.01$; mean = 17.23, $p<0.01$) with regard to the study variables. The study has made few recommendations to the students and faculty members to facilitate the teaching and learning process.

Keywords: E-learning, English language, teaching-learning process, English proficiency, digital teaching

Introduction

Around four hundred million people speak English as their primary language and it is an official language in fifty three nations (Xu et al., 2019). Additionally, it is the greatest extensively spoken second language in the entire world. Knowing English increases the likelihood of getting a garbed job at a global corporation in one's own country or in another. Learning English is decisive for socializing, leisure, and employment because it is the language of the internet, media, and worldwide communication. Geoffrey Chaucer is regarded as the "Father of the English Language" because of his contributions to the development of the language (Slim and Hafedh, 2019). He was a London-born author, poet, and diplomat. Four elements that represent the learning goals for junior-level are: (i) understanding spoken language; (ii) speaking the language; (iii) writing the language; and (iv) reading the language in both written and spoken form. This decision resulted in the English Education Act of 1835, a legislative act of the Council of India (Nawaila et al., 2020; Lengkoan et al., 2022; Van et al., 2021). The grammar-translation technique is the oldest way of teaching English in India, according to earlier studies. With this approach, speaking and listening are both emphasized, and proper grammar and pronunciation are valued. English has such a wide range that it can be used in any subject of study or profession. English proficiency is required in every field of job or study, including science, commerce, business, tourism, politics, media, internet, Google, computers, and others. Wiki, blog, podcast, social network, and video conferencing are some of the most popular Web 2.0 technologies that have shown the potential of the most recent technology in language teaching and learning (Gayed et al., 2022). Through active interaction, technology also enables students to practice their language skills outside of the classroom. Schools employ a number of digital resources to aid in the learning process (Liang, 2022; Tarrayo et al., 2023). Technology makes learning more meaningful and fulfilling by converting students from passive recipients to active learners. Students use feature-rich apps that sync even without the internet to study for their English classes. Undoubtedly, the extent to which conventional language had been broken up got tongues wagging. A methodology connected to the communicative approach is task-based teaching. The cognitive-code method and the aural-oral method (also known as the audio-lingual method) are further approaches (Afzal, 2019; Halim et al., 2021; Pradana, 2023). Previous studies confined the fact that modern strategies of learning English language are huge. However, only few studies have focused on identifying the problems of learners. The present study investigates the impact of technology and the biggest problems that the learners face in English language learning.

The present study has the following hypotheses:

- There is no significant difference between group A (25 score and above) and group B students (24 and less score) who come across problems while learning English.

[a]jainabzareena@gmail.com, jainabzareena@saveetha.ac.in

DOI: 10.1201/9781003587538-16

- There is no significant difference between group A and B students with regard to the study variable, "Individual Students' perception towards the utility of technology to learn English language".

Methodology and interpretation

The secondary source of data is obtained through the literature review. Primary data is collected from two hundred engineering students who are in the first year of their study in India. A self-designed questionnaire consisting of two sections: Part A and B are used. The demographic details (questions 1 – 5) are included in Part A. The part B consisted of twenty questions measuring the study variables. Ten questions assessed the variable, "individual students' perception towards the utility of technology to learn English language" and the remaining ten questions assessed, "problems encountered by an individual student while learning English language". The respondents are given dummy number starting from R1 to R200. Inter item correlation was calculated to study the association amongst the questions that are used to measure each of the study variable and results are tabulated in Appendix A and B. The scores for each respondent were added to get a single score. The minimum score was 10 and the maximum score was 50 for each of the study variables. "Problems encountered by an individual student while learning English language" variable is assessed by 10 questions. Among which Q1, Q2, Q3, Q4, Q5, Q7, Q8, and Q9 are the negative statements. Hence, the rating marked as "5" by an individual student for the negative statements are entered as "1" in the data sheet, "5" entered as "1", "4" entered as "2", and vice versa. Further, the cut off score of 25 out of 50 was fixed to the variable that assessed the problems faced by the students while learning English language. Students who have scored 25 and above are categorized as group A and the students who have obtained 24 and below score were put in group B. The demographic details of the sample are tabulated in Table 16.1.

First hypothesis was tested using independent sample T-test. The result revealed huge difference between group A (Mean = 42.92, Std. deviation = 3.465) and group B students (Mean = 14.98, Std. deviation = 1.00) with T-value 67.197 and p-value significant (p = 0.000, df = 198). Therefore, null hypothesis is rejected and it is inferred that group B students come across many problems while learning English when compared to group A students and the results are tabulated in Table 16.2.

Through percentage analysis, the problems encountered by group B students while learning

Table 16.1 Demographic details of the sample.

S. No.	Description			In numbers	Percentage %
1	Gender	Male	First year	113	56.5
		Female	First year	87	43.5
2	Age	17–19 years		200	100
3	Medium of instruction in school	English language		111	55.5
		Other South Indian language (Tamil, Telugu, Malayalam, Kannada)		89	44.5
4	Year of study	First year		200	100
5	Branch of study	Computer Science		50	25
		Electronics		50	25
		Mechanical		50	25
		Civil		50	25

Source: prepared by author

Table 16.2 Problems encountered by an individual student while learning English.

Categorization of students	Total no. of students	Mean	SD	T-value	Sig.
Group A	127	42.92	3.465	67.197	0.000**
Group B	73	14.98	1.00		

**p<0.01

Source: prepared by author

English language are identified and summarized as follows:

- Pronunciation problem (100%).
- Grammatical patterns to be difficult while communicating a message (100%).
- Learning a (foreign) English language is one of the hardest things (100%).
- Not knowing when to use which article (100%).
- Difficulty in understanding, as the medium of instruction in schooling is not English (100%).
- Parents do not know English (100%).
- Speaking poor English, as they do not get enough opportunity to communicate (100%).
- Getting confused in using words like him/his in different sentences (100%).
- Psychological factors (poor confidence, fear, shyness, anxiety) stop to learn English (100%).
- English teacher not focusing on the slow learners in the class (100%).

The second hypothesis test score revealed that there this significant difference between group A (Mean=42.97, Std. deviation=3.47) and group B students (Mean=17.23, Std. deviation=2.34). The calculated T-test value was 56.315 and p-value significant (p=0.000, df=198). The results are given in Table 16.3. It is inferred that group B students have poor perception towards the technology that are available to learn English language.

Through the self-assessment, group B students agree that the technologies such as digital language lab, video conferencing software, e-books, audio visual aids, and interactive games for English fluency have not played a significant role in enhancing their English language skills.

Conclusions

Knowing English has now become mandatory to sustain in this competitive world. The study variables clearly prove the importance of the study. Comparatively, group B students face many problems to communicate in English than the group A students. The present study undoubtedly proves the fact that the group B students find English as a difficult language to use in practice and communicate with others. It is also inferred that group A students have positive perception towards technology and they acknowledge that it facilitates them to improve their proficiency level. On the contrary, group B students are found to have negative discernment towards English learning technologies. The study recommends group B students to give up shyness, nervousness, and fear of criticism. A significant problem that Indian English speakers face is that they translate phrases directly from their respective native language which makes pronouns, and article learning a big struggle. In order to facilitate the teaching–learning pedagogy, the present study suggests teachers to have one-to-one counseling with those students falling under the group B category. The study recommends providing informative feedback, and giving verbal praise for successful accomplishments.

References

Afzal, N. (2019). A study on vocabulary-learning problems encountered by BA English majors at the university level of education. *Arab World English J. (AWEJ)*, 10, 27–32.

Gayed, J. M., Carlon, M. K. J., Oriola, A. M., and Cross, J. S. (2022). Exploring an AI-based writing assistant's impact on English language learners. *Comp. Edu. Artif. Intel.*, 3, 100055.

Halim, N. A., Ariffin, K., and Darus, N. A. (2021). Discovering students' strategies in learning English online. *Asian J. Univer. Edu.*, 17(1), 261–268.

Lengkoan, F., Andries, F. A., and Tatipang, D. P. (2022). A study on listening problems faced by students of higher education. *Globish English-Indonesian J. English Edu. Culture*, 11(1), 41–50.

Liang, W. (2022). Towards a set of design principles for technology-assisted critical-thinking cultivation: A synthesis of research in English language education. *Think. Skills Creat.*, 101203.

Nawaila, M. B., Kanbul, S., and Alhamroni, R. (2020). Technology and English language teaching and learning: A content analysis. *J. Learn. Teach. Digit. Age*, 5(1), 16–23.

Pradana, A. (2023). An analysis of students' difficulties in English conversation practice. *J. Corner Edu. Ling. Lit.*, 2(3), 215–222.

Slim, H. and Hafedh, M. (2019). Social media impact on language learning for specific purposes: A study in English for business administration. Teach. English Technol., 19(1), 56–71.

Tarrayo, V. N., Paz, R. M. O., and Gepila Jr, E. C. (2023). The shift to flexible learning amidst the pandemic: The case of English language teachers in a Philippine state university. *Innov. Lang. Learn. Teach.*, 17(1), 130–143.

Van, L. K., Dang, T. A., Pham, D. B. T., Vo, T. T. N., and Pham, V. P. H. (2021). The effectiveness of using technology in learning English. *AsiaCALL Online J.*, 12(2), 24–40.

Appendix A

Interitem correlation (individual students' perception towards the utility of technology to learn English language).

	Q2	Q3	Q4	Q5	Q6	Q7	Q8	Q9	Q10
Q1	0.940**	0.727**	0.978**	0.981**	0.985**	0.981**	0.984**	0.981**	0.975**
Q2		0.842**	0.894**	0.898**	0.898**	0.899**	0.903**	0.939**	0.907**
Q3			0.691**	0.691**	0.709**	0.693**	0.700**	0.705**	0.727**
Q4				0.997**	0.977**	0.998**	0.964**	0.975**	0.981**
Q5					0.977**	0.998**	0.964**	0.978**	0.981**
Q6						0.978**	0.986**	0.964**	0.985**
Q7							0.964**	0.979**	0.981**
Q8								0.965**	0.971**
Q9									0.968**

**p<0.01

Appendix B

Interitem correlation (problems encountered by an individual student while learning English language).

	Q12	Q13	Q14	Q15	Q16	Q17	Q18	Q19	Q20
Q11	0.944**	0.758**	0.996**	0.996**	1.000**	0.996**	1.000**	0.979**	1.000**
Q12		0.862**	0.919**	0.919**	0.944**	0.919**	0.944**	0.964**	0.944**
Q13			0.695**	0.695**	0.758**	0.695**	0.758**	0.718**	0.758**
Q14				1.000**	0.996**	1.000**	0.996**	0.978**	0.996**
Q15					0.996**	1.000**	0.996**	0.978**	0.996**
Q16						0.996**	1.000**	o.979**	1.000**
Q17							0.996**	0.978**	0.996**
Q18								0.979**	1.000**
Q19									0.979**

**p<0.01

17 Identity crisis in generation Z students: Understanding its effect on social skill development

Vishakha Mandrawadkar[a], *Jayanti Shinge*[b], *Naveenkumar Aigol*[c] *and Khezia Olagundi*[d]

Department of Humanities and Social Sciences, KLE Technological University, Hubli, Karnataka, India

Abstract

Gen Z engineering students struggle with self-perceptions of personality, hindering social skill development. Misunderstandings of traits like extroversion and introversion prevent students from learning and growing socially. This study highlights a gap between self-perception and reality, urging for interventions to improve social skills. This identity crisis is illustrated by a comparative analysis of the perceived personality traits and actual personality traits using a condensed version of the big five personality scale.

Keywords: Gen Z, identity, labeling, personality, social skills

Introduction

We spend our lives learning about ourselves and how we interact with the world. This is especially important during young adulthood, as we deal with social pressures, school, and future careers. Our unique personality traits, as defined by McCrae and Costa (1999), play a big role in this journey of self-discovery. Personality is complex, shaped by a mix of how we think, feel, and act (Bieri, 1966; Powell, 1982). Understanding these nuances is key for Gen Z (1997–2012), known for their digital ease, entrepreneurial drive, and social awareness (Twenge, 2010). In today's global world, understanding Gen Z's multifaceted personalities is crucial. Especially Gen Z, engineering students, who show a wide range of personality traits. Social skills aren't just for popularity. Research shows they boost grades, friendships, and job prospects (Stone and Daig, 2010). In fields like engineering, where teamwork matters, social skills are even more crucial for success (Goleman et al., 2002). Incorrect labels hurt Gen Z engineering students. This research explores how misjudging their personalities hinders their social skill development. This research tackles the challenges faced by Gen Z engineering students due to misconceptions about personality. It highlights that personality is a spectrum, not restricted to introversion or extroversion.

Literature review

Studies show understanding personality is key to social skills (Kiderle et al., 2021). Personality and emotions are linked, making it crucial for designing effective social interactions. Sarkar (2020) highlighted the development of personality through effective communication. Gen Z's social skills are complex. Technology connects them but can hurt in-person interactions (Aziz et al., 2019; Ajmain, 2020). Weak social skills can lead to stress, loneliness, and even dropping out (Segrin, 2017; Jitaru et al., 2023). Community service can help by teaching emotional control, responsibility, and teamwork (Afzal and Hussain, 2020). Schools should make community service mandatory, not optional. This builds social skills crucial for today's global workforce, where employees need to understand others (Premack and Woodruff, 1978) and collaborate effectively (Thomas et al., 2015).

Investigative focus

Gen Z engineering students often misjudge themselves, especially on the introvert–extrovert scale. This can hurt their identity, social skills, and future success (McCrae and Costa, 1999; Stone and Daig, 2010). This study explores this link to help students understand themselves better and develop strong social skills.

Methodology

This study was conducted in three phases to fully understand the extent of the observed discrepancy in the perceptions of the personality of Gen Z engineering students (Table 17.1).

[a]vishakhamandrawadkar24@gmail.com, vishakha.mandrawadkar@kletech.ac.in, [b]jayanti_s@kletech.ac.in, [c]naveenkumar.aigol@kletech.ac.in, [d]khezia.olagundi@kletech.ac.in

DOI: 10.1201/9781003587538-17

Table 17.1 Research design.

Phase 1	*Students are asked two questions in a semi-structured interview:* • *Which label is applicable to you: introvert or extrovert?* • *Give 2 instances to support your claim* *After completing the personality test, students participated in a community service project for "Design Thinking in Social Innovation" offered by KLE Technological University. They collected data on social issues, then shared their findings in class presentations*
Phase 2	TIPI Scale is administered to the students and their actual personality traits are quantified. This comprehensive information is shared with them through one-on-one discussion
Phase 3	Students are once again sent to engage with the community and share their findings in the form of a classroom presentation

Source: Author

Table 17.2 Phase I.

Label	*Frequency* (%)
Extrovert	80
Introvert	20

Source: Author

Sample and instrument

Twenty fresh engineering students from KLE Technological University, Hubli with diverse backgrounds participated. The ten item personality inventory, a condensed version of the big five personality test (Gosling, 2003) was used to see how these students

Table 17.3 Phase II.

S. No.	*Openness*	*Conscientiousness*	*Extraversion*	*Agreeableness*	*Neuroticism*
1	6	5	7	8	4
2	8	7	6	6	3
3	4	6	5	9	2
4	7	8	4	5	6
5	5	4	8	7	5
6	9	3	6	6	2
7	6	6	7	7	3
8	3	7	8	5	4
9	8	5	4	8	2
10	4	8	5	6	7
11	7	6	7	5	4
12	6	7	4	6	8
13	5	8	6	7	4
14	9	4	7	5	3
15	3	6	8	7	4
16	7	5	6	8	3
17	5	7	6	6	5
18	6	6	5	6	6
19	8	4	7	5	4
20	4	5	8	6	7

Source: Author

view their own personalities. This helps identify any gaps between their self-perception and reality, potentially impacting social skills.

Data analysis

In phase 1, most Gen Z students (80%) thought they were extroverts (talkative, friendly) while 20% thought they were introverts (shy) (Table 17.2). But their social skills seemed weak overall, regardless of label. This mismatch between self-perception and reality hurt their ability to engage with the community and collect information.

In phase 2, students took a personality test (TIPI scale) to measure actual traits. Individual results were shared for discussion (means and interpretations based on norms). Table 17.3 shows the individual scores of the students on the following personality traits. Students mark their responses against a Likert scale of 1–7, where 1 is strongly disagree and 7 is strongly agree. The traits are evaluated as per TIPI norms and shared with the students:

1. Openness to experience: This dimension showed moderate to high openness (3–9 on the scale) in the students. This means they are generally curious, creative, and open to new ideas. People lower in openness tend to prefer routine and tradition.
2. Conscientiousness: The students also scored moderately high in conscientiousness (3–8 on the scale), meaning they are likely organized, responsible, and goal-oriented. Lower scores suggest a more relaxed approach to tasks.
3. Extraversion: The students scored moderately high in extraversion (4–8 on the scale). This means they are generally outgoing and social, while lower scores suggest a preference for quieter settings.
4. Agreeableness: Agreeableness (5–9 on the scale) was also moderate to high. This means the students tend to be cooperative and empathetic, while lower scores suggest a more competitive nature.
5. Neuroticism: The students showed a range in neuroticism (2–8 on the scale). This means some experience more anxiety and stress than others. Lower scores indicate higher emotional stability.

The actual personality traits are disclosed to the students in a one-on-one discussion. Each trait and how mislabeling themselves might have hurt their social skills during the community project. For phase 3, students returned to the community for a second round of engagement. Their classroom presentations showed a clear improvement, reflecting their newfound understanding of their personalities. They were more confident and gathered richer data for their projects.

Conclusions

The study found a gap between how Gen Z engineering students saw themselves and reality. This mismatch between self-perception and actual personality traits suggests inaccurate labels can hurt development of social skills. Phase 2 moved beyond self-labels and used the TIPI test to measure actual personality traits. This gave a clear picture of each student's personality (openness, etc.) and closed the gap between how they saw themselves and reality, boosting self-awareness. The final phase showed a big change in the students. Learning their real personality traits (not just introvert/extrovert), they understood personality is flexible. Their presentations showed success in connecting with the community, proving they could interact skillfully. This study shows closing the self-awareness gap empowers Gen Z students to navigate social situations, leading to success in both academics and life. Personality is complex, but self-awareness unlocks its potential. Self-awareness is key. Understanding their true personalities helps Gen Z engineering students succeed in today's competitive world.

Future scope

It would be beneficial to design effective classroom interventions focused on personality development and building self-awareness in the Gen Z engineering students to guarantee success in a social situation. The study could also explore other influencing factors like emotional intelligence, group dynamics and communication styles and measure its impact on social skills in the future.

References

McCrae, R. R. and Costa, P. T. (1999). A five-factor theory of personality. L. Pervin and O. John (Eds.). Handbook of personality: Theory and research (2nd ed.). 139–153. Guilford Press.

Bieri, J. (1966). The challenge of the personal construct theory. *Ann. Rev. Psychol.*, 17(1), 262–282.

Powell, J. H. (1982). Why the personal construct theory of personality deserves wider recognition. *Br. J. Clin. Psychol.*, 21(4), 321–332.

Gompers, P. A. and Metrick, A. (2001). Institutional investors and equity prices. *Q. J. Econ.*, 116(1), 229–259.

Twenge, J. M. (2010). Generation me: Why today's young Americans are more confident, assertive, entitled—and more miserable than ever before. Atria Books.

Stone, D. A. and Daig, A. (2010). The importance of soft skills in the workplace. *J. Gen. Manag.*, 36(2), 35–51.

Goleman, D., Boyatzis, R., and McKee, A. (2002). Primal leadership: Unleashing the power of emotional intelligence. *Harvard Business Review Press.*

Kiderle, T., Ritschel, H., Janowski, K., Mertes, S., Lingenfelser, F., and Andre, E. (2021). Socially-aware personality adaptation. *2021 9th Internat. Conf. Affec. Comput. Intel. Interac. Workshops Demos (ACIIW)*. IEEE. https://doi.org/10.1109/aciiw52867.2021.9666197.

Sarkar, S. (2020). Communication skills and personality development. *Ind. J. Appl. Res.*, 10.

Aziz, N., Hashim, H., and Yunus, M. Md. (2019). Using social media to enhance ESL writing skill among Gen-Z learners. *Creat. Edu.*, 10(12), 3020–3027. https://doi.org/10.4236/ce.2019.1012226.

Ajmain, T. (2020). Impacts and effective communication on generation Z in industrial revolution 4.0 era. *J. English Teach. Appl. Ling.*, 2(1), 37–42. https://doi.org/10.36655/jetal.v2i1.204.

Segrin, C. (2017). Indirect effects of social skills on health through stress and loneliness. *Health Comm.*, 34(1), 118–124. https://doi.org/10.1080/10410236.2017.1384434.

Jitaru, O., Bobu, R., and Bostan, C. M. (2023). Social and emotional skills in adapting students to the academic environment. *Revista Romaneasca pentru Educatie Multidimensionala*, 15(1), 590–605. https://doi.org/10.18662/rrem/15.1/713.

Afzal, A. and Hussain, N. (2020). The impact of community service learning on the social skills of students. *J. Edu. Edu. Dev.*, 7(1), 55. https://doi.org/10.22555/joeed.v7i1.2988.

Premack, D. and Woodruff, G. (1978). Does the chimpanzee have a theory of mind? *Behav. Brain Sci.*, 1(4), 515–546.

Thomas, A., DiPrete, Jennifer, L'., and Jennings (2015). Study of academic achievement among social skill deficient and non-deficient school students.

Gosling, S. D., Rentfrow, P. J., and Swann, W. B., Jr. (2003). A very brief measure of the big five personality domains. *J. Res. Personal.*, 37, 504–528.

McCrae, R. R. and Costa, P. T. (1999). A five-factor theory of personality: The NEO-PI. In P. J. Rogers, D. P. Heninger, and M. H. Kehle (Eds.). Handbook of personality psychology. 179–215. Academic Press.

Stone, J. and Daig, A. (2010). Social and emotional learning curriculum and instruction in the classroom. *Preven. School Fail. Altern. Edu. Prog. Child. Youth*, 54(3), 141–151.

18 Comparative analysis of technology-enhanced teaching methods in engineering mathematics using TOPSIS method

N. Kavitha[1,a], *G. Uthra*[2,b] *and V. Anandan*[3,c]

[1,3]Department of Mathematics, Saveetha Engineering College, Chennai, Tamilnadu, India

[2]Department of Mathematics, Pachaiyappa's College, Chennai, Tamilnadu, India

Abstract

Prelude: Learning methods that engage students and attain the course outcomes are in higher demand in engineering education, especially in the branch of mathematics. The modern teaching methods usually incorporate technology to improve student performance and learning opportunities. **Methodology:** Students are evaluated and grouped based on various teaching approaches as follows: adaptive learning (Group A), collaborative learning (Group B), flipped classroom (Group C), interactive simulations (Group D), and gamified learning (Group E). The Technique for order of preference by similarity to ideal solution (TOPSIS) method is used to examine the efficacy of technology-enhanced teaching methods in engineering mathematics education. **Result:** The results demonstrated remarkable advancements in every evaluated category. This study handles the use of technology in education analytically and offers insightful information to educators and educational institutions, looking for optimal and efficient methods to improve engineering mathematics instruction.

Keywords: Teaching methodology, TOPSIS, effective teaching

Introduction

Effective and interesting teaching techniques are in high demand in the field of engineering education, particularly in mathematics and other related subjects. Technology is frequently used in these fields to increase student performance and learning opportunities (Johnson, 2021). It is becoming very difficult to fix and determine the best approach because of different skill level among students. The core of engineering education is mathematics, which gives students the fundamental knowledge and problem-solving abilities they need. However, conventional teaching approaches often have trouble annoying students' attention or satisfying their range of learning requirements. Students did not completely understand the fundamental ideas which they are supposed to apply in the core engineering, as a result, they perform below their ability.

Several teaching methodologies were proposed in the recent years to solve the above-mentioned issues and create self-centric and dynamic learning environments (Jackson, 2021). The methods includes 1) flipped classroom – where students should go through the study materials at home and learn and the doubts, if any, will be cleared in the classroom, 2) collaborative learning – where the peer group learning is motivated through team works, 3) adaptive learning – where contents are adjusted based on the learners ability and strengths, 4) gamified learning - where games are included and 5) interactive simulations – where students are allowed to get hands-on training (Adams, 2019; Brown, 2018).

MCDM and TOPSIS

The method of obtaining the best alternative among the multiple conflicting criteria is referred as multi-criteria decision-making (MCDM). It provides a sequential algorithm in decision making, specifically in tough situations with multiple variables (Wang, 2015). TOPSIS is the technique of making preferences depending on the correlation between the positive and negative ideal solution, a popular MCDM technique that supports in dealing complicated decision-making methods (Lee, 2020).

Research gap

The best approach to enhance learners' performance using technology-enhanced technique continue uncertain at times, due to its development, particularly in the content delivery of engineering mathematics. Few works have investigated the efficiency of certain techniques, but the evaluation and ranking

[a]kavithan@saveetha.ac.in, [b]uthragopalsamy@yahoo.co.in, [c]anandanworld@gmail.com

DOI: 10.1201/9781003587538-18

are not comprehensive because of the multiple criteria (Gokhale, 1995; Johnson, 2020). TOPSIS uses a systematic evaluation procedure, which will fulfill the research gap in this work (Renato et al., 2015; Subrata, 2022). The purpose of this work is to provide a useful and precise technique for the institutions and teachers to the enhance the teaching-learning practices. The study systematically assesses and rank the various technology-enhanced teaching methods based on the student performance in engineering mathematics.

Methodology

Study design and teaching methods selection

The evaluation of the efficiency of the technology-enhanced teaching practices in engineering mathematics education should be done using a perfect experimental design (Smith, 2020). The five key strategies in teaching-learning are categorized as follows:

Group A: Adaptive learning
Group B: Collaborative learning
Group C: Flipped classroom
Group D: Interactive simulations
Group E: Gamified learning

These reason for choosing these techniques are due to its prevalence in educational research and their significance to improve the learning outcomes.

Criteria for evaluation

The norms that are used to determine the technique's efficiency are:

1) Improvement in pre- to post-test scores: Assessment of learner's improvement in pre-tests to post-tests because of the strategy adopted.
2) Mid-term exam performance: Assessment of learner's performance in mid-term exam based on the reflection in the halfway period.
3) Final exam performance: Assessment of learner's performance in final exams based on the attainment of the course outcomes.

These norms will provide a comprehensive evaluation to determine the use of teaching strategy adopted.

Data collection process

Learners in all groups are assessed based on the measure of pre, post, mid-term and final exams output. Prior to the implementation of the instructional strategies, the pre-tests provide a baseline assessment of the students' understanding. Exams for post-tests, midterms, and finals are given to students to assess their performance at various points during the course.

TOPSIS algorithm

The TOPSIS algorithm is described as follows.

Step 1. Normalization: $n_{ij} = \frac{x_{ij}}{\sqrt{\sum_{i=1}^{m} x_{ij}^2}}$, i = 1, ….., m; j = 1,….,n

Step 2. Weighted normalization: $v_{ij} = w_j n_{ij}$, i = 1,…., m; j = 1,…., n

Step 3. Positive and negative ideal solutions: $A^+ = \{v_1^+, \ldots, v_n^+\}$ $A^- = \{v_1^-, \ldots, v_n^-\}$

Step 4. Separation measures: $D_i^+ = \left\{\sum_{j=1}^{n} (v_{ij} - v_j^+)^2\right\}^{\frac{1}{2}}$, i = 1, …., m

Step 5. Closeness coefficient: $= C_i = \frac{D_i^-}{D_i^+ + D_i^-}$

Step 6. Alternatives are ranked based on the logic: Higher C_i is the Best

Course and assessment pattern

Course: Calculus and Laplace transforms

Credit: 4

Pre- and post-assessment: Pre-assessment was done on the first day of the course. All students have attended 50 MCQ questions. The test contains prerequisite contents which will assess the fundamental ability of the learners

Formative assessment: All other tests such as midterm and end semester will contain **Part A** questions to test the knowledge levels remember to apply; **Part B** will test application level and **Part C** will test apply and evaluation level

Duration: 90 minutes; Total marks: 50 marks

Question pattern: Part A (5 * 2 marks). Part B (2 * 13 marks), Part C (1 * 14 marks)

Learners group formation

Heterogeneous groups with learners having different learning skills were formed based on the pre-assessment score. Each group will contain learners from all 4 years, various departments, boys, and girls. In the heterogeneous group, learners work with other students who are at different reading levels (Subrata, 2022).

Problem statement

The study aims to evaluate and rank the effectiveness of different technology-enhanced teaching methods in engineering mathematics education using the TOPSIS method. The data collected includes pre-test scores, post-test scores, midterm exam scores, and final exam scores for students across five teaching method groups (Group A – Adaptive learning, Group B – Collaborative learning, Group C – Flipped classroom,

Table 18.1 Groupwise Student Performance.

S. No.	Group	Register No.	Student name	Pre-test	Post-test	Mid-term	Final test
1	Group A	212223060001	Aadhithya S. V.	51	60	86	100
2	Group A	212221060001	Aakash G.	15	50	50	100
39	Group E	212222060225	Catherine I.	5	88	40	75
40	Group E	212223020024	Sherena Jenice J.	63	94	98	86

Source: Author

Table 18.2 Groupwise TOPSIS Score.

	Group A	Group B	Group C	Group D	Group E
Pre-test to Post-test Improvement	0.45	0.78	0.65	0.58	0.72
Mid-term exam performance	0.63	0.82	0.74	0.68	0.79
Final exam performance	0.78	0.89	0.75	0.72	0.85
TOPSIS score	0.55 Grade B	0.83 Grade A	0.70 Grade B+	0.65 Grade B	0.79 Grade A-

Source: Author

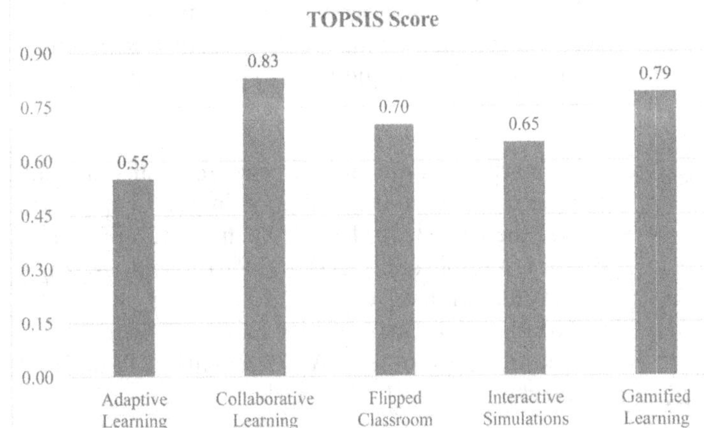

Figure 18.1 Groupwise TOPSIS Score
Source: Author

Group D – Interactive simulations and Group E – Gamified learning) as shown on Table 18.1. The research seeks to determine which teaching method demonstrates the highest effectiveness in improving student performance based on pre-defined evaluation criteria.

The TOPSIS grade is obtained and shown in Table 18.2 represents the effectiveness of different technology-enhanced teaching methods in engineering mathematics education.

Results and discussion

On assessing the technology-enhanced teaching practices, the TOPSIS method provides the following results:

The best teaching-learning practices are ranked as follows: 1) collaborative learning (Group B), 2) gamified learning (Group E), 3) flipped classroom (Group C), and 4) interactive simulations (Group D), whereas Group A's adaptive learning exhibited moderate efficiency level.

The engagement and motivation provided by collaborative and gamified components considerably improved the learning outcomes. Student learning was positively enhanced by the hands-on activities of interactive simulations and the flipped classroom method (Choi, 2018), (White, 2020).

As seen in Figure 18.1, the study's overall findings highlight the significance of creative teaching techniques in the teaching of engineering mathematics and provide practical advice for enhancing student learning.

Significance of TOPSIS grade

The TOPSIS method has played a significant role in evaluating the effectiveness of various technology-enhanced teaching methods in engineering mathematics education, as observed by the TOPSIS and corresponding grades. Collaborative learning (Group B) stands better among all other groups with a TOPSIS score of 0.83 enhancing student engagement and learning outcomes, following closely is gamified learning (Group E) with a TOPSIS score of 0.79 projecting its importance. Group C's flipped classroom received a TOPSIS score of 0.70, indicating that it is successful in encouraging student participation and active learning. The TOPSIS method provides a systematic approach for the comparative study which permits to rank the teaching techniques based on learner performance (Mohamed, 2015). Based on the results, Group B's collaborative learning and Group E's gamified learning exhibited better efficiency in in increasing performance. Moreover, flipped classroom (Group C) and interactive simulations (Group D) also secured decent marks. These observations provide complete idea to the educators to improve engineering mathematics teaching.

Conclusions

The TOPSIS method was proposed in this work to determine the efficiency of the technology-enhanced teaching practices for engineering mathematics education. The observations are concluded as follows:

1) The TOPSIS method provides a systematic framework for the educators to analyze and compare the outcomes of different instructional strategies adopted.
2) Collaborative learning and gamified learning are found to be effective with TOPSIS grade 0.83 and 0.79 respectively and secures top rank.
3) The incorporation of collaborative and gamified components in the teaching creates positive impact on student engagement and improves learning outcomes.
4) The importance of incorporating collaborative and interactive tools optimize teaching practices in engineering mathematics education.

References

Adams, E. (2019). Interactive simulations in STEM education: A systematic review of their effectiveness. J. Interac. Learn. Res., 25(4), 200–215.

Brown, K. (2018). Flipped classroom models: A meta-analysis of their impact on student achievement. *Edu. Psychol. Rev.*, 14(2), 90–105.

Choi, H. (2018). Comparative analysis of teaching methods: A quantitative approach. *J. Edu. Res.*, 25(1), 50–65.

Gao, F. and Liu, Y. (2022). Research on classification evaluation system based on Topsis evaluation model. *Proc. 7th Internat. Conf. Intel. Inform. Proc.*, 1–5.

Gokhale, A. A. (1995). Collaborative Learning Enhances Critical Thinking. Journal of Technology Education, 7(1), 22–30.

Jackson, R. (2021). Evaluating technology-enhanced teaching methods: A comparative study. *J. Edu. Technol.*, 15(3), 102–115.

Johnson, L. (2021). Technology in the classroom: Exploring its impact on student performance in mathematics. *Edu. Technol. Res. Dev.*, 68(3), 245–260.

Johnson, M. (2020). The role of gamification in enhancing learning experiences: A review of current literature. *J. Gamified Edu.*, 7(1), 30–45.

Lee, S. (2020). Application of TOPSIS in educational research: A systematic review. *Edu. Sci.*, 8(4), 75–88.

Mohamed, A. (2015). Innovative pedagogical practices in engineering education. *Edu. Prac. Innov.*, 2, 9–15.

Smith, J. (2020). Enhancing engineering mathematics education: A review of effective teaching strategies. *J. Engg. Edu.*, 10(2), 45–58.

Renato, A. K. and Pacheco, A. G. C. (2015). A-TOPSIS – An approach based on TOPSIS for ranking evolutionary algorithms. *Proc. Comp. Sci.*, 55, 308–317. https://doi.org/10.1016/j.procs.2015.07.054.

Subrata, C. (2022). TOPSIS and modified TOPSIS: A comparative analysis. *Dec. Anal. J.*, 2, 100021. https://doi.org/10.1016/j.dajour.2021.100021.

Taylor, L. (2020). Adaptive learning systems: A comprehensive review of their application in higher education. *Comp. Edu.*, 36(2), 180–195.

Wang, C. (2019). Multi-criteria decision making in education: A comprehensive guide. *Edu. Res. Rev.*, 12(3), 150–165.

White, S. (2020). Effectiveness of technology-enhanced learning methods in STEM education: A meta-analysis. *J. STEM Edu.*,15(2), 78–93.

Gokhale, A.A. (1995). Collaborative learning enhances critical thinking. Journal of Technology education. 7(1), Retrieved 5 Nov. 2011, from:
http://scholar.lib.vt.edu/ejournals/JTE/v7n1/gokhale.jte-v7n1.html.

Gokhale, A.A. (1995). Collaborative learning enhances critical thinking. Journal of Technology education. 7(1), Retrieved 5 Nov. 2011, from:
http://scholar.lib.vt.edu/ejournals/JTE/v7n1/gokhale.jte-v7n1.html.

Gokhale, A.A. (1995). Collaborative learning enhances critical thinking. Journal of Technology education. 7(1), Retrieved 5 Nov. 2011, from:
http://scholar.lib.vt.edu/ejournals/JTE/v7n1/gokhale.jte-v7n1.html.

Gokhale, A.A. (1995). Collaborative learning enhances critical thinking. Journal of Technology education. 7(1), Retrieved 5 Nov. 2011, from:

http://scholar.lib.vt.edu/ejournals/JTE/v7n1/gokhale.jte-v7n1.html.

Gokhale, A.A. (1995). Collaborative learning enhances critical thinking. Journal of Technology education. 7(1), Retrieved 5 Nov. 2011, from:
http://scholar.lib.vt.edu/ejournals/JTE/v7n1/gokhale.jte-v7n1.html.

Gokhale, A.A. (1995). Collaborative learning enhances critical thinking. Journal of Technology education. 7(1), Retrieved 5 Nov. 2011, from:
http://scholar.lib.vt.edu/ejournals/JTE/v7n1/gokhale.jte-v7n1.html.

Gokhale, A.A. (1995). Collaborative learning enhances critical thinking. Journal of Technology education. 7(1), Retrieved 5 Nov. 2011, from:
http://scholar.lib.vt.edu/ejournals/JTE/v7n1/gokhale.jte-v7n1.html.

Gokhale, A.A. (1995). Collaborative learning enhances critical thinking. Journal of Technology education. 7(1), Retrieved 5 Nov. 2011, from:
http://scholar.lib.vt.edu/ejournals/JTE/v7n1/gokhale.jte-v7n1.html.

Gokhale, A.A. (1995). Collaborative learning enhances critical thinking. Journal of Technology education. 7(1), Retrieved 5 Nov. 2011, from:
http://scholar.lib.vt.edu/ejournals/JTE/v7n1/gokhale.jte-v7n1.html.

19 Assessing the flipped classroom approach for management students

Niranchana Shri Viswanathan[1,a] *and Delecta Jenifer Rajendren*[2,b]

[1]Assistant Professor, School of Management studies, Sapthagiri NPS University, Chikkasandra, Hesarghatta Main Road, Bengaluru, 560057

[2]Assistant Professor, Department of Management Studies, Saveetha Engineering College, Chennai, Tamil Nadu, India, 602105

Abstract

FLIP is an acronym created by Bergmann and Sams, who mentioned that the four fundamental principles of foreign language education are a flexible learning environment, a culture that is favorable to learning, instructional content that is purposeful, and educators who are highly trained and informed. A total of 216 undergraduate and graduate students from management studies were asked for their opinions on eight distinct characteristics as part of this research. To gain a better understanding of the topic, secondary and primary data were used. The researcher was able to demonstrate, based on the data and association tests, that there is a connection between the liking of FL and the demand for more work and material to implement, the effectiveness of interventions on knowledge and skills, and the support of the FL approach that assisted me in learning better, as well as the demand for more work and material when it comes to implementation and support for the goal.

Keywords: Active learning, educational technique, 5-point Likert scale, pedagogical teaching strategy, flipped classroom

Introduction

In a flipped classroom, it is expected that students will complete preparation work outside of class and that they will actively participate and interact with one another during coaching. As soon as pre-recorded lectures became accessible on the Internet, two management teachers at UG and PG programme started implementing the flipped classroom method of instruction. To provide pupils who are absent from school with the opportunity to benefit from this, Woodland Park High School in Colorado might use this system. This not only addresses the high percentage of absenteeism but also indirectly addresses the poor performance of students and the decline in their academic achievement. The 2012 Bergmann Sams says FC is established as a diverse organization of courses in which applications of teaching methods (such as pre-recorded sessions, reading online courses, and demonstrations) are distributed as readings to be contemplated prior to class, and the time spent in class is devoted to discussion and practice of material that is progressively more difficult. Participating in a study session with others is an appealing idea. This method of teaching, which requires the provision of theoretical support materials prior to instruction, was something that Phillips (2015) found to be problematic.

Literature review

Using pre-recorded movies, the classroom was turned upside down, and active learning was encouraged while the classroom was in operation. Hashemifardnia et al. (2021) stated that on February 20, 2019, the Minister of Human Resource, Aalbert (2014), Arbaugh (2014), and Davies et al. (2013) are the authors of published works. Only four out of thousands of students investigated the effect of FC on their final grades.(Walsh, 2024a) which frees teachers to focus on more difficult topics and abilities. The 2019 study by Cheng et al. Numerous studies have shown that classroom settings can be flipped. Ferreri and O'Connor (2013), Papadopoulos and Roman (2010), and Albert and Beatty (2014) evaluated the impact of FC versus lecture on grades in an introductory management course. This approach is characterized by its hybrid nature (Tucker, Jr.) The flipped classroom, often known as "flipped learning," is a method of instruction that places significant emphasis on the utilization of content-sharing platforms and technological tools (Wang, 2023) that are accessible online. Rivera (2015), Garver Roberts (2013), and Kim Khera Get Man (2014) are included in this list. ***In this article, the potential advantages of the flipped classroom paradigm are studied, and the findings of research that compares it to regular classroom settings are provided.***

[a]niranchanaphd@gmail.com, [b]delectajenifer@saveetha.ac.in

DOI: 10.1201/9781003587538-19

Research methodology

This paper presents an outcome-based survey designed for UG and PG students in management studies. Total 216 students' perception was recorded on 5point Likert scale for 8 different attributes: 1. I like FL, 2. FL approach helped me to learn better, 3. FL required more work,4. I would want more of the material in the course thought this way, 5. I would want other course to use this approach learning,6. The effectiveness of interventions on knowledge,7. The effectiveness of interventions on skills,8. The effectiveness of interventions on self-learning skill. The researcher analyzed the data using the Chi-square test and p-value (Tables 19.2 and 19.3) and the association between the FL tool and better understanding, requirement of more work and material for implementation, and effectiveness of interventions on knowledge and skills. Both males and females were considered to avoid gender bias.

Objective

1) *Management examines undergraduate and graduate students' FL satisfaction, the need for more effort and material for implementation, and how interventions affect knowledge and capacities.*
2) *How do undergraduates and graduates see FL? This would enhance my learning and provide more material for implementation. I would also learn how knowledge and skills affect intervention success.*

Hypothesis

$1H_0$: There is no association between liking FL and the requirement of more work and material for implementation and effectiveness of interventions on knowledge and skills.

$1H_1$: There is an association between the liking of FL and the requirement of more work and material for the implementation and effectiveness of interventions on knowledge and skills.

$2H_0$: There is no association between the FL approach to help me learn better and the requirement of more work and material for implementation and effectiveness of interventions on knowledge and skills.

$2H_1$: There is an association between the FL approach to help me learn better and the requirement of more work and material for implementation and effectiveness of interventions on knowledge and skills.

Data analysis and interpretation

Demographic data is based on qualification and gender of respondents. There are 216 total respondents and 114 are females and 102 are male. UG students are more in number as compared to PG students from management studies.

Interpretation: Students' perceptions about FL on 8 different parameters were studied on 216 students of UG and PG level and it reveal that 118 students like FL approach, but it required more work. One hundred and fifty-one students stated that it helped them to learn in better way.

$1H_0$: There is no association between liking FL and the requirement of more work and material for implementation and effectiveness of interventions on knowledge and skills. [Alternate Hypothesis] $1H_1$: There is an association between the liking of FL and the requirement of more work and material for the implementation and effectiveness of interventions on knowledge and skills. Degree of freedom = (c-1) * (r-1) = (3-1) * (2-1) = 2*1= 2. At 0.05 level of significance = 5. 991.Result: The calculated chi-square statistic is 40.0499= 40.05 and which is higher than table value of Chi-square i.e. 5. 991.Interpretation: Therefore, we reject the null hypothesis. This means Researcher accepts the alternative hypothesis that there is an association between liking of FL and requirement of more work and material

Table 19.1 Author own work (Questionnaire)with the respondents response.

	Students 'perceptions about...	Agree	Nil	Disagree
1	I like FL	118	5	93
2	FL approach helped me to learn better	151	5	60
3	FL required more work	118	57	41
4	I would want more of the material in the course thought this way	129	39	48
5	I would want other course to use this approach learning	124	45	47
6	The effectiveness of interventions on knowledge	138	17	61
7	The effectiveness of interventions on skills	139	42	35
8	The effectiveness of interventions on self-learning skill	149	12	55

Source: Author

Table 19.2 Includes all calculated Chi-square statistic values.

Statement no.	Calculated Chi-square statistic	p-Value
1 and 3	63.792	0.00001
1 and 4	41.1243	0.00001
1 and 5	47.263	0.00001
1 and 6	14.7573	0.000624
1 and 7	57.1249	0.00001
1 and 8	16.2384	0.000298
Average Chi-square value = 240.2999/6=40.0499		

Source: [Null Hypothesis]

for implementation, effectiveness of interventions on knowledge and skills and supports objective.

$2H_0$: There is no association between FL approach to help me to learn better and requirement of more work and material for implementation, effectiveness of interventions on knowledge and skills.$2H_1$: There is an association between FL approach to help me to learn better and requirement of more work and material for implementation, effectiveness of interventions on knowledge and skills. Degree of freedom = (c-1) * (r-1) = (3-1) * (2-1) = 2*1= 2.At 0.05 level of significance = 5.991.**Result:** The calculated Chi-square statistic is 27.2092= 27.21 and which is higher than table value of Chi-square i.e. 5.991.**Interpretation:** Therefore, we conclude that H_0 must be false which means researcher accepts the alternative hypothesis that there is an association between FL approach to help me to learn better and requirement of more work and material for implementation, effectiveness of interventions on knowledge and skills and it also supports objective.

Findings

By carefully considering all factors before investing resources, based on the findings, we fail to reject the null hypothesis that the intervention of incorporating the $2H_0$ programme was ultimately discontinued and did not achieve its intended goals of enhancing student learning. By carefully analyzing the reasons behind the failure of the $2H_0$ programme, educators can learn valuable lessons for future interventions. Through this approach, schools can Moreover, through this approach ,schools can use formative assessment, educators and support staff can regularly monitor the impact of interventions implemented in provision to enable informed judgements on which strategies are proving most effective. At the end of the day, collaboration and sharing information make for a more supportive environment in which students can succeed. Each and every student would benefit from having a structure of radical inclusion through continuous improvement in practice, within the context most accessible to children: schools. Using data to guide decision-making, educators can personalize interventions for each student - providing targeted support that reads students where they are in both their academic and social-emotional development improves the overall experience of school and ensures our children are prepared for future success.

Table 19.3 Includes all calculated chi-square statistic values.

Statement no.	Calculated chi-square statistic	p-Value
2 and 8	3.1131(slightly low)	0.2108
2 and 7	36.2032	0.00001
2 and 6	7.1385	0.028177
2 and 5	36.2303	0.00001
2 and 4	29.3346	0.00001
2 and 3	51.2355	0.00001
Average Chi-square value = 163.2552/6= 27.2092		

Source: [Null Hypothesis] H01:

Discussion

The study is supported (Samadi et al., 2024a) by evidence that has been demonstrated through theoretical research. This highlights the necessity of aligning intervention strategies (Zhou et al., 2024a) with available resources to optimize student learning experiences and outcomes.

Conclusions

It has already proven itself to be an effective technique of instruction that is held in high regard by students at both the undergraduate and graduate levels of education. Additionally, the researcher can demonstrate, through the utilization of statistical methods (Aljermawi et al., 2024a). This is in addition to the fact that it acknowledges (Shen et al., 2024a) their existence.

Limitation

One of the limitations of this research article on the flipped classroom is that it is by addressing issues (Walsh, 2024b) such as the utilization of films Simply put, it is a situation in which both parties benefit (Hong, 2024b).for instance, researchers have the potential to raise the dependability and effectiveness of their study (Wu et al., 2024b). (Aljermawi et al., 2024b; Pratiwi et al., 2024b) classroom model, but it also paves the way for research results that are not only more significant but also more cost-effective.

Future directions for flipped classroom research

Research on the flipped classroom should be conducted in the future by selecting students, such as those in the fields of medicine and engineering,(Shen et al., 2024b; Zhou et al., 2024b). During the process of conducting research on flipped classrooms, it is feasible to lessen the number (Karimian et al., 2024b; Shen et al., 2024c) of objections that are addressed towards it. Among these criticisms are films of less than satisfactory quality and the need for additional preparation time. It is for those who are fortunate enough to have this opportunity, it will be possible to gain experience that is (Ni et al., 2024b; Samadi et al., 2024b) more beneficial in general. The fact that it is possible to reduce the amount of criticism that it receives is one of the reasons why this is something that has the potential to be achieved.

Abbreviations

FLIP (Flexible learning environment), UG (Undergraduate), PG (Postgraduate), FL (Flipped learning), FC (Flipped classroom), HRD (Human resource development)

Declaration

Availability of data and material: Not applicable,Funding: Not applicable

References

Ahmed, A. and Ibrahim, A. A. (2023). The effectiveness of adaptive learning methods in mathematics compared to flipped learning analyzed by ANN and IoT. *Wireless Person. Comm.*, 1–21. https://doi.org/10.1007/S11277-023-10500-6/METRICS.

Andujar, A. and Çakmak, F. (1 C.E.). Research Anthology on Applying Social Networking Strategies to Classrooms and Libraries Foreign language learning through Instagram: A flipped learning approach. 278–299. https://Services.Igi-Global.Com/Resolvedoi/Resolve.Aspx?Doi=10.4018/978-1-6684-7123-4.Ch016. https://doi.org/10.4018/978-1-6684-7123-4.CH016.

Ay, K. and Dağhan, G. (2023). The effect of the flipped learning approach designed with community of inquiry model to the development of students' critical thinking strategies and social, teaching and cognitive presences. *Edu. Inform. Technol.*, 28(11), 15267–15299. https://doi.org/10.1007/S10639-023-11809-2/TABLES/10.

Bıyık Bayram, Ş., Gülnar, E., Özveren, H., and Çalışkan, N. (2023). The effect of flipped learning on blood pressure knowledge and self-directed learning skills of first-year nursing students: A randomized controlled trial. *Nurse Edu. Prac.*, 67, 103557. https://doi.org/10.1016/J.NEPR.2023.103557.

Chai, A. and Hamid, A. H. A. (2023). The impact of flipped learning on students' narrative writing. *Internat. J. Adv. Res. Edu. Soc.*, 4(4), 159–175. https://doi.org/10.55057/ijares.2022.4.4.15.

Huang, H., Hwang, G. J., and Chang, S. C. (2023). Facilitating decision making in authentic contexts: an SV-VR-based experiential flipped learning approach for professional training. *Interac. Learn. Environ.*, 31(8), 5219–5235. https://doi.org/10.1080/10494820.2021.2000435.

Hwang, G. J. and Chang, C. Y. (2023). Facilitating decision-making performances in nursing treatments: a contextual digital game-based flipped learning approach. *Interac. Learn. Environ.*, 31(1), 156–171. https://doi.org/10.1080/10494820.2020.1765391.

Lazzari, E. (2023). Flipped learning and affect in mathematics: Results of an initial narrative *analysis. Eur. J. Sci. Mathemat. Edu.*, 11(1), 77–88. https://doi.org/10.30935/scimath/12435.

Lo, C. K. (2023a). How can flipped learning continue in a fully online environment? Lessons learned during the COVID-19 pandemic. *PRIMUS*, 33(2), 175–185. https://doi.org/10.1080/10511970.2022.2048929.

Jurnal Pendidikan. (n.d.). Meta-anal ysis: The effectiveness of Iot-based flipped learning to improve students' problem-solving abilities. *Edumaspul.* https://ummaspul.e-journal.id/maspuljr/article/view/6195(1)

Ng, H. K. Y. and Lam, P. (2023). How the number of lessons flipped influence the overall learning effectiveness and the perceptions of flipped learning experiences? *Interac. Learn. Environ.*, https://doi.org/10.1080/10494820.2020.1826984.

Vitta, J. P. and Al-Hoorie, A. H. (2020). The flipped classroom in second language learning: *Language Teaching Research* A meta-analysis. 27(5), 1268–1292. https://doi.org/10.1177/1362168820981403.

20 Improvisation of reading efficacy among the technical students through phygital learning model

K. Velmurugan[a]

Assistant Professor, Department of English, Anurag University, Telangana, India

Abstract

Reading in students tends to activate their inquisitive ability with an interest in learning new and unfamiliar vocabulary, concepts applied in different structures, theoretical form of the context that stabilizes their prior knowledge in an advanced manner to compete the ability of reading in their day-to-day life. Language teachers and experts seek the samples from the application of language skills by the second language or L2 learners. The phygital learning is developing a domain that helps the learners in acquiring the language skills through learning domains especially the mind and body connections where the digital platform only helps to categorize visual and audio elements. The interest in phygital reading is inculcated among the students to acquire the fundamental knowledge about the concern topic and creates an awareness to harness the technology-oriented strategies to comprehend the text quickly. Phygital Reading as a strategy develops the technical students' and triggers their interest towards learning a concept without much considering of time consuming at the initial stage. The following sentence and response symbolizes with score of 26.66 "*I couldn't match the reading pace of other students*" where the students could not sustain their reading either only with text book or digital reading alone. The embedded form of reading plays a significant role where the students can sustain their pace of reading and comprehend the content.

Keywords: Language acquisition, phygital reading, focused task, technology-oriented, experimental learning

Introduction

Language acquisition plays vital role within oneself in the listening, speaking, reading, and writing skills. These skills are divided under two different sections as receptive skills and productive skills, where listening and reading skills are considered as receptive skills and speaking and writing skills are considered to be productive skills. Apart from this section, language skills also direct the learners to understand the importance and features in language acquisition. From the study, it is proposing that there is a decline in improvising the language skills especially reading skills among the students in recent times. Apart from these four language skills, the importance of language skills targets the learners in attaining the productive skills through the receptive skills.

Reading in students tends to activate their inquisitive ability with an interest in learning new and unfamiliar vocabulary, concepts applied in different structures, theoretical form of the context that stabilizes their prior knowledge in an advanced manner to compete the ability of reading in their day-to-day life. This also anticipates the readers' ability in reading skills competent towards the task they take up either in reading or incorporating new strategies that are applicable in their reading a text material. The text material would change depending on the situation and circumstances they encounter in their daily work. This assists the reader in implementing novel strategies for synchronizing different events in the text, different linguistic elements used, and various situations interpreted by the writer or author in the text. The novelty of employing reading strategies motivates the reader to assess their level of comprehension while reading as well as comprehend the text quickly. This even prompts their experience in various aspects to evaluate the authenticity of the text materials they choose and read.

Language elicitation

Language teachers and experts seek the samples from the application of language skills by the second language or L2 learners which are being categorized accordingly in different demographical form. This also instigates the teachers and researchers to assess the samples collected from the L2 learners in order to identify the hindrances faced by the L2 learners as well as teachers. This intends to the necessary changes to be carried by the researchers to recognize the level of competency in language skills and the strategies applied to comprehend the passage in reading, understanding and decoding the message received during conversation especially focus on the receptive skills of language which directly and/or indirectly promotes

[a]velmuruganhs@anurag.edu.in

DOI: 10.1201/9781003587538-20

the progress level in the productive skills as well. Second language learners need to emphasize their requirement consistently in terms of "literacy" and vocabulary development that relates to the transformations through different digital influences (Matzner et al., 2018; Nadkarni and Prügl, 2021; Priyono et al., 2020).

The purpose of reading

The reader's dependability in the form of where, when, what, how, and why is reflected in the many purposes of reading. These are the broad objectives that describe the content of the reading. The ultimate goal of reading is to build conceptual ideas through cognitive processes in reading in order to comprehend the text materials used to read as well as knowledge resources that tend to activate the readers knowledge towards the relevant resources available at the time of reading. Learning additional concepts concerns more of their occupations or concerning their hobbies. At the side of digital devices, reading tends to be the first supply of data on foreign problems, urban drawback and culture values, reading satisfying intellectual and political demands, student use reading to accumulate information that is said to scholastic success, so youngsters view as the primary motive for reading (Velmurugan and Smruti, 2021). The purpose also tries to emphasize their prior knowledge through schemata in reading. The students naturally intend to learn language skills by applying different strategies they are comfortable with and ready to instigate new experience in different approach and/or pedagogies carried out by the teachers in second language teaching and learning. In such cases, the teachers are incidentally concentrating on the task-based language teaching that motivates and encourages the students to focus on meaning and form as well to comprehend the text while reading. The teachers should be delicate in identifying the interest of the students in acquiring the second language skills and the strategies they apply to compete the difficulty in language acquisition. English grammar is taught through numerous methods by various teachers. Some teach implicitly, while others in an explicit manner (Smruti et al., 2021).

Tasks in English as a second language

Teacher development is another domain that focuses on the important responsibilities which activates language skills for those English is a second language or L2 learners, where both teachers and learners are noted to be reliant on one another to achieve the competency level of language abilities from receptive to productive. This serves as a foundation for learning language skills by instilling the connections between classroom activity, social attachment, and religious bias, as well as the cultural variations that L2 learners encounter when dealing with content focused on native speakers. L2 learners must demonstrate expertise in language strategies. Learning English as a L2 plays a vital portion as it is being played in the main domain of communication for everyone in their routine life. This domain of language moves horizontally, interrupting any discipline viz., education, science, commerce, history, geography, politics, trade, and technology. Pupil's practice takes away their own strategies according to the circumstances for the betterment of fluency (Velmurugan et al., 2022).

A task in second language acquisition focuses on language use in language learning and language pedagogy, which focuses on the learner's participation in the language classroom. The tasks practiced in a language class or activity naturally focus on performing incidental learning, which is intended to test the level of competency in language abilities for second language learners as well as language teachers. The intensity with which language abilities are assessed is determined by the language used by second language learners as well as the proficiency of language teachers working with second language learners. The task for the second language learners here would vary with the ability of the students who are aware of the language skills and different strategies they try to apply in their reading comprehension to comprehend the text. As a result, pupils with the tasks they have been assigned can deduce the meaning of the text with a clear knowledge of the text.

1. The nature of language
2. Student centered approach
3. Focus on syntax
4. Focus on meaning
5. Focused and unfocused task
6. Rejection of the traditional approach

Objective of the study

At the end of the course, students would be able to read proficiently and enhance the ability to comprehend any type of text. The present generation read the content with much influence of the digital devices and social media without recognizing the authenticity of the information in a text they read. First, they need to confirm with the authenticity of the information, second properly comprehend the text through different sentence structure they come across in a passage. Third, understanding the text what the writer or author has expressed in his or her perspectives. These factors can be achieved by the students through careful reading a news from the newspaper, articles from journals, abstracts, reports and so on. The practice of speaking or rehearsal for more than five times for presentation or for mock interview conducted in the institute campus. This helps and supports as focused self-learning materials. The sample size for this study were around 180 technical students who were from different branches of B. Tech. course like IT, ECE, DS, and CS.

Instructions for the students

Students were given assignment to be carried out at their home or room where they try to collect the articles, newspaper cuttings for improving their reading skills and recognize their way of strategies they practice in their self-reading. The way they read before their classmates or before some may get varied due to some anxiety in making mistakes like mispronouncing, more conscious about right way of reading, missing the punctuations while reading. To avoid certain misunderstanding the content, they asked to practice in the home. Once they come to the traditional classroom, they try performing their reading and respond to the questions followed by the passage. The new passage or new context is being given to the students to analyze and evaluate their level of understanding through reading. Reading comprehensive passages are given to perform their test. Students are tested with different types of reading like intensive reading or extensive reading a particular passage. They are asked to identify the passages for intensive reading and extensive reading which can stimulate in understanding their own strategies not only to infer the meaning but also to interpret the text without any misconception. The output of this activity triggered students to focus on their reading with different sub-activity designed made them little more concentrate than the traditional reading. The instructions are witnessed through classroom activity in the Figure 20.1.

Phygital reading

This learning is an astute perspective that encourages learners to participate in both physical and digital learning in a smart learning environment that sought to create the ability to sustain digital information or technology-driven activities with physical learning of concepts. The phygital learning is developing domain that helps the learners in acquiring the language skills through learning domains especially the mind and body connections where the digital platform only helps to categorize visual and audio elements where there are no adequate components in learning process.

The physical and digital learning which is blended in the recent classroom for the L2 learners that depict the idea of learning process in realistic manner to acquire the language elements from digital skilling as well as by experiencing through the physical learning. Phygital learning and reading are correlated with the influence of many factors such as environment, subject taken to read, background knowledge of the content, text material chosen as per the age group, technology focused with the physical bases, as well as the opportunities can be thought to be effective means of adoptive reading habits. The tremendous growth in the technology with sufficient, updates and upgradation especially in the academic side. This enables the growth in the reading style and habit among all people invariant to the age and gender where adolescents mostly influenced and develops much impact of the digital devices in different forms such as mobile, tab, laptop, smart watches which triggers the human cognition which adapts towards the different applications and relevant software to focus on digitalized reading (Andrade and Dias, 2020; Veer and Dobele, 2021).

Pleasure in phygital learning

Enjoyment is not that what has been instructed from schools. There is a need and requirement of enjoyment while reading the text either from printed document or digital file which varies the implementation of the strategies. The reader may experience pleasure or pressure while reading the printed document or digital file which categorizes the mood and pace of the reader. The pace of the reader is being monitored and analyzed in this paper through different responses they have registered that they have experienced in this

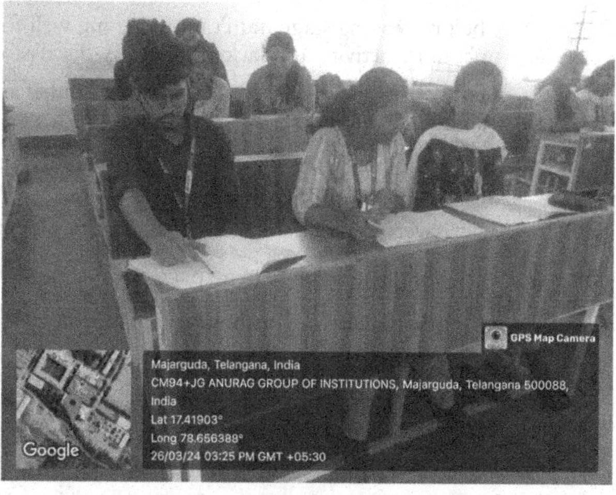

Figure 20.1 Technical student's response to the physical reading
Source: Author

phygital learning. The phygital learning not play the role blended or embedded with two different aspects of reading in the form printed document and digital file but also expose the reader to gain compounded experience that they undergo while processing and comprehending the text through physical touch of the printed document and visualizing the digital files. Mele and Russo-Spena (2022, p.14) opined that a phygital learner expedition as "the path a customer get trained through interaction with the concern in a synergetic physical and digital context to have a seamless and personalized experience which inculcates the innovative experiences in reading with pleasure."

Findings and analysis

The findings of the study tend to concentrate on some suggestions for the betterment of students and attitudes towards the use and application of educational technology during the pandemic period for online learning. The teachers and students reacted positively to this emergency online learning as it would not be permanent. The need to perform and act towards the explanations which infer through the comprehension level reading to be progressive where mental modeling deliberately dwells around to improvise the attitude at this pandemic. Reading comprehension through digital creates an impact of input which can be equated with receptive skills and productive skills through producing more like output skills (Velmurugan et al., 2024), where it highlights the growth of upgradation among the technical students in their way and style of reading.

Table 20.1 extracts the statement *"I felt conscious to seek clarification for doubts as I would be highlighted"* scored as **28.33%** as the students were nervous to get clarification. The statement *"There were internet connectivity and technological issue"* scored more in strongly disagreed parameter as **30%** by the students. The statement *"I couldn't match the reading pace of other students"* responded more in strongly agreed by the students as **26.66%** where the students find it difficult in reading pace. The students were conscious about the learning reading where models try to track the missed concepts and strategies to be implemented in the part of reading in comprehending a text material.

The word embedding and blended learning especially reading focus on the application used by the technical students that targets contemporary vector sequence training algorithm for the limited corpus target. The word embedding is processed based on the pre-trained billion-word corpus text for the different digital reading (Deepak Kumar et al., 2023).

Conclusion and further study

Phygital reading as a strategy develops the technical students' interest towards learning a concept without much considering of time consuming at the initial stage. As the time verves, they do get interested in participating the activity on reading hosted by the

Table 20.1 Technical student's response to the efficiency in phygital reading (in percentage).

S. No.	Questionnaire	SA	A	D	SD	CS
1	The reading session was dull and monotonous.	16.66	31.66	15	23.33	13.33
2	The pre-reading stage didn't motivate me well for the upcoming activity and lacked the lively discussion	25	20	13.33	25	16.66
3	I felt conscious to seek clarification for doubts as I would be highlighted	15	28.33	18.33	16.66	21.66
4	The session lacked warmth of one-to-one teacher student interaction	18.33	23.33	25	20	13.33
5	There were internet connectivity and technological issues	16.66	11.66	26.66	30	15
6	I couldn't match the reading pace of other students	26.66	15	25	20	13.33
7	I found difficulty in vocabulary recognition	28.33	23.33	20	13.33	15
8	Time constraint in pre- and post-reading sessions were hindrances	15	25	18.33	21.66	20
9	Continuous exposure to technology without a human presence demotivated me	18.33	26.66	20	20	15
10	I missed "Experimental learning" or team-based learning	16.66	18.33	23.33	21.66	20

(Strongly agree - SA, Agree - A, Disagree - D, Strongly disagree - SD, CS - Can't say)

Source: Author

facilitator. The interest in phygital reading is inculcated among the students to acquire the fundamental knowledge about the concern topic and creates an awareness to harness the technology-oriented strategies to comprehend the text with less time consumption to make it ease in their classroom than learning the content only by text book reading to attain the reading criteria. This helps to determine more conventional way of comprehending a text that they have been practicing for longer time which consumes more time as well. A great mindset with passion in their way of reading concepts dwells their etiquettes they follow with impact in comprehending a text in their reading and a complete mediator among behavioral intention and attitudes is much important personality trait for enhancing the successfulness in the readership quality (Velmurugan et al., 2023).

References

Andrade, J. G. and Dias, P. (2020). A phygital approach to cultural heritage: Augmented reality at Regaleira. *Virt. Archaeol. Rev.*, 11(22), 15–25.

Ahmadi, A. and Bajelani, M. R. (2012). Barriers to English for specific purposes learning among Iranian University students. *Proc. Soc. Behav. Sci.*, 47, 792–796.

Deepak Kumar, L. Vertivendan, Velmurugan, K., Kumarasamy, M., Dhanashree, and Vajid Nabilal, K. (2023). Semantic marginal autoencoder model for the Word embedding technique for the marginal denoising in the different languages. *Internat. J. Intel. Sys. Appl. Engg.*, 11(3s), 204–210.

Matzner, M., Büttgen, M., Demirkan, H., Spohrer, J., Alter, S., Fritzsche, A., Ng, I. C., Jonas, J. M., Martinez, V., Möslein, K. M., and Neely, A. (2018). Digital transformation in service management. *SMR J. Ser. Manag. Res.*, 2(2), 3–21.

Mele, C. and Russo-Spena, T. (2022). The architecture of the phygital customer journey: A dynamic interplay between systems of insights and systems of engagement. *Eur. J. Market.*, 56(1), 72–91.

Nadkarni, S. and Prügl, R. (2021). Digital transformation: A review, synthesis and opportunities for future research. *Manag. Rev. Quart.*, 71(2), 233–341.

Poedjiastutie, D., Darmaji, D., Musrina, M., and Novikasari, R. (2018). Task-based language teaching: An alternative approach in teaching reading comprehension in Indonesia. *J. Asia TEFL*, 15(3), 856–863. http://dx.doi.org/10.18823/asiatefl.2018.15.3.22.856.

Priyono, A., Moin, A., and Putri, V. N. A. O. (2020). Identifying digital transformation paths in the business model of SMEs during the COVID-19 pandemic. *J. Open Innov. Technol. Market Comp.*, 6(4), 104.

Smrutisikta, M., Jeyasakthi, V., and Velmurugan, K. (2021). The intervention of physical games in teaching English grammar to secondary school students: A review with special reference to the secondary school pupil in Puducherry, India. *Psychol. Edu. J.*, 58(2), 176–185. https://doi.org/10.17762/pae.v58i2.1547.

Tsai, Y. R., Chang, Y. K., and Ouyang, C. S. (2012). Proposing a recommender system through mining English reading comprehension errors of Engineering students. *Proc. 2012 Internat. Conf. Mac. Learn. Cybernet. (ICMLC '12)*, 1571–1575.

Veer, E. and Dobele, A. (2021). Big boys don't cry [Ofine]: The phygital disconnect between online and ofine mental wellness engagement. *J. Strat. Market.*, 3(2), 1–21.

Velmurugan, K. and Smrutisikta, M. (2021). Enhancing the reading skills of the technical students through mental modelling. *Psychol. Edu. J.*, 58(3), 1302–1312. https://doi.org/10.17762/pae.v58i3.3862.

Velmurugan, K., Smrutisikta, M., Jeyasakthi, V., and Mahapatra, G. S. (2022). Reading skill as a receptor of language acquisition tool among the technical engineering students of India: A strategic study and model with multi-attribute decision-making. *J. Pos. School Psychol.*, 6(2), 2748–2758.

Velmurugan, K., Kedia, N., Dhiman, A., Shaikh, M., and Chouhan, D. S. (2023). Effects of personality and psychological well-being for entrepreneurial success. *J. ReAttach Ther. Dev. Diver.*, 6(10s), 481–485.

Velmurugan, K., Smrutisikta, M., and Mahapatra, G. S. (2024). Multi-criteria decision-making interventions to reinforce the reading skills enhancement of the technical students. *J. Engg. Edu. Transform.*, 37, 544–552. DOI: https://dx.doi.org/10.16920/jeet/2024/v37is2/24086

21 Technology-based pedagogical approaches for teaching and learning

Kalpana Pawase[1,a], *Mahesh Goudar*[1,b], *Usha Verma*[1,c] *and Bhagyashri Alhat*[2,d]

[1]School of Electronics and Telecommunication Engineering, MIT Academy of Engineering, Alandi (D), Pune, Maharashtra, India

[2]School of Computer Engineering, MIT Academy of Engineering, Alandi (D), Pune, Maharashtra, India

Abstract

The goal of education is to help students actively developing their potential will enable them to have the kind of moral fiber, intellect, and personalities that are advantageous to both themselves and society. Education is a conscious effort to realize the learning process. In order to teach computer programming, emerging technology, natural science, and prototyping in undergraduate engineering degree programs, this paper provides a technology driven teaching and learning approach. Teachers were employing only tedious lecture techniques in the traditional teaching and learning process, which lowers student engagement. The conventional approach also has cost, location, and time restrictions. Computers, information, and communication technology help to overcome these by replicating thrilling learning experiences that could potentially provide more educational chances. The learning management system (LMS), e-content, artificial intelligence (AI), and other information and computer technology (ICT) tools are only a few examples. A typical learning management system supports debates, training sessions, and communication between other LMS users. LMS features include discussion boards, teacher and instructor-facilitated learning, and rubrics. E- content is a very powerful teaching tool. This most modern method of instruction has the interest of both students and teachers across all educational systems. It possesses important resources for the growth of an information-rich society. This method makes it possible to deliver lectures that are both dynamic and interesting, giving students the best opportunity to widen their knowledge paradigm and learn more efficiently. ICT must be integrated into the educational system at every level if we want to create a society that is informed.

Keywords: Information and computer technology (ICT), learning management system (LMS), artificial intelligence (AI)

Introduction

Learning with electronic technology, such as the Internet, intranets, satellite broadcasts, audio and video conferencing, bulletin boards, chat rooms, webcasts, and CD-ROM, is referred to as technology-based teaching and learning (TBL). TBL also includes terminologies that are connected to it, such as online learning is limited to learning that takes place via computers. Since it can handle bigger numbers of learners at minimal additional expense it is easily scalable to both large and small groups (Koller, n.d.). From the perspective of the learners, TBL can be self-paced, tailored to the needs of the learners, and built upon pedagogy. It highlights the benefits of discovery learning; because it clearly allows for the hands-on manipulation of course materials as well as the usage of simulations and games, it has the potential to foster improved comprehension and retention, particularly for difficult content. Perhaps as a result of these factors, TBL has seen a significant increase in the demand for training from the government, business, and educational sectors. There are eight ways to adopt technologically oriented instructional methods (Molenda, 2013).

Pedagogical approaches

In order to encourage student acquisition of ideas and principles, the teaching strategy known as problem-based learning (PBL) replaces challenging real-world problems with the direct presentation of facts and concepts. Along with learning the course information, PBL can help students improve their communication, critical thinking, and problem-solving abilities. Additionally, it might offer opportunities for research source discovery and evaluation, group cooperation, and lifetime learning (Waight and Wang, 2000).

Student-created content

More and more technology-based methods give students the chance to produce information that can be shared in the learning management systems (LMS), online learning platforms, and classrooms across schools.

[a]kalpana.pawase@mitaoe.ac.in, [b]director@mitaoe.ac.in, [c]uyverma@mitaoe.ac.in, [d]bralhat@mitaoe.ac.in

DOI: 10.1201/9781003587538-21

Collaborative learning
Technology has made it possible for us to learn anywhere. As a result of technological advancements like Skype and FaceTime, online learning platforms may now be accessed from anywhere at any time (Dhavale and Paliwal, n.d.).

Competency-based education
Competency-based education is another learning strategy. Instead of emphasizing time-based learning, this alternate strategy wants to emphasize learning that sticks.

In essence, it's an effort to adapt to how the younger generations work. Spending hours on a subject when the listener is not receptive is frequently ineffective. Stress is lessened and learning may be done more effectively by dissolving the bonds between time and place.

Active Learning
Active learning, also known as hands-on learning, is when students experiment with various teaching techniques, encourage one another's ideas, and, most importantly, are given the freedom to think and act independently.

Another crucial component of the active learning process is letting students suggest enjoyable and helpful tech tools for learning (Mugenyi, n.d.).

Blended learning
The majority of the time, blended learning is defined as the fusion of conventional classroom instruction and online digital media (Sivakumar and Singaravelu, 2016). While it is necessary for teachers and students to be present physically, the content and student assignments are broken up into sections that may be completed digitally on laptops, tablets, etc. Together, these educational opportunities strengthen and complete one another (Alrushi and Olfman, n.d.).

Flipped learning
A hybrid learning approach called "flipped learning" flips the way that lessons are typically taught. It emphasizes using knowledge in the real world and increasing degrees of student–teacher interaction in the classroom (University of Mumbai). It operates outside of the classroom, students learn about a topic using digital resources and intermittent micro-learning techniques, such watching brief films on a certain topic at home.

Integrated subjects
Additionally, it moves education away from strictly specialized and subject-focused methods and toward integrated, cross-curricular study that more accurately reflects the connected world in which we live.

Proposed work

This strategy of technology-based approach is used by authors while teaching to 1st and 2nd year B. Tech students in the courses like C programming, cloud computing, Science of Nature, and Digital prototyping (University of Mumbai). Authors employed various technology-based approaches to improve the effectiveness of the course delivery of subjects like cloud computing, Science of Nature, and Digital prototyping. Due to which authors were successful in optimizing the learning experience for students which enables active and interesting lessons. Figure 21.4 shows the plot of student satisfaction survey. It also aids in increasing students' knowledge paradigms. Following are the examples of activities conducted in this approach.

Problems were given to students to design an application/idea by applying knowledge gained from their track (as shown in Figure 21.1a and b) or related technology in group. As we are celebrating international year of millets 2023 (IYM 2023). Students have to go through question statement provided on LMS and present a Poster/PowerPoint presentation/web page. Submission guidelines were also given on LMS. The International Year of Millets (IYM) 2023, proposed and sponsored by the Government of India, was accepted by the United Nations General Assembly (UNGA). The theme for IYM 2023 is "Harnessing the untapped potential of millets for food security, nutrition, and sustainable agriculture". The glimpses of the activities under collaborative learning are as shown in Figures 21.2 (a, b) and 21.3a.

In active learning activity is assigned in which students need to prepare a locomotion model of any animal. Animals in the wild move with different kinds of locomotion styles, some hop, some run, some swim. The assignment focuses on analysis, breaking down a complex phenomenon into simpler parts as shown in Figure 21.3b for the sake of better understanding of a phenomenon.

Effectiveness of implementation

Result and discussion
Technology-driven teaching and learning help students learn more efficiently and makes it easier for them to locate relevant information and knowledge for their studies. It was also noted that this method produces better results as shown in Figure 21.6. Additionally, students are now ready for the initial step of campus interview.

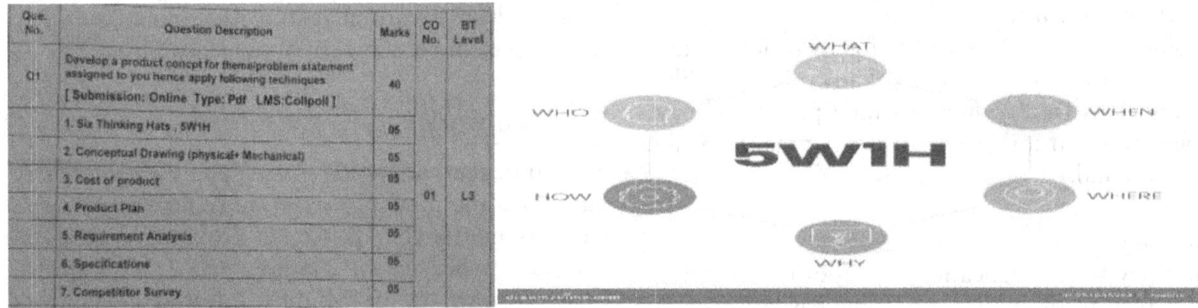

Que. No.	Question Description	Marks	CO No.	BT Level
Q1	Develop a product concept for theme/problem statement assigned to you hence apply following techniques [Submission: Online Type: Pdf LMS:Collpoll]	40	01	L3
	1. Six Thinking Hats , 5W1H	05		
	2. Conceptual Drawing (physical+ Mechanical)	05		
	3. Cost of product	05		
	4. Product Plan	05		
	5. Requirement Analysis	05		
	6. Specifications	05		
	7. Competititor Survey	05		

Figure 21.1 (a and b): Examples of problem-based learning
Source: https://online.visual-paradigm.com/diagrams/templates/5w1h/5w1h-questions

Figure 21.2 (a and b): Student uploaded project on LMS for utilizing technology to drive sustainable millet production for climate balance (project-based)
Source: Author

Figure 21.3 (a) Glimpses of collaborative learning. (b) Glimpses of active learning
Source: Author

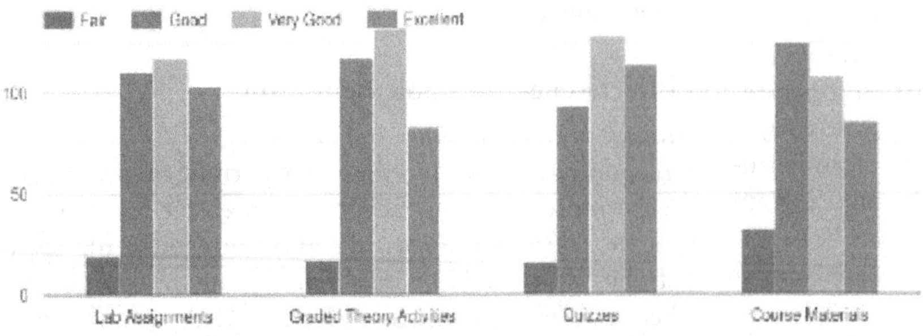

Figure 21.4 Students' satisfaction survey
Source: Author

```
#include<stdio.h>
#include<stdlib.h>

struct batsman
{
  char name[25];
  int runs,score,balls,toruns,tobal,ones,twos,threes,fours,sixes;
  int max_six,max_run,max_four;
  float str;
```

Figure 21.5 Project showing Olympic game details using C
Source: Author

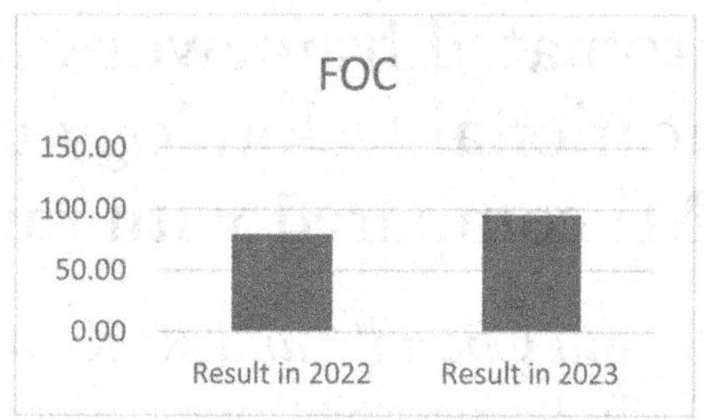

Figure 21.6 Observed improvements in final result of subject foundation of computing
Source: Author

The survey of student happiness, which has a higher percentage than earlier approaches, is shown in Figure 21.4 (Ojukwu, 2021). Figure 21.5 shows the glimpses of active learning where students were engaged and motivated with a passion for writing a program code. Here learners were engaged with concepts, solve problems and explore new ideas practically. Program code can be used to simulate or model real-world scenarios, allowing learners to apply their knowledge and skills in practical contexts. This bridges the gap between theory and practice, helping learners develop the skills needed to tackle real-world challenges in their field of study or profession (Figure 21.6).

Conclusions

The improvement of educational quality and student growth are the main focus of our endeavor. It is feasible to reach the greatest number of students and keep track of each student's progress using ICT tools. From the above experimentation it can be concluded that by using technology driven teaching learning we can make teaching more effective and increase the involvement of the students in learning. It is intended for students to have easy access to study materials within the system.

References

Koller, V., Harvey, S., and Magnotta, M. (n.d.). Technology-based learning strategies. *U.S. Depart. Labor Creat. Edu.*, 5(11). Pg no.1–39

Waight, C. and Wang, X. C. (2000). E-learning: A review of literature. ResearchGate Pg no.1–74

Dhavale, S. M. and Paliwal, A. S. (n.d.) . Development of e-content for effective teaching and learning. *Nat. Conf. Adv. Elec. Interdis. Appl. [NCAEIA-2014]. Pg. No. 363–365*

Mugenyi, J. K., Zhu, C., and Kagambe, E. (n.d.) . Blended learning effectiveness: the relationship between student characteristics, design features and outcomes. *Internat. J. Edu. Technol.*, 14(1). DOI:10.1186/s41239-017-0043-4. Pg. No.1–20

Sivakumar, A. and Singaravelu, G. (2016) . Enabling teachers on technology driven teaching head. *Depart. Edu. J. Soc. Sci.* Pg No. 82–84

Alrushi, N. and Olfman, L. (n.d.). Aiding participation and engagement in a blended learning environment. *J. Inform. Sys. Edu.*, 24(2), 133–145. *https://www.education.gov.in/statistics.*

Institute of Distance and Open Learning, University of Mumbai. *B.A. Semester VI paper: International Relations: India in World Politics.* Mumbai, India.

Raharjo, F. D., Raharjo, T. J., and Sumsudi. (2016). The implementation of multiple intelligence-based school management. *J. Edu. Dev.*, 4(1), 31–36.

Kirkwood, A. and Price, L. (2006). Adaptation for a changing environment: Developing learning and teaching with information and communication technologies. *Internat. Rev. Res. Open Dis. Learn.*, 7(2), 1–14.

Ojukwu, N., Nwawih, N., Agim, E. C., and Ameh, C. (2021). International Journal of Research and Review DOI: https://doi.org/10.52403/ijrr.20210713 Vol.8; Issue: 7; July 2021, Pg No.90–97

Educational technology for teaching and learning in the post Covid-19 era: A case study of tertiary institutions in Imo state, Nigeria. 8(7).

Briggs, L. J. (1977). *Instructional Design.* Englewood Cliffs, New Jersey: Educational Technology Publications Inc. Pages 61–75

Wicklein, R. C. (1998) . Designing for appropriate technology in developing countries. *Technol. Forecast. Soc. Change.* Pg No. 202–206

Sudarsana, I. K. (2013). Pengaruh Model Pembelajaran Cooperatif Terhadap Prestasi Belajar Siswa Dengan Tipe J. Panamanian Mutu. *J. Penjaminan Mutu*, 4(1), 20. DOI:10.25078/jpm.v4i1.395.

Molenda, J. (2013). *Educational Technology: A Definition with Commentary.* Taylor & Francis Group, LLC. https://doi.org/10.4324/9780203054000, Second Edition Sept 2007. Pages:384

Arsyad, A. (2018). Pengaruh Model Pembelajaran Kooperatif Terhadap Peningkatan Mutu Hasil Belajar Siswa. *J. Penjaminan Mutu*, 4(1), 20. DOI:10.25078/jpm. v4i1.395.

Rajarapollu, P., Bansode, N . V., and Katkar, V. (2022). CT - A tool to enhance teaching learning activity in technical education. 35(Special Issue-1), 14–18. DOI: 10.16920/jeet/2022/v35is1/22002. Pages:14–18

22 An automated framework for improving paper correction in educational technology using long brief-term memory (LBTM) compared with random forest calculation

Thota Bhanu Chand[1,a] *and R. Kesavan*[2,b]

[1]Research Scholar, Department of Computer Science and Engineering, Saveetha School of Engineering, Saveetha Institute of Medical and Technical Sciences, Saveetha University, Chennai-602105, Tamil Nadu, India

[2]Research Guide, Department of Computer Science and Engineering, Saveetha School of Engineering, Saveetha Institute of Medical and Technical Sciences, Saveetha University, Chennai-602105, Tamil Nadu, India

Abstract

This aim to explore and looks at the adequacy of long brief term memory (LBTM) organizations and irregular woodland estimations in further developing paper adjustment processes inside instructive innovation. The point is to upgrade amendment accuracy and proficiency by utilizing artificial intelligence (AI) procedures for blunder recognition. **Materials and methods:** To comprehensively evaluate the potential of LBTM networks and random forest calculations in paper correction, we compiled a diverse dataset comprising various types of student papers across different subjects and grade levels. This dataset includes both error-free papers and those containing errors, essential for training and evaluating the models effectively. **Results:** The comparative analysis between LBTM networks and random forest calculations provides valuable insights into their effectiveness in paper correction. LBTM networks demonstrate strong performance in capturing contextual dependencies and linguistic nuances, resulting in accurate error detection and correction. **Conclusion:** In conclusion, this study highlights the potential of LBTM networks and random forest calculations in enhancing paper correction processes within educational technology. Through a comprehensive comparative analysis, we gain valuable insights into their strengths, limitations, and applicability in improving correction precision and efficiency.

Keywords: Paper correction, long brief-term memory (LBTM), random forest calculation, comparison, accuracy, educational technology, error detection

Introduction

The integration of artificial intelligence (AI) into educational technology has opened new avenues for enhancing various processes, including the correction and grading of student papers. This study aims to explore the adequacy of long brief-term memory (LBTM) networks and random forest (RF) algorithms in improving paper correction processes. The objective is to enhance the accuracy and efficiency of these processes by utilizing machine learning techniques for error detection and correction B. Yang, et al. (2021). AI's ability to process and analyze large volumes of data rapidly and accurately makes it an ideal tool for educational applications. In particular, machine learning models like LBTM networks and RF algorithms offer promising approaches for automating and refining the paper correction process. LBTM networks, a variation of long short-term memory (LSTM) networks, are designed to capture and utilize contextual dependencies and linguistic nuances, which are crucial for accurate error detection and correction in written texts. On the other hand, RF algorithms, known for their robustness and versatility, offer a reliable method for handling various types of data and can be effectively trained to identify and correct errors in student paper S. Wang et al. (2021). To thoroughly evaluate the potential of these AI techniques, a diverse dataset was compiled, comprising student papers from various subjects and grade levels. This dataset includes both error-free and error-containing papers, which are essential for training and evaluating the models' performance. By comparing the performance of LBTM networks and RF algorithms, the study aims to determine which method provides superior accuracy and efficiency in the context of educational paper correction A. Solyman et al. (2021). The results of this comparative analysis provide valuable insights into the effectiveness of LBTM networks and RF algorithms. LBTM networks, with their advanced capabilities in understanding linguistic context, demonstrate strong performance in identifying and correcting errors. In contrast, RF algorithms also show significant potential, particularly due to their ability to handle diverse datasets and provide consistent results

[a]thotabhanuchand0744.sse@saveetha.com, [b]kesavanr.sse@saveetha.com

DOI: 10.1201/9781003587538-22

H. Yang (2023). This study highlights the potential of integrating LBTM networks and RF algorithms into educational technology to enhance paper correction processes. Through a comprehensive comparative analysis, we gain valuable insights into the strengths and limitations of these models, paving the way for their application in improving correction precision and efficiency in educational settings. This research underscores the importance of continuous innovation in educational technology, leveraging AI to support educators and improve student learning outcomes (Aggarwal, 2004; Criminisi and Shotton, 2013), with the overarching goal of enhancing learning experiences for students (Shah et al., 2016).

Materials and methods

The primary goal of this research is to enhance paper correction processes in educational technology by exploring the adequacy of LBTM networks and RF algorithms Q. F. Zhao and Y. F. Li (2013). To achieve this, we compiled a diverse dataset comprising various types of student papers across different subjects and grade levels, including both error-free papers and those containing errors. This dataset is essential for training and evaluating the models effectively. The preprocessing of the dataset involved several critical steps. Text normalization was performed to standardize the text by converting it to lowercase, removing special characters, and correcting common misspellings. Tokenization split the text into individual tokens, facilitating analysis by the models Z. Yuan (2021). Additionally, part-of-speech tagging was conducted to annotate each token with its grammatical role, helping the models understand the sentence structure. Errors in the papers were manually annotated to provide ground truth data for model training M. Yagci and M. Üünal (2014).

Two primary models were trained and evaluated: LBTM networks and RF algorithms. The LBTM networks, a variant of LSTM networks, are designed to capture long-range dependencies in text through input, hidden, and output layers with recurrent connections that maintain contextual information B. Sadayapillai and K. Kottursamy (2022). The training process involved optimizing the LBTM model for error detection and correction using back propagation through time (BPTT) to adjust the network's weights, with various hyper parameters tuned for optimal performance. In contrast, the RF model consists of an ensemble of decision trees, each trained on a random subset of the data. The final prediction is derived from the averaged predictions of all individual trees. The RF model was trained using the same annotated dataset, with a focus on feature extraction from the text to classify and correct errors. Important features such as word frequency, part-of-speech tags, and n-grams were engineered to enhance the model's accuracy A. Zemliak (2022).

The performance of both models was evaluated using metrics like accuracy, precision, recall, and F1 score. Accuracy measured the percentage of correctly identified and corrected errors, precision assessed the proportion of true positive error detections, recall evaluated the proportion of true positive error detections out of all actual errors, and the F1 score provided a balanced measure of the model's performance. A comparative analysis was conducted to evaluate the effectiveness of LBTM networks versus RF algorithms H. Yu et al. (2022). The LBTM networks were assessed for their ability to capture contextual dependencies and linguistic nuances, which are crucial for accurate error detection and correction. The robustness of RF

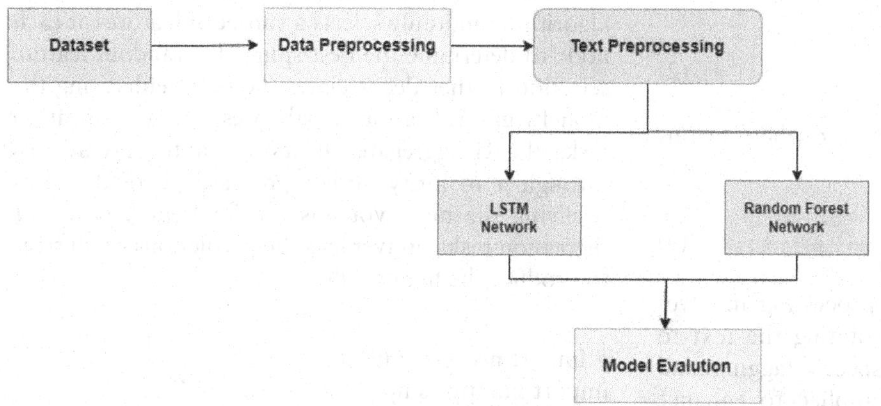

Figure 22.1 Flow chart in paper correction in educational technology for illustrating the process described in your abstract, focusing on the flow of data and operations involving LSTM networks and random forest algorithms for paper correction in educational technology

Source: Author

algorithms was evaluated based on their ability to handle diverse datasets and provide consistent results. The computational efficiency and scalability of both models were compared, considering training time and resource requirements.

The results of this analysis provided valuable insights into the strengths and limitations of each model. LBTM networks demonstrated strong performance in capturing contextual information, leading to accurate error detection and correction. Meanwhile, RF algorithms showed robustness in handling diverse data but required extensive feature engineering to achieve comparable accurace.

A. Long brief-term memory

The long brief-term memory (LBTM) algorithm is a variant of LSTM network, designed specifically to handle and utilize long-range dependencies in sequential data. This model is particularly useful for tasks requiring an understanding of sequences over extended periods while retaining crucial contextual information. LBTM combines the strengths of LSTM with specific optimizations to enhance performance in applications such as error detection and correction in educational technology. Key concepts of LBTM include memory cells and gating mechanisms, such as the input gate, forget gate, and output gate, which regulate the flow of information and manage how long past data is retained. The architecture of LBTM involves an input layer that accepts sequential data, hidden layers composed of LBTM cells that process the data, and an output layer that generates the final prediction, classification, or error detection result.

$$i_t = \sigma(W_i \blacksquare x_t + U_i . h_{t-1} + b_i) \tag{1}$$

Input gate i_t: Controls the extent of new information added to the cell state.

$$f_t = \sigma(W_f \blacksquare x_t + U_i . h_{t-1} + b_i) \tag{2}$$

Forget gate f_t: Determines how much of the previous cell state to retain.

$$o_t = \sigma(W_o \blacksquare x_t + U_i . h_{t-1} + b_i) \tag{3}$$

In the training process, data preprocessing involves normalizing, tokenizing, and annotating the text to prepare it for training. Part-of-speech tagging and other linguistic annotations are applied to enhance the model's understanding. The model is trained using BPTT, unrolling the network for a fixed number of time steps, and updating weights based on error gradients. Hyper parameters such as learning rate, batch size, and the number of hidden units are tuned for optimal performance. The LBTM model is specifically trained to identify and correct errors in text by learning from annotated examples, leveraging its ability to understand context to make accurate corrections.

Applications of the LBTM model extend beyond educational technology to various natural language processing (NLP) tasks like machine translation, sentiment analysis, and text summarization, where long-term dependencies are crucial. In speech recognition, the LBTM model improves the accuracy of transcriptions by maintaining context over long sequences of spoken words. The benefits of LBTM models include high accuracy in tasks requiring long-term context understanding, optimized gating mechanisms that reduce computational overhead, and flexibility in applying to a wide range of sequential data tasks.

B. Random forest algorithm

The random forest algorithm is a robust ensemble learning method utilized for both classification and regression tasks. It operates by constructing multiple decision trees during the training phase and synthesizing their predictions to enhance overall accuracy and mitigate the risk of overfitting. The key strength of RF lies in its ability to combine the outputs of numerous weak learners (decision trees) to create a strong predictive model.

To generate a RF, the algorithm employs a technique known as bootstrap aggregating, or bagging. This involves creating several subsets of the original training data through random sampling with replacement. Each subset is then used to train a different decision tree. By using different data subsets, the algorithm ensures that each tree is trained on unique data points, which promotes diversity among the trees and reduces the likelihood of overfitting to any single dataset. During the construction of each decision tree, the algorithm randomly selects a subset of features at each node to determine the best split. This random feature selection further decorrelates the trees, enhancing the model's generalization capabilities. For classification tasks, the RF aggregates the individual tree predictions through a majority voting process, where the class receiving the most votes is the final prediction. For regression tasks, it averages the predictions of all trees to produce the final result.

```
# Import necessary libraries
import numpy as np
from sklearn.datasets import load_iris
from sklearn.model_selection import train_test_split
from sklearn.ensemble import RandomForestClassifier
from sklearn.metrics import accuracy_score
```

```
# Load dataset (example using Iris dataset)
iris = load_iris()
x = iris.data
y = iris.target

# Split data into training and testing sets
X_train, X_test, y_train, y_test = train_test_split(X, y,
test_size=0.2, random_state=42)
# Initialize Random Forest classifier
clf = RandomForestClassifier(n_estimators=100,
max_depth=2, random_state=42)
# Train the model
clf.fit(X_train, y_train)

# Predict on the test data
y_pred = clf.predict(X_test)

# Evaluate accuracy
accuracy = accuracy_score(y_test, y_pred)
print(f'Accuracy: {accuracy}')
```

Random forest offers several advantages, including improved accuracy due to the reduction of variance through ensemble learning, and robustness to overfitting due to the use of bagging and random feature selection. It also handles high-dimensional data effectively, making it suitable for a wide range of applications. Additionally, RF can provide valuable insights into feature importance, identifying the most relevant features for prediction tasks. In the context of educational technology, RF can be leveraged to enhance paper correction processes. By training the model on a diverse dataset of student papers annotated with various types of errors, the algorithm can learn to detect and correct these errors accurately. Its robustness and accuracy make it well-suited for handling the complex and varied nature of textual data in educational assessments.

Statistical analysis

Measurable examination was led to thoroughly assess the presentation of both LBTM organizations and arbitrary woodland estimations in paper revision. Elucidating insights were processed to sum up the attributes of the dataset, including proportions of focal propensity and scattering. Also, inferential measurable tests, for example, T-tests or ANOVA, were utilized to evaluate the meaning of contrasts in adjustment accuracy between the LBTM and arbitrary woods drawn near. These measurable examinations gave experiences into the adequacy of every technique and worked with informed dynamic in choosing the most reasonable methodology for paper rectification in instructive innovation.

Results

The image you sent is a screenshot of a T-test output, likely from a statistical software program. It displays the results of a T-test, which is a statistical test used to determine if there is a significant difference between the means of two groups. The data in the image is

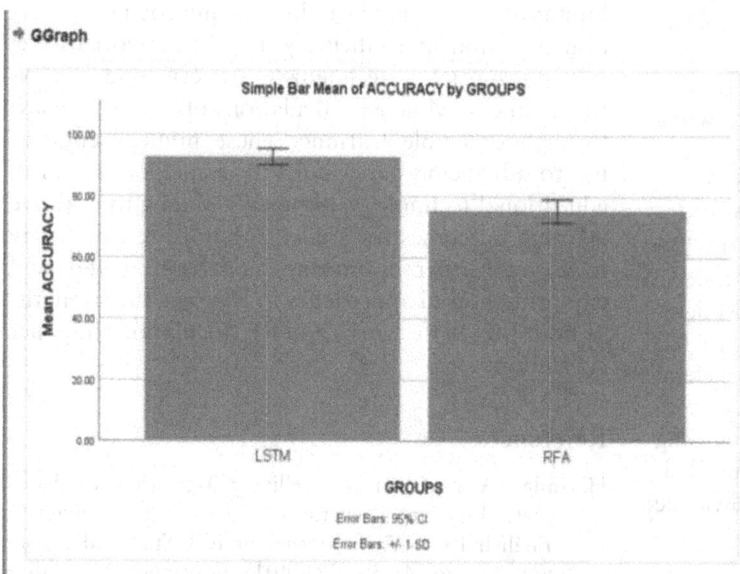

Figure 22.2 The image you sent is a graph showing the mean accuracy of two groups, likely created with software like SPSS. The x-axis shows the groups, labeled "LBTM" and "RF." The y-axis shows the mean accuracy, but the values are not displayed in the part of the image you sent

Source: Author

Table 22.1 The mean accuracy LBTM versus RF.

	Algorithm	N	Mean	Std. dev	Std. error mean
Accuracy	LBTM	10	80.50	5.91	1.32
	RF	10	75.10	5.88	1.31

Source: Author

Table 22.2 Independent sample T-test calculation.

Accuracy				95% Credible interval	
	Sig	F	T	Lower bound	Upper bound
Equal variances assumed	0.017	38	0.211	6.68487	9.63113
Group=VGG 16		95.113	0.211	6.67441	9.64157

Source: Author

labeled "ACCURACY" and compares two groups, "LBTM" and "RANDOM FOREST". The table shows that the mean accuracy for the "LBTM" group is higher than the "RANDOM FOREST" group, with a mean difference of 10.58.

The bars in the graph are both blue, making it difficult to visually compare their heights. It's also important to note that error bars, which are used to show the variability of the data, are not displayed in the image. Without the specific values and error bars, it is impossible to say definitively which group has a higher mean accuracy or if the difference is statistically significant.

The study evaluates the performance of algorithms Long Brief Term Memory (LBTM) and Random Forest (RF) algorithms for improving paper correction in educational technology. The results indicate that LBTM outperforms RF in terms of accuracy.

Table 22.1 presents the mean accuracy results: LBTM achieved an average accuracy of 80.05%, while RF achieved 75.10%. The standard error of the mean for RF is 1.322, slightly higher than LBTM 1.315.

Table 22.2 shows that there is no statistically significant difference in mean accuracy between LBTM and RF, with a p-value of 0.984 (p>0.05, two-tailed). Despite the lack of statistical significance, LBTM demonstrates a higher mean accuracy compared to RF, suggesting its potential superiority in musculoskeletal detection tasks.

This comparison highlights LBTM as a promising algorithm for enhancing the accuracy of musculoskeletal detection in real-time physiological knowledge prediction applications.

Discussion

Envision a heat map outwardly addressing the consequences of a measurable test, maybe looking at the method for two gatherings across various factors. The variety force could demonstrate the meaning of the distinctions, with more profound tones addressing genuinely huge varieties. Examining this heat map could include This situation could include a heat map showing the connections between different highlights in a dataset.

Conclusions

In conclusion, this study sheds light on the potential of LBTM networks and RF calculations in enhancing paper correction processes within educational technology. Through a comprehensive comparative analysis, we gain valuable insights into their strengths, limitations, and applicability in improving correction precision and efficiency. LBTM networks excel in capturing intricate linguistic patterns and contextual nuances, while RF calculations offer effectiveness through ensemble learning. These findings contribute to advancing paper correction methodologies in educational technology, ultimately aiming to improve learning outcomes for students. Future research may focus on further optimizing model parameters and exploring hybrid approaches to leverage the strengths of both LBTM networks and RF calculations in paper correction.

References

H. Yang, "A Study on an Intelligent Algorithm for Automatic Test Paper Generation and Scoring in University English Exams," in Journal of ICT Standardization, vol. 11, no. 4, pp. 391–401, November 2023, doi: 10.13052/jicts2245-800X.1144.

A. Solyman, Z. Wang, Q. Tao, A. A. M. Elhag, M. Toseef and Z. Aleibeid, "Synthetic data with neural machine translation for automatic correction in arabic grammar", Egypt. Inform. J., vol. 22, no. 3, pp. 303–315, 2021.

S. Wang, N. Shrestha, A. K. Subburaman, J. Wang, M. Wei and N. Nagappan, "Automatic Unit Test Generation for Machine Learning Libraries: How Far Are We?", 2021 IEEE/ACM 43rd International Conference on Software Engineering (ICSE), pp. 1548–1560, 2021.

B. Yang, H. Xie, K. Ye, H. Qin, R. Zu and A. Liu, "Analysis of intelligent test paper generation method for online examination based on UML and particle swarm optimisation", Int. J. Inform. Commun. Technol., vol. 18, no. 3, pp. 317–333, 2021.

D. Wang, Y. Zhao, H. Lin and X. Zuo, "Automatic scoring of Chinese fill-in-the-blank questions based on improved P-means", J. Intell. Fuzzy Syst., vol. 40, no. 3, pp. 5473–5482, 2021.

Z. Yuan, "Interactive intelligent teaching and automatic composition scoring system based on linear regression machine learning algorithm", J. Intell. Fuzzy Syst., vol. 40, no. 2, pp. 2069–2081, 2021.

Q. F. Zhao and Y. F. Li, "Research and development of online examination system", Adv. Mater. Res., vol. 756-759, pp. 1110–1113, 2013.

M. Yagci and M. Üünal, "Designing and Implementing an Adaptive Online Examination System", Proc. Soc. Behav. Sci., vol. 116, pp. 3079-3083, 2014.

B. Sadayapillai and K. Kottursamy, "A blockchain-based framework for transparent secure and verifiable online examination system", J. Uncer-tain Syst., vol. 15, no. 03, 2022.

A. Zemliak, "A modified genetic algorithm for system optimization", COMPEL, vol. 41, no. 1, pp. 499–516, 2022.

H. Yu, M. Zheng, W. Zhang, W. Nie and T. Bian, "Optimal design of helical flute of irregular tooth end milling cutter based on particle swarm optimization algorithm", P. I. Mech. Eng. C. J. MEC, vol. 236, no. 7, pp. 3323–3339, 2022.

S. Vaziri, J. Abbatematteo, M. Fleisher, A. B. Dru, D. T. Lockney, P. S. Kubilis, et al., "Correlation of perioperative risk scores with hospital costs in neurosurgical patients'", J. Neurosurg., vol. 132, no. 3, pp. 818–824, 2019.

23 Leveraging machine learning for paragraph-based answer generation

Maridu Bhargavi[a], Cherukuri Sowndaryavathi[b], Kshama Kumari[c], Ankit Kumar Prabhat[d] and Manish Kumar[e]

Dept of CSE Vignan's Foundation for Science, Technology, and Research (Deemed to be University), Vadlamudi, Guntur, Andhra Pradesh, India

Abstract

This paper introduces an intuitive system designed to generate answers to questions based on a given paragraph. Extracting answers from lengthy texts or paragraphs can be arduous, particularly for researchers or individuals seeking specific information. Therefore, the ability to automatically generate answers from large texts, especially for educational purposes, would be highly beneficial. This system accepts a paragraph as input along with a question. It produces an answer tailored to the question's type, which could include MCQ's (Tan 2011), FB, T/F, binary questions, or WH-type questions. The model operates by tokenizing the input text, converting the tokens into numerical representations the model understands. These tokens are then fed into multiple layers of transformer models, which capture the meaning of each token in context. For answer generation, the system identifies relevant portions of the text by comparing embedding of the question with the text, aiming to provide a pertinent answer.

Keywords: BERT model, text understanding, tokenization, numerical encoding, transformer architecture, contextualized embeddings

Introduction

Automated answer generation systems have significant applications in education, research, and professional environments, providing quick and accurate information retrieval. Our system employs bidirectional encoder representations from transformers (BERT) to comprehend and respond to questions by processing contextual embeddings from text inputs.

Literature review

Past research highlights various methodologies for question generation and answering. Many techniques were implemented to generate questions from the given text and generate answers. The first implemented paper was by Kumar et al. (2019) present two hierarchical models for generating questions from paragraphs, addressing the challenge of dealing with long contextual information. The first model is a hierarchical BiLSTM with selective attention, while the second model was by Kumar et al. (2021) who introduce an innovative approach to question generation called an "Answer Driven Model for Paragraph-level Question Generation." The primary focus is on generating questions that are directly related to a given context and answers. The paper by Parthasarathy et al. (2021) introduces a novel approach for a web-based question answering (QA) system aimed at enhancing e-learning experiences. By utilizing template mapping and content clustering techniques, the system efficiently detects question types, constructs tailored search queries, and extracts detailed answers from relevant web pages, offering a comprehensive and user-friendly solution for e-learning needs. Another paper by Kumar et al. (2021) proposes a deep learning-based approach for automatic question-answer pair generation, leveraging BERT to combine answer extraction (AE), question generation (QG), and question answering (QA) models. By fine-tuning BERT, the system aims to create WH-type questions from paragraphs and accurately retrieve corresponding answers, facilitating time-saving and efficient question creation across various domains. Another paper by Chen et al. (2021) introduces an innovative answer-driven model for paragraph-level question generation, which dynamically integrates answer information into the decoder to focus on relevant question aspects. Through reinforcement learning, the model effectively utilizes answer distribution differences to outperform existing attention-based seq2seq models, enhancing question-answer coherence. Varathan, Sembok, and Kadir (2010) developed an automatic lexicon generator to improve the accuracy

[a]bhargaviformal@gmail.com, [b]cherukurisowndaryavathi@gmail.com, [c]kshamakumari894@gmail.com, [d]ankitprabhat5@gmail.com, [e]manishkrydv4212@gmail.com

DOI: 10.1201/9781003587538-23

and efficiency of logic-based question-answering systems by analyzing large text corpora to create comprehensive lexicons. Abd Rahim et al. (2017), who developed an automated exam question generator using a genetic algorithm to efficiently create exam questions, and Dhami et al. (2015), who presented an automatic question generator in the International Journal of Science and Technology, emphasizing the growing utility of machine learning in educational assessments.

Methodology

The methodology outlined involves a systematic approach to implementing an answer generator using BERT for natural language understanding tasks, particularly question answering.

Initially, the text undergoes preprocessing where it is tokenized using the BERT tokenizer. This involves breaking down the input paragraph into individual tokens, with special tokens like [CLS] marking the beginning and [SEP] marking the end. This tokenization process employs libraries like Hugging Face's transformers to facilitate the conversion of text into BERT-compatible tokens. Mathematically, this can be represented as

$$T = [CLS] + \text{BERTTokenizer}(P) + [SEP]$$

where T represents the tokenized sequence and P represents the input paragraph.

Following tokenization, the tokens are encoded into numerical representations using the BERT tokenizer's vocabulary, resulting in a sequence of token IDs. These token IDs are then formatted as input to the BERT model, typically comprising token IDs, segment IDs, and position embeddings. The input sequence traverses through multiple layers of BERT's Transformer encoder architecture, incorporating self-attention mechanisms and feed-forward neural networks. This can be symbolized as

$$EBERT = \text{BERT Model}(IBERT)$$

where $EBERT$ represents the output embeddings from BERT's layers and $IBERT$ represents the formatted input sequence.

As the input sequence traverses BERT's layers, each token's representation is updated based on its context within the sequence, resulting in contextualized embeddings. The final layer's output embeddings contain rich contextual information about each token, capturing semantic meanings and relationships within the input text. For question-answering tasks, additional layers or attention mechanisms are applied to focus on relevant parts of the input paragraph to generate an answer to the question. This involves comparing BERT embeddings for the paragraph and the question to identify the most relevant answer option, typically using similarity metrics like cosine similarity. Mathematically, this process can be expressed as

$$Aselected = \text{Select Answer}(O, Q, P, Econtext)$$

where $Aselected$ represents the selected answer, O represents answer options, Q represents the question, and

$Econtext$ represents the contextualized embeddings (Figure 23.1).

Results

The evaluation of the model's performance is crucial to assess its accuracy and effectiveness (Dalianis and Dalianis, 2018). By examining the calculated values of evaluation metrics, we can gauge the model's performance. Higher values indicate that the model is performing well, while lower values signify the need for further improvements to ensure the relevance of the answers provided. The system's performance can be calculated by using recall, precision, and F1 score are computed. These performance metrics offers intuition into the system's ability to generate precise answers

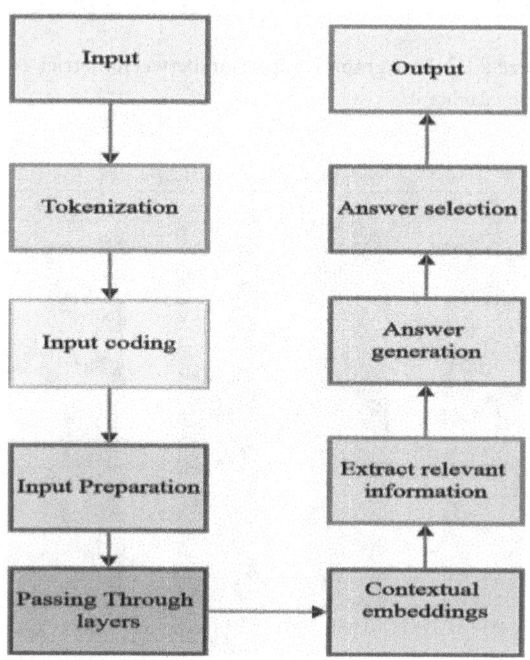

Figure 23.1 Model diagram
Source: Author

and its overall effectiveness in question-answering tasks (Table 23.1).

Table 23.1 Performance metrics for different question types.

Question type	Precision	Recall	F1 score
MCQs	0.90	0.92	0.91
Fill-ups	0.95	0.97	0.96
True/false	0.99	0.92	0.90
Binary	0.92	0.90	0.91
Wh-questions	0.95	0.90	0.92

Source: Author

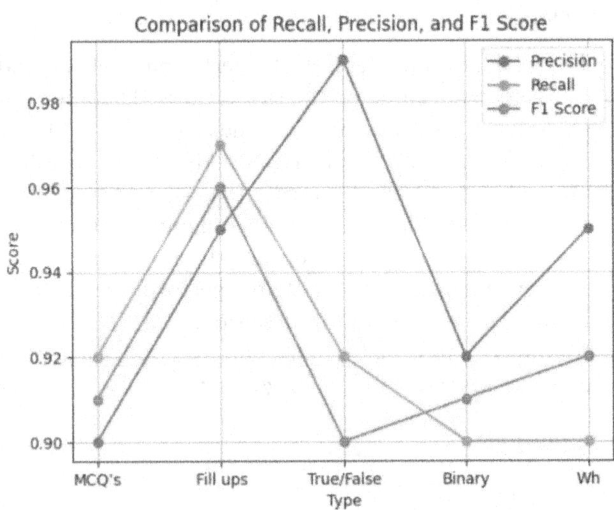

Figure 23.2 Line graph comparison between metrics
Source: Author

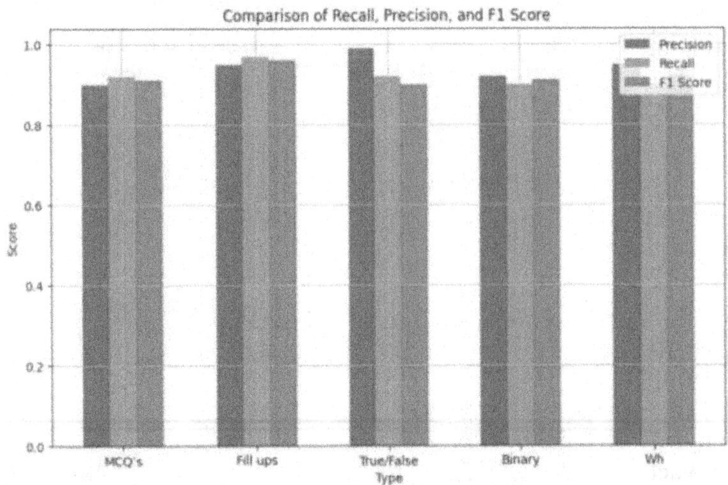

Figure 23.3 Bar graph comparison between metrics
Source: Author

The graphs seen in Figure 23.2 and 23.3 compares the performance of a model on different question types (multiple choice, true/false, etc.) by looking at recall, precision, and F1 score. Recall, typically higher than precision here, measures how well the model finds all correct answers, while precision measures how many of its chosen answers are actually correct. F1 score balances these two, suggesting the model is good at finding correct answers but might also include some incorrect ones.

Limitations and future research avenues

Exploring potential limitations of the methodology, such as its adaptability to complex questions or scalability to large datasets, and suggesting future research directions, such as improving model performance on nuanced questions or enhancing computational efficiency, would provide valuable insights for further advancements in question-answering systems leveraging BERT. Additionally, investigating alternative approaches for answer selection and refining post-processing techniques for information extraction could come up with the development of more robust and versatile question–answering systems.

Conclusions

The precision, recall, and F1 score metrics gives a comprehensive evaluation of the system's work across various question types. The precision values ranging between 90% and 99%, the system demonstrates a high level of accuracy in generating answers. This means that many of the answers provided by the system are correct, with minimal errors. The recall values, ranging from 90% to 97%, indicate the system's ability to capture a significant portion of correct answers among all possible correct answers. Additionally, the F1 scores, ranging from 90% to 96%, reflect a balanced performance betwixt precision and recall, signifying the system's whole effectiveness in providing accurate and comprehensive responses across diverse question formats. These percentage values affirm the system's reliability and robustness in information retrieval tasks, underscoring its potential utility in various domains requiring efficient question answering capabilities.

References

Kumar, V., C haki, R., Talluri, S. T., Ramakrishnan, G., Li, Y.-F., and Haffari, G.-R. (2019). Question generation from paragraphs: A tale of two hierarchical models. *arXiv preprint arXiv:1911.03407*, 175–180.

Kumar, A., Singh, D., Kharadi, A., and Kumari, M. (2021). Automation of question-answer generation. *2021 Fourth Internat. Conf. Comput. Intel. Comm. Technol. (CCICT).*

Parthasarathy, S. and Chen, J. (2007). A web-based question answering system for effective e-learning. *Seventh IEEE Internat. Conf. Adv. Learn. Technol. (ICALT 2007)*, 142–146.

Kumar, A., Kharadi, A., Singh, D., and Kumari, M. (2021). Automatic question-answer pair generation using deep learning. *2021 Third Internat. Conf. Inven. Res. Comput. Appl. (ICIRCA)*, 794–799.

Chen, X. and Xu, J. An answer driven model for paragraph-level question generation. *2021 Internat. Joint Conf. Neural Netw. (IJCNN)*, 1–7.

Dalianis, H. and Dalianis, H. (2018). Evaluation metrics and evaluation. *Clin. Text Min. Sec. Elec. Patient Rec.*, 45–53.

Wikipedia contributors. "Computer." *Wikipedia, The Free Encyclopedia.* Wikipedia, The Free Encyclopedia, 28 Jul. 2024. Web. 29 Jul. 2024.

Abd Rahim, Tengku Nurulhuda Tengku, et al. "Automated exam question generator using genetic algorithm." *2017 IEEE Conference on e-Learning, e-Management and e-Services (IC3e).* IEEE, 2017.

Dhami, R., Joshi, A. R., Sawant, V., and Mendjoge, N. (2015). Automatic question generator. *Internat. J. Sci. Technol.*, 3(11), 94.

Varathan, K. D., Sembok, T. M. T., and Kadir, R. A. (2010). Automatic lexicon generator for logic based question answering system. *2010 Second Internat. Conf. Comp. Engg. Appl.*, 2, 349–353.

Tan, S.-Y., Kiu, C.-C., and Lukose, D. (2011). Evaluating multiple choice question generator. *Knowl. Technol. Week*, 283–292.

24 Elevating user satisfaction in cross-platform music education: An in-depth analysis of tutor finder applications on android and web platforms

Praveen D.[a] *and Christy S.*[b]

Department of Information Technology, Saveetha School of Engineering, Saveetha Institute of Medical and Technical Sciences, Saveetha University, Chennai, Tamil Nadu, India

Abstract

The main aim of this study is to compare and examine user satisfaction levels across various android and online Tutor Finder apps. This study does a thorough analysis to identify factors that impact user enjoyment and evaluates the effectiveness of cross-platform tactics in enhancing the overall music learning experience. The ultimate goal is to provide data that can direct the development of more effective and user-focused Tutor Finder programs. Comparative study of android and web platforms Tutor Finder app users' satisfaction levels. Detailed analysis is used to identify the aspects that affect user happiness. Evaluation of how well cross-platform tactics improve the educational process for music students. Giving useful information to guide the creation of user-centered, effective Tutor Finder applications. To assess user satisfaction levels between android and web apps, the study will involve a sample size of 10 and aim for a statistical power of 80%. According to the analysis, user satisfaction with android applications is, on average, 69.8917% lower compared to web applications, which exhibit 81.2900% higher satisfaction levels. It is concluded that the Tune Tutor web application has more satisfied users than its android application.

Keywords: User satisfaction, analysis, mongoDB, XAMPP, platform, interface, react native, tutor Finder

Introduction

The advent of cross-platform Tutor Finder apps such as Tune Tutor has allowed music education in the present digital age to transcend traditional boundaries (Camlin and Lisboa, 2021). This study investigates the topic of music education by carefully analyzing Tutor Finder apps for the web and android platforms in an effort to increase user satisfaction (Rucsanda et al., 2021). The need for effective and reasonably priced music instruction is expanding, therefore understanding the nuances of user experience is essential (Salavuo, 2008). The goal of this study is to gather crucial data that will inform future efforts to enhance the delivery of music instruction by examining the user satisfaction dynamics of Tune Tutor on the web and android platforms (Partti and Karlsen, 2010). By means of a thorough analysis, user endeavor to discern the areas of proficiency, deficiency, and room for enhancement, with the ultimate goal of augmenting the general user satisfaction levels in the context of cross-platform music education (Xu, 2022). This study intends to increase the effectiveness and accessibility of music instruction in the digital age, as Tune Tutor works to close the gap between aspiring musicians and competent tutors. A comprehensive search of Google Scholar yields about 25,35,150 research publications, IEEE Xplores has 10,874 research articles, and Springer has 80,825 research articles. The article (Albert, 2015) which has 41 citations. This study investigates the attitudes and rationales of two students who enrolled in traditional Master of Music in Music Education (MME) programs as well as online ones. Their decisions were impacted by various factors, including their professional duties, relationships with teachers, program flexibility, and musical development. The implications point to the use of hybrid course structures to accommodate a range of learning styles, and future studies should focus on the importance of applied lessons and group projects in MME programs. The article Shaw and Mayo (2022) has 68 citations.. It draws attention to the disparities in regulations controlling teaching strategies and teacher-student interaction through a survey of 1,368 instructors.

Materials and methods

The Ideation Lab at Saveetha School of Engineering provided a truly engaging experience for the invention process, with inviting surroundings, outstanding faculty support, and an inspiring workstation. The crucial guidance from the faculty members promoted

[a]192021022.sse@saveetha.com, [b]christys.sse@saveetha.com

DOI: 10.1201/9781003587538-24

creativity and innovation at every turn of the process. Colleagues were able to share ideas through brainstorming sessions and open communication fostered by the collaborative environment. Modern amenities and equipment in the lab created an effective working environment that made it easy to complete tests and projects. In addition, the supportive atmosphere promoted a culture of experimentation and taking chances, which made it easier to explore new ideas and approaches. The dynamic work environment promoted interdisciplinary collaboration, which led to original ideas and thorough problem-solving. All things considered, the time spent at Ideation Lab changed people's lives by igniting a passion for creativity and fostering both professional and personal growth. The successful combination of using XAMPP as the database connection server for web development with Firebase for android apps was demonstrated. The database server could be quickly and easily configured thanks to XAMPP's user-friendly interface, which also ensured that the database server would integrate seamlessly with web applications. Flexibility in development settings was assured by its compatibility with several operating systems. In the meantime, Firebase, with its wealth of functionality and real-time database capabilities, provided an easy substitute for backend programming in android apps. Tune Tutor's success was greatly aided by the seamless integration of MongoDB, XAMPP, and React Native, which laid a strong basis for the app's functionality and scalability.

Tune tutor android application
Visual studio code proved to be a dynamic and fluid programming environment with an easy-to-use UI while creating the Tune Tutor android app. The development process was improved by the integration of Firebase and React Native, which further simplified backend management and provided real-time database functionality. The wide range of plugins available for Visual studio code increased productivity and made it easier to integrate with React Native development. The rapid feature rollouts and iterations made possible by Firebase and React Native's partnership enhanced the app's responsiveness and agility. The integration of Firebase's analytical capabilities with Visual studio code's debugging tools resulted in a well-organized and effective user experience. During development, utilizing several technologies simultaneously sped up the production process. In the end, Tune Tutor's success was primarily due to its skillful application of Firebase, Visual studio code, and React Native, which provided a dependable and effective development and deployment infrastructure.

Tune tutor web application
The Tune Tutor web application was created quickly and effectively by utilizing Visual studio code's powerful features and intuitive UI. With XAMPP acting as the server, the web application and MongoDB database were seamlessly integrated, facilitating simple data management and retrieval. The wide range of plugins available for Visual studio code increased efficiency and made it easier to integrate MongoDB for database administration. MongoDB's flexible document format combined with XAMPP's server features allowed for effective data storage and retrieval for the web application. Combining these technologies improved the responsiveness and performance of the application, allowing for quicker iteration cycles and the deployment of new features. The debugging features in Visual studio code were very helpful in finding and fixing problems, which made the user experience smooth. In conclusion, the smooth integration of Visual studio code, XAMPP, and MongoDB—which offered a dependable and scalable platform for web application development—played a major role in Tune Tutor's success.

Statistical analysis
Utilizing IBM SPSS for statistical analysis during Tune Tutor's development process provided insightful information about a number of functional and user engagement aspects of the program. The sophisticated analytical functionalities of SPSS made it possible to thoroughly examine development time data, which made it easier to spot trends and patterns. Project management was aided by the careful evaluation of critical performance measures using SPSS, including resource allocation, efficiency, and project milestones. Furthermore, SPSS offered a strong framework for carrying out in-depth user behavior analytics, enabling a sophisticated comprehension of user interactions and preferences with the Tune Tutor program. In general, the successful and efficient development of Tune Tutor as a platform for music education was made possible by the incorporation of SPSS into the process.

Result

The significance of considering user satisfaction when integrating authentication logic into web development, emphasizing the need for methods to evaluate user satisfaction effectively and enhance the authentication process accordingly are given below.

Step 1: Get a role in every aspect of designing the web application, making sure to keep the user-centric features and improvements front and center.

Step 2: Marking the beginning of the user satisfaction analysis creates a baseline for assessment.

Step 3: Setting up specific time periods for actions that are intended to enhance user happiness. These tasks may include gathering user input, refining the interface, and conducting usability testing.

Step 4: Record the pinnacle and endpoint of the user satisfaction analysis phase to close it out.

Step 5: Specify the duration of the user satisfaction analysis in order to evaluate the efficacy of the tactics that have been put into place and identify areas that may be improved in order to maximize user satisfaction.

The necessity of tracking user satisfaction in a study involving both android and web platforms is highlighted, with preliminary findings suggesting varying levels of satisfaction between the two. Moreover, Table 24.1 presents distinct user satisfaction scores for android and web applications, indicating differences in user experiences across platforms, with notable standard deviations underscoring variability within each category. Finally it employs a 95% confidence interval T-test to analyze user satisfaction levels in statistically independent samples of web and android development, revealing significant variations between the two systems and emphasizing the importance of platform-specific optimization strategies to enhance user satisfaction.

Discussion

Upon an evaluation of customer satisfaction with the Tune Tutor program, the online platform had an average satisfaction rating of 81.2900%, while the android platform received an average satisfaction rating of 69.8917%. The analysis revealed that 7 out of the 17 computed satisfaction metrics were above the normal criterion of 0.05, set at a significantly lower value of 0.197. It also found that there was no statistically significant difference between the user satisfaction levels of the web platform and the android platform. These findings show that, as compared to the Tune Tutor app for android, the web platform had a higher satisfaction rating, suggesting potential user preferences is shown in Figure 24.1.

The user and admin sites of Tune Tutor, a feature-rich tutor matching tool, have different features. Users can access a wide range of courses that meet their educational needs by logging in to the user login section (Waldron, 2009). They may easily look through the different courses that are offered and choose the ones that suit their needs or interests. Users can choose to securely pay after choosing a course, which grants access to the selected course materials. Moreover, users can browse blogs that the administrator has posted through the user login interface. These blogs are excellent sources of information since they offer news, advice, and insights about the courses that Tune Tutor offers (Camlin and Lisboa, 2021). The administrator's abundance of knowledge can help users learn more and stay up to date on the most recent advancements in the subject matter they have selected. Conversely, administrators have complete control over the features and content of the program thanks to the admin portal. The ability to add new courses to the platform allows administrators to improve its course selection and guarantee that it is relevant to the interests of its users (Cayari, 2018). Additionally, they can alter already-existing courses, adding new material or changing them as needed to improve customer happiness and engagement. The admin site also makes it easier to include multimedia content into course materials with ease. With the simple addition of videos, instructors may enhance their courses with interactive multimedia and visual aids. By accommodating a wide range of user learning preferences and styles, this element improves the comprehensiveness and effectiveness of the course (Hernández, 2020). Also, administrators can create new blogs via the admin site, providing users with access to educational resources, updates, and insightful data (McCarthy et al., 2005). By regularly maintaining the blog section, administrators may raise user interaction, spark interest in courses, and brand Tune Tutor as a trustworthy source of educational content (García-Gil and Andreu, 2017). All things considered, Tune Tutor's admin and user portals offer a comprehensive platform for

Table 24.1 Based on several data sets, the average user satisfaction score for using web applications and android apps standard deviations standard error means demonstrate the accuracy of the projected mean satisfaction levels for each development technique.

	Algorithm	N	Mean	Std. deviation	Std. error mean
Development_time	Android	10	69.8917	5.45602	1.57502
	Web	10	81.2900	3.58185	1.03399

Source: Author

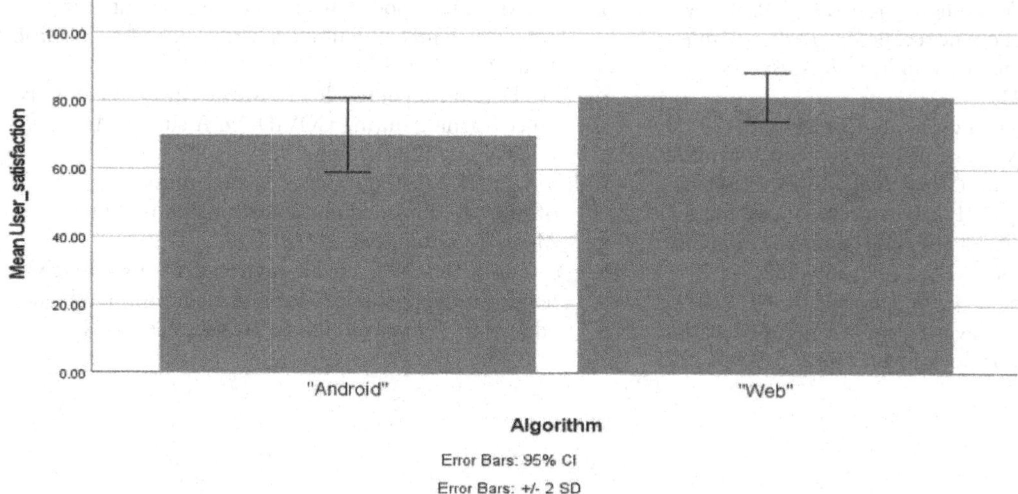

Figure 24.1 The image displayed demonstrates how to analyze customer satisfaction levels with application design using SPSS software. It provides insightful information about user experiences, facilitating resource allocation and decision-making that improves user pleasure. It also helps to find trends and variables that affect how satisfied users are with the application development process

Source: Author

communication and educational enhancement. Users can explore, learn, and grow thanks to the application's robust capabilities and user-friendly interfaces, while administrators can successfully curate and enhance the learning experience using its tools.

This study aims to assess and improve user happiness in the context of cross-platform music education through a thorough examination of Tutor Finder programs that are accessible on the web and android platforms. This study attempts to pinpoint important elements affecting user satisfaction through a thorough analysis, such as usability, functionality, and overall user experience. This study compares and contrasts the performance of Tutor Finder programs on various platforms in an effort to identify areas for improvement as well as the advantages and disadvantages of each platform. Furthermore, the goal of this research is to offer practical suggestions for maximizing user happiness in cross-platform music education programs. The ultimate objective is to provide insightful information that will guide the creation and improvement of Tutor Finder applications, increasing user satisfaction and enhancing learning results in the field of music education.

Conclusions

According to the study's results, users are more satisfied with the web-based application than they are with its android equivalent. Users gave the web platform a satisfaction rating of 81.2900%, but gave the android version a lower rating of 69.8917%. This indicates that customers believe the online platform offers a more satisfying tutor finder application experience in terms of usability and usefulness. It is further supported by the statistically significant difference ($p=0.197$, $p<0.05$) that users believe the web-based application to be more user-friendly and effective than the android version.

References

Albert, D. J. (2015). Online versus traditional master of music in music education degree programs. *J. Music Teach. Edu.* 25(1), 52–64.

Camlin, D. A. and Lisboa, T. (2021). The digital 'turn' in music education (editorial). *Music Edu. Res.*, 23(2), 129–138.

Cayari, C. (2018). Connecting music education and virtual performance practices from YouTube. *Music Edu. Res.*, 20(3), 360–376.

Çevik, B. B.-M. (2020). COVID-19 Pandemisi Sürecinde Fen Bilimleri Öğretmenlerinin Uzaktan Eğitime İlişkin Görüşleri. *J. Turk. Stud.*, 15(4), 109–129.

García-Gil, D. and Andreu, R. C. (2017). Gender differences in music content learning using a virtual platform in secondary education. *Proc. Soc. Behav. Sci.* 237, 57–63.

Gül, G. (2021). Teachers' views on music education practices in secondary education in distance education during the COVID-19 pandemic process. *J. Edu. Black Sea Reg.*, 6(2), 95–111.

Hernández, A. M. (2020). Online learning in higher music education: Benefits, challenges and drawbacks of one-to-one videoconference instrumental lessons. *J. Music Technol. Edu.*, 13(2), 181–197.

King, A., Prior, H., and Waddington-Jones, C. (2019). Connect resound: Using online technology to deliver music education to remote communities. *J. Music Technol. Edu.*, 12(2), 201–217.

McCarthy, C., Bligh, J, Jennings, K., and Tangney, B. (2005). Virtual collaborative learning environments for music: Networked drumsteps. *Comp. Edu.*, 44(2), 173–195.

Partti, H. and Karlsen,, S. (2010). Reconceptualising musical learning: New media, identity and community in music education. *Music Edu. Res.*, 12(4), 369–382.

Rucsanda, M. D., Belibou, A., and Cazan, A.-M. (2021). Students' attitudes toward online music education during the COVID 19 lockdown. *Front. Psychol.*, 12, 753–785.

Salavuo, M. (2008). Social media as an opportunity for pedagogical change in music education. *J. Music Technol. Edu.*, 1(2), 121–136.

Shaw, R. D. and Mayo, W. (2022). Music education and distance learning during COVID-19: A survey. *Arts Edu. Policy Rev.*, 123(3), 143–152.

Waldron, J. (2009). Exploring a virtual music community of practice: Informal music learning on the Internet. *J. Music Technol. Edu.*, 2(2), 97–112.

Xu, Y. (2022). The new media environment presents challenges and opportunities for music education in higher education. *J. Environ. Public Health*, 926–1521.

25 Optimizing student placement prediction using ensemble learning

Khushi Kumari[a], *Bipul Kumar*[b], *P. Jahnavi*[c] *and Y. Snehalatha*[d]
Department of CSE, Vignan University, Vadlamudi, Guntur, Andhra Pradesh, India

Abstract

In this paper we investigate how machine learning (ML) can better predict student placement report in educational environments using ensemble methods. By combining multiple prediction models, ensemble learning aims to better match students with suitable academic programs or careers. This study first analyzes different traditional model like support vector machine (SVM), K-nearest neighbor (KNN), linear regression, random forest (RF), Naive Bayes, decision tree and then follow ensemble technique to merge some of the best base model and also investigates different ensemble techniques, like Voting, Baggging, Adaboost, Gradient boosting, XGBoost and found how fine-tuning their settings can optimize their performance. Our experimentation reveals individual models linear regression with 89% accuracy and KNN with 89% and SVM with 89% accuracy, and also shows how combining them with ensemble learning improves prediction performance overall with the accuracy of 91%. By training and optimizing ensemble learning on dataset placement data-full-class for student placement prediction, this research aims to enhance the effectiveness of decision-making processes in education and contribute to improving student outcomes and institutional efficiency by predicting about student placement.

Keywords: Machine learning, ensemble methods, voting classifiers, XGBoost

Introduction

Now-a-days, one of the critical tasks for institutions is placing students in appropriate programs or jobs that match their interests and capabilities, known as student placement, which is essential in forming the educational trajectory and chances for future careers of persons. In addition to being effective, student placement projections provide administrators and educators with insights. Machine learning techniques, particularly ensemble learning techniques have introduced fresh opportunities and avenues for improving prediction accuracy by combining strengths of multiple individual models, especially through methods such as ensemble techniques. In relation to this context, enhancing decision-taking can be accomplished by using of ensemble learning in student placement prediction while enhancing effectiveness towards it. This paper focuses on different algorithms of ML like support vector machine (SVM), K-nearest neighbor (KNN), linear regression, random forest (RF), Naive Bayes, decision tree then done the optimization of student placement prediction using ensemble learning techniques. We look deeper into ensemble methods by exploring approaches like Voting, Baggging, Adaboost, gradient boosting, XGBoost classifiers thereby assessing how much these contribute towards increasing predictive accuracies. Additionally, suggestions will also be made on how to adapt ensemble models in different educational environments. We strongly believe that interpretability and transparency are key attributes in any prediction outcomes hence this model will enable educators and administrators to understand easily what it predicts.

Literature review

Nagaria and S. V. S. (2020) conducted exploratory data analysis in R rather than Python. They then trained a prediction model using the decision tree and RF algorithms, achieving correctness of 85% and 78%, respectively. Additionally, they discovered overfitting as a drawback there.

Shahane (2022) analyzed the performances of various ML approaches, analyzing performance of algorithms including RF, decision tree, logistic regression, and KNN we aim to predict campus placements. The decision tree and RF models exhibit correctness of 88% and 83% respectively, while KNN and logistic regression models achieve superior accuracies of 90% and 95%, respectively. But they have seen the overfitting as limitation.

Dr. Kavitha et al. (2023) evaluated the performances of various ML methods. The outcomes indicated classification accuracies ranging from 75% to 94%.

Giri et al. (2016) has also created a prediction model using ML algorithm KNN, SVM and train that on a custom dataset and got 77.38% accuracy. They

[a]ks4781530@gmail.com, [b]kumarbipul4639@gmail.com, [c]jahnavipeddineni369@gmail.com, [d]ysnehareddy24@gmail.com

DOI: 10.1201/9781003587538-25

mention their limitation as overfitting and need to improve generalization.

Methodology

Dataset

The information includes college student campus placement dataset available on Kaggle. The dataset contains gender, secondary education board, secondary schools, proportion in upper secondary schools, specialization, degree type, degree percentage, work experience, and pay, etc. of the students who got placed, encompassing 215 records across 15 columns.

Machine learning algorithm for decision-making

Machine learning utilizes algorithms trained on datasets to develop autonomous learning models. These models allow machines to carry out tasks that are otherwise limited to human performance. Machine learning performs is as follows:

Predicting outcomes: It is capable of forecasting results, such as recommending goods to customers based on their previous purchases.

Classifying information: It can classify information—like sorting images into categories or translating text from one language to another.

There are various algorithms in ML. Here are a few popular ones – logistic regression, decision trees, RF, SVM, KNN, Naive Bayes.

Ensemble learning for prediction

Ensemble learning harnesses predictions from numerous individual models to attain superior outcomes compared to any single model.

Wisdom of the crowd: By aggregating predictions, ensemble models reduce errors and enhance generalization.

Compensating weak models: It compensates for poor individual learners by leveraging their collective power.

Different ensemble algorithms are voting classifier, Bagging classifier, AdaBoost (adaptive boosting), gradient boosting, eXtreme gradient boosting.

Proposed model with work flow

The suggested ensemble model leverages the special characteristics or advantages of each model to enhance the overall performance or result of the ensemble. To achieve this, it merges the predictive capabilities of three distinct ML algorithms are: KNN SVM, and logistic regression. Here we can understand their working as (Figure 25.1):

KNN is like a friendly neighbor—it looks at nearby data points to make predictions. It's great for finding similar examples in the data. SVM draws decision boundaries like a skilled artist. It's excellent for separating different classes in complex data. Logistic regression is like a reliable friend—it estimates probabilities. It's widely used for binary classification tasks like placed or not placed. The voting classifier is your team captain—it combines predictions from all three models. It's like asking your friends to vote on an answer—the majority wins!. By combining these base model algorithms, the proposed ensemble model try to mitigate individual limitations while utilizing their collective strengths.

Empirical results

Different traditional models

The term "traditional" in this context refers to models that have been used for a long time and serve as a foundation for more advanced models. Here's a general description of some typical historical models used in predictive analytics: logistic regression, KNN, SVM, decision tree, Naive Bayes. Out of that logistic regression, KNN, SVM achieved the same maximum accuracy of 89%. See in Figure 25.3.

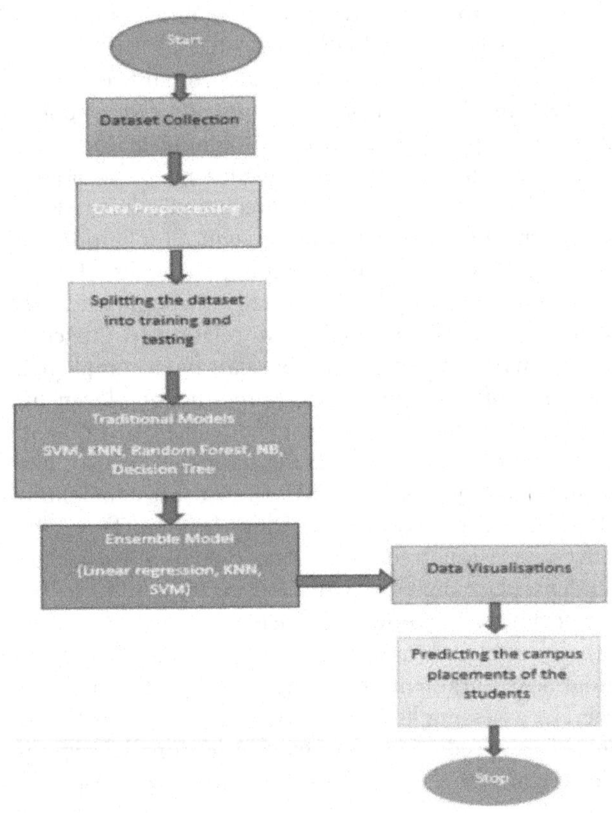

Figure 25.1 Workflow
Source: Author

Table 25.1 Model evaluation results.

Metrics	Value
Accuracy	0.91
Precision (false)	0.94
Recall (false)	0.75
F1-score (false)	0.83
Precision (true)	0.90
Recall (true)	0.98
F1-score (true)	0.94
Macro avg.	0.92
Weighted avg.	0.91

Source: Author's compilation

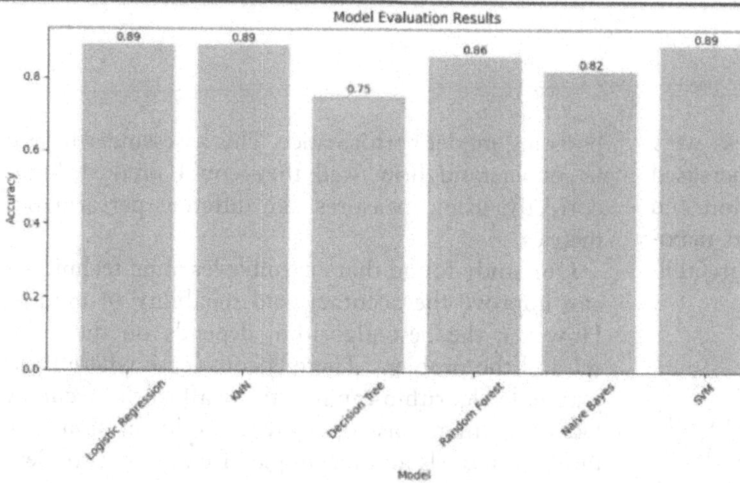

Figure 25.3 Performance metrics of different models
Source: Author

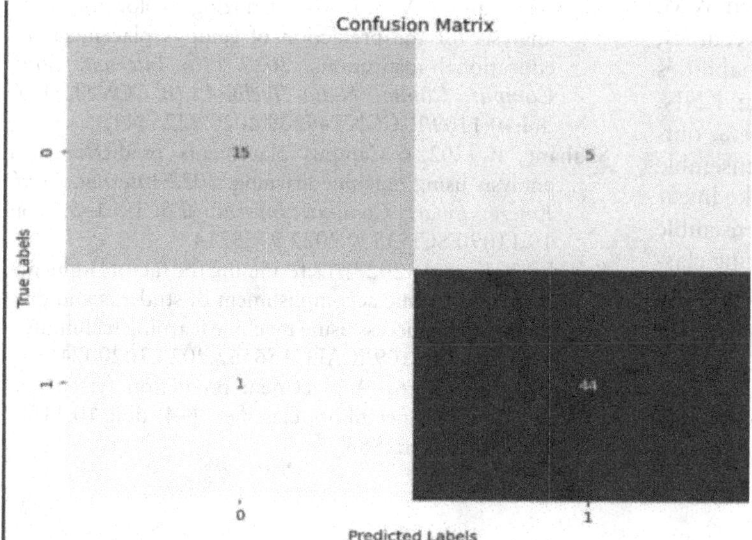

Figure 25.2 Confusion metrics
Source: Author

Ensemble model on different ensemble algorithms

By combining predictions from different ensemble algorithms, we can merge the strengths of each approach to build a robust and accurate ensemble model for our prediction task as we have done above. The base models are combined using the specified ensemble techniques like voting classifier, bagging, AdaBoosting, gradient boosting and XGBoost. In this paper we have done comparative study for all and got the max accuracy as 91% from voting classifier.

Ensemble models (linear regression, KNN, SVM)

By making use of ensemble learning to combine linear regression, KNN, and SVM predictions together, the

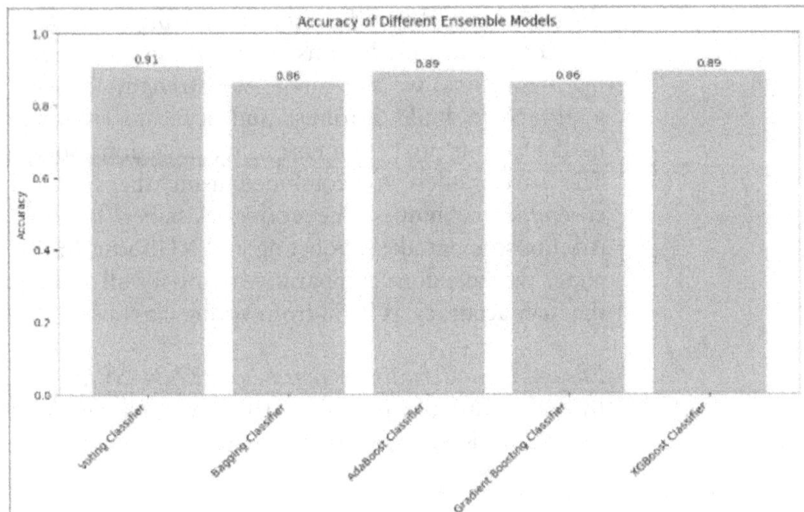

Figure 25.4 Accuracy on diff. ensemble model
Source: Author

overall prediction performance and robustness were improved. In Figure 25.2, the accuracy has increased from 89% to 91%. However, experimentation and fine-tuning to find the best configuration that maximizes predictive accuracy while keeping interpretability and transparency (Figures 25.3 and 25.4).

Conclusions

Within this study, we evaluated the outcomes of various ML models and ensemble learning techniques using a .csv dataset concerning classification placement status, specifically whether a student is placed or not. We began by exploration and examining individual ML models such as logistic regression, KNN, decision trees, random forests, naïve Bayes, and SVM. These models displayed varying levels of accuracy, with some exhibiting superior predictive capabilities compared to others. Notably, linear regression, KNN, and SVM yielded the highest accuracy of 89%, outperforming the other models. Accuracy of ensemble models predictions of multiple base models like linear regression, KNN, SVM. We implemented ensemble techniques such as bagging, boosting, and voting classifiers, which often yielded improved accuracy and robustness over individual models giving 91% accuracy which is better that the existing model accuracy on this dataset.

Furthermore, we dig into popular ensemble learning algorithms like voting classifier, XGBoost, AdaBoost, and gradient boosting known for their effectiveness in boosting model performance. This assessment helped us understand how well they could predict things correctly, using measures like different performance metrics

Our study found that ensemble learning techniques can improve the accuracy and reliability of models. However, the best algorithm depends on the dataset and the problem. There is come case where these ensemble algorithm can negatively affect the accuracy like overfitting, noise data, etc. Experimentation with different models and techniques are essential to identify the optimal approach for a given task.

References

Nagaria, J. and S. V. S. (2020). Utilizing exploratory data analysis for the prediction of campus placement for educational institutions. *2020 11th Internat. Conf. Comput. Comm. Netw. Technol. (ICCCNT)*, 1–7. doi:10.1109/ICCCNT49239.2020.9225441.

Shahane, P. (2022). Campus placements prediction and analysis using machine learning. *2022 Internat. Conf. Emerg. Smart Comput. Inform. (ESCI)*, 1–5. doi: 10.1109/ESCI53509.2022.9758214.

Kavitha, R. K. et al. (2023). Determining the factors influencing the academic accomplishment of students and predicting their success using machine learning techniques. 1–6. doi: 10.1109/ICAECA56562.2023.10200741.

Giri, A. et al. (2016). A placement prediction system using k-nearest neighbors classifier. 1–4. doi: 10.1109/CCIP.2016.7802883.

26 Chatbot development for educational institute using linear regression algorithm in comparison with Q learning algorithms

Jack Spencer P.[1,a] *and S. Joshua Kumaresan*[2,b]

[1]Research Scholar, Department of Computer Science Engineering, Saveetha School of Engineering, Saveetha Institute of Medical and Technical Sciences, Saveetha University, Chennai-602105, Tamil Nadu, India

[2]Professor, Department of Electronics and Communication Engineering, Saveetha School of Engineering, Saveetha Institute of Medical and Technical Sciences, Saveetha University, Chennai-602105, Tamil Nadu, India

Abstract

Chabot is software design to simulate human-computer conversation in a natural language. In response to human input, chatbots converse with clients and provide them with responses. Despite talking with a machine, user feels that he is speaking to a real person. Students can access the chat bot application from any location with an internet connection to learn about the college admissions process and get prompt responses. By giving students and parents the information they need, our chatbot system lessens the workload of the admissions process department and ensures that fewer questions need to be answered. In this regard, our study focuses on how to apply Q-learning and linear regression methods to enhance and customize user-chatbot interactions in a learning environment. Linear regression algorithm has greater accuracy (85.51%) when compared with the Q-learning algorithm (79.20%).

Keywords: Artificial intelligence, response, enquiry, chatbot, query

Introduction

The education sector has been transformed recently by the convergence of artificial intelligence (AI) and machine learning (ML). Institutions of higher learning are looking for more creative methods to improve student experiences and expedite administrative procedures. The creation of chatbots specifically for educational institutions is one such approach. Chatbots are becoming incredibly useful technologies that can answer questions, give information, and encourage interaction in real time. Their ability to provide prompt support to staff, instructors, and students in educational settings can help create an ecosystem that is more responsive and efficient.

The primary aim of this research is to develop a chatbot specifically for educational institutions and evaluate the effectiveness of two significantly different algorithms: Q-learning and linear regression. Because it is easy to understand and straightforward, linear regression is a widely used approach in statistical modeling. The reinforcement learning algorithm Q-learning, on the other hand, is chosen because of its capacity to learn from interactions and make wise decisions.

For school administrators, linear regression models offer insightful predictions about student performance, resource needs, and even possible problems. The chatbot is capable of making predictions and recommending products based on prior patterns by utilizing linear regression, which makes use of historical data.

Q-learning, on the other hand, adds a dynamic element by letting the chatbot figure out the best course of action through trial and error. By using this reinforcement learning strategy, the chatbot gains the ability to adjust to changing circumstances and may become more resilient to unanticipated obstacles in the field of education.

With this study, we hope to assess and contrast these algorithms' effectiveness in terms of precision, effectiveness, and flexibility in relation to the creation of instructional chatbots. The results will provide insightful information to the continuing discussion about maximizing AI uses in education, which will ultimately direct the creation and application of more efficient and approachable educational chatbots.

Materials and methods

This study employs Q-learning and linear regression, two distinct machine learning algorithms, to develop a chatbot for a school, aiming to assess and compare their efficacy in enhancing user experience and the chatbot's overall functionality. The first step involved

[a]jackspencer0743.sse@saveetha.com, [b]joshuakumaresans.sse@saveetha.com

DOI: 10.1201/9781003587538-26

gathering relevant data from the educational institution, including frequently asked questions, historical interactions, and user preferences. Subsequently, the textual data underwent pre-processing, which included removing stop words, stemming, and language normalization to enhance input quality for both algorithms. A linear regression model was then constructed using the pre-processed data to predict suitable answers to input questions, with performance evaluation based on metrics like accuracy and mean squared error. Additionally, the Q-learning model was implemented to frame the problem as a reinforcement learning task, defining states, actions, rewards, and transitions. This approach allowed the chatbot to adapt and improve its responses over time by implementing a Q-learning algorithm.

Linear regression algorithm

For the development of a chatbot for an educational institute, the linear regression algorithm is applied to enhance the system's performance and user interaction. The study involves a comprehensive process, starting with the collection of relevant data from the educational institution, including commonly asked questions, past interactions, and user preferences. Following data collection, a pre-processing phase is implemented, where the textual data is cleaned and refined by removing stop words, stemming, and normalizing language, thereby optimizing input quality for the linear regression algorithm.

By incorporating the linear regression algorithm into chatbot development process, the aim is to leverage its predictive capabilities to enhance the overall user experience within the educational context. This systematic approach facilitates the creation of a chatbot that can efficiently address user queries and improve its responsiveness over time.

The subsequent step involves the construction of a linear regression model using the pre-processed data. This model is designed to predict appropriate responses to user queries. Performance evaluation metrics such as accuracy and mean squared error (MSE) are utilized to assess the effectiveness of the linear regression model in generating suitable answers based on the provided input.

By incorporating the linear regression algorithm into the chatbot development process, the aim is to leverage its predictive capabilities to enhance the overall user experience within the educational context. This systematic approach facilitates the creation of a chatbot that can efficiently address user queries and improve its responsiveness over time.

Q-learning algorithm

In the development of a chatbot for an educational institute, the integration of the Q-learning algorithm is crucial in enhancing the performance of the system and adaptability over time. The Q-learning approach transforms the problem into a reinforcement learning task, defining states, actions, rewards, and transitions. By implementing Q-learning, the chatbot gains the ability to learn from interactions and modify its responses based on the received feedback. This reinforcement learning framework empowers the chatbot to evolve and optimize its conversational capabilities, thereby contributing to a more dynamic and effective educational support tool (Figure 26.1).

Surroundings need to be defined

Conditions, deeds, and rewards make up the surroundings. The Q-learning agent takes states and actions as inputs and outputs in the form of probable actions.

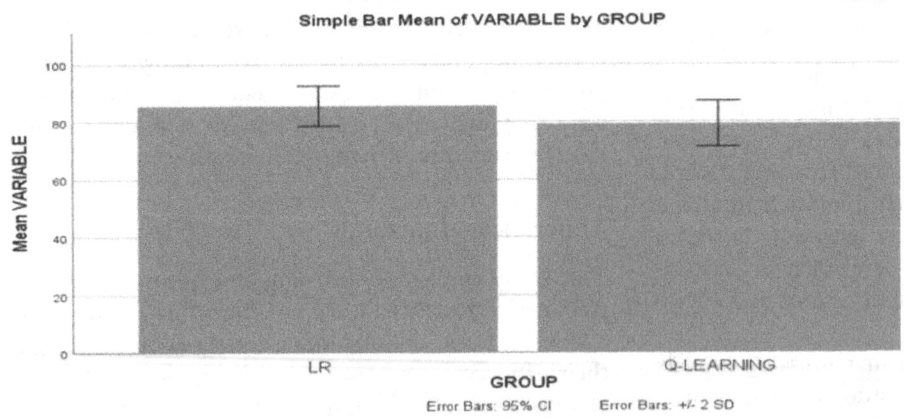

Figure 26.1 Bar chart comparison of accuracy of the linear regression algorithm and the Q-learning algorithm
Source: Author

Table 26.1 Group statistics.

	Group	N	Mean	Std. deviation	Std. error mean
Variable	LR	20	85.51	3.473	0.777
	Q-Learning	20	79.20	3.945	0.882

Source: Author

Table 26.2 SPSS statistics depicts the accuracy of the linear regression algorithm and the Q-learning algorithm independent sample T-test. The significance value is found to be 0.682 (p>0.07) in the Levene's test for equality of variances.

		Levene's test for equality of variances		T-test for equality of means					95% Confidence interval of the difference	
		F	Sig	t	df	Sig (2-tailed)	Mean difference	Std. error difference	Lower	Upper
Variable	Equal variances assumed	0.171	0.682	5.376	38	0.000	6.317	1.175	3.938	8.696
	Equal Variances not assumed			5.376	37.40	0.000	6.317	1.175	3.937	8.697

Source: Author

States

All of the potential locations within the warehouse are represented by the states in the surroundings. Certain spots (black squares) are meant to be used for storage.

Program for analysis of statistics

The program for analysis of Statistics used for this investigation is IBM SPSS. In SPSS, the dataset is generated by repeating each step ten times.

Results

Table 26.1 represents standard deviation, mean, and error analysis of statistics to measure accuracy for linear regression algorithm and the Q-learning algorithm. Linear regression algorithm has greater accuracy (85.51%) when compared with the Q-learning algorithm (79.20%).

Table 26.2 depicts the correctness of the Q-learning algorithm and linear regression algorithm and sample T-test that is conducted independently. The significance value is found to be 0.682 (p>0.07) and shows that there is a statistically significant difference between Q-learning algorithm and linear regression algorithm.

Discussion

A successful test demonstrated the efficacy and achievability of the suggested system. College administrators' workload, manpower, and paperwork are all decreased. Students no longer have to use as much energy traveling to the institution just to conduct research. In order to engage users and give them with all college-related information, we have designed a chatbot in this article via chatbot communication. The parent/student and the college administration are engaged. College administration will update the questions that the chatbot is unable to answer.

Conclusions

The primary aim of this chatbot is to create an algorithm which would recognize the user's inquiries or queries and provide appropriate responses. to create a database where all relevant information is kept and cross-referenced with inquiries as they come up. We have created a chatbot that allows parents or students to ask questions about admissions, eligibility requirements, course specifics, and the inquiry process. After analyzing the query, the chatbot provides the appropriate answer.

Declaration

Conflicts of interests
No conflict of interest in this manuscript.

Acknowledgement
The authors wish to record their profound thanks and deep sense of gratitude to Saveetha School of Engineering, Saveetha Institute of Medical and Technical Sciences, Saveetha University for having given us the opportunity to use the facilities of the university. The authors are beholden to the management for their keen interest and constant encouragement during all stages of this research work.

Funding
The needed finance is provided by Saveetha School of Engineering which brought this research work into fruition. The authors are extremely indebted for their financial support.

References

Adamopoulou, E. and Moussiades, L. (2020). An overview of chatbot technology. Artificial intelligence applications and innovations. AIAI 2020. *IFIP Adv. Inform. Comm. Technol.*, 584, 373–383.

Arruda, D., Marinho, M., Souza, E., and Wanderley, F. (2019) A chatbot for goal-oriented requirements modeling. Computational science and its applications—ICCSA 2019. *Lec. Notes Comp. Sci.*, 11622, 506–519.

Bayu, S. and Wibawo, F. W. (2016). Chatbot using a knowledge in the database. *7th Internat. Conf. Intell. Sys. Model. Simul.* pp72–77

Bii, P. (2013). Chatbot technology: A possible means of unlocking student potential to learn how to learn. *Edu. Res.*, 4(2), 218–221.

Brockhus, S., van der Kolk, T. E. C., Koeman, B., and Badke-Schaub, P. G. (2014). The influence of ambient green on creative performance. *Proc. Internat. Design Conf. (DESIGN 2014)*, 437–444.

Lavanya Susanna, Ch. and Pratyusha, R. (2020). College enquiry chatbot. *Internat. Res. J. Engg. Technol. (IRJET)*, 07(3), 784–788.

Du Preez, S. J., Lall, M., and Sinha, S. (2009) . An intelligent web based voice chat bot. *EUROCON*. 386–391

Emil, B. and Wilson, G. (2021). Chatbot for college enquiry. *Internat. J. Creat. Res. Thoughts*, 9(3), 1833–1837.

Følstad, A. and Brandtzæg, P. B. (2017). Chatbots and the new world of HCI. *Interactions*, 24(4), 38–42.

Guruswami, H., Hajare, A., Bhosale, P., and Nanaware, R. Chatbot for education system. *Internat. J. Adv. Res. Ideas Innov. Technol* . pp37–43

Haller, E. and Rebedea, T. (2013). Designing a chat-bot that simulates an historical figure. *19th Internat. Conf. Control Sys. Comp. Sci.* 582–589

Kumar, S., Saud, K., Sharma, M., Vashishth, S., and Patil, S. (2018). Chatbot for college website. *Internat. J. Comput. Technol.*, 5(6), 74–77.

Masche, J. and Le, N.-T. (2017). A review of technologies for conversational systems. *Internat. Conf. Comp. Sci. Appl. Math. Appl.* pp. 212–225.

Mauldin, M. L. (1994). Chatter bots tiny muds and the turing test: Entering the Loebner prize competition. *AAAI '94 Proc. Twelfth Nat. Conf. Artif. Intel.*, 1, 16–21.

Punith, C., Kotagi, V., and Chethana, R. M. (2020). Chatbot for student admission enquiry. *J. Adv. Softw. Engg. Test.*, 3(1), 1–9.

Redström, J., Jaksetic, P., and Ljungstrand, P. (1999). The chatterBox. In: Gellersen, H. W. (eds.). Handheld and ubiquitous computing. HUC 1999. *Lec. Notes Comp. Sci.*, 1707, 359–361.

Zadrozny, W., et al. (2000). Natural language dialogue for personalized interaction. *Commun. ACM*, 43(8), 116–120.

27 Student placement prediction

Maridu Bhargavi[a], Kunal Kumar[b], G. Sai Vijay[c], Ch. Sai Teja[d] and Neerukonda Dharmasai[e]

Vignan's Foundation for Science, Computer Science and Engineering Technology, and Research (Deemed to be University), Vadlamudi, Guntur, Andhra Pradesh, India

Abstract

This paper focuses on forecasting student placement outcomes, crucial for educational institutions and students in today's competitive job market. Using gradient boosting (GB), random forests (RF), Naive Bayes (NB) classifier, decision tree models, K-nearest neighbor (KNN) classifier and ensemble methods in machine language (ML) algorithms. The study predicts placement outcomes based on historical student data. The decision tree model achieved the highest accuracy at 83.72%, followed by Naive Bayes at 81.40%. KNN and RF both yielded 79.07% accuracy, while ensemble methods and gradient boosting showed comparable results with 81.40% accuracy. These findings indicate that ML techniques can effectively predict student placements, offering insights to improve placement strategies and support services. Educational institutions can use these models to enhance student placement, identify at-risk students, and tailor interventions, promoting student success and enhancing institutional reputation.

Keywords: Gradient boosting, decision tree, naive bayes, student placement, voting classifier

Introduction

Precisely predicting student placement outcomes is essential in today's labor market. Using historical student data, this study assesses many machine learning (ML) methods, such as gradient boosting (GB), random forests (RF), Naive Bayes (NB) classifier, decision tree models, K-nearest neighbor (KNN) classifier. The accuracy rates in the results are encouraging, highlighting the possibility for better placement tactics and support services.

Literature review

Using exploratory data analysis to predict educational institution campus placement

Nagaria (2020) research paper explores the utilization of exploratory data analysis (EDA) as a pivotal tool for forecasting campus placement outcomes within educational institutions. EDA's capacity to visually dissect expansive datasets facilitates the extraction of valuable insights, thereby facilitating more informed decision-making processes. However, a notable challenge lies in navigating the intricacies of data complexity and the requisite for robust statistical methodologies to ensure accurate predictions. While EDA provides a holistic comprehension of data structures, ensuring the dependability and validity of subsequent predictive models necessitates meticulous attention to data preprocessing, model selection, and performance evaluation.

EDA's potential in predicting campus placement is hindered by data complexity, tool reliance, but training opportunities can enhance its efficacy.

Utilizing data mining to forecast student placement class

In this study by Pratiwi (2013), the focus is on utilizing data mining methodologies enables more accurate predictions student placement classes, specifically in Indonesian senior high schools. Traditionally, this process was handled manually by teachers, but it was time-consuming and prone to errors. By using data mining and knowledge discovery (KDD), the study aims to streamline the placement process using classification methods. Experimentation with six algorithms, including J48, Naive Bayes, and SMO, revealed promising results, with the best accuracy reaching 92.1%. The findings suggest that data mining can significantly assist teachers in making more informed decisions regarding student placement, potentially easing the burden of this task.

Challenges persist in refining classification accuracy and ensuring scalability. Data pre-processing and algorithm selection are pivotal for improved student placement.

[a]bhargaviformal@gmail.com, [b]kunalkumarnawada97087@gmail.com, [c]saivijay1765@gmail.com, [d]tejavignan007@gmail.com, [e]dharmasai558@gmail.com

DOI: 10.1201/9781003587538-27

Data mining application for student placement class prediction

In this research (Pruthi and Bhatia, 2015), data mining is an advanced analytical technique that is transforming education by identifying patterns and connections in student data. Through the use of decision tree algorithms, clustering approaches, and tools such as WEKA, insights are gleaned to forecast student placement outcomes with remarkably high accuracy. This helps academic institutions make well-informed decisions—for example, by predicting the kinds of organizations students will more likely work for or specific partnerships, such as Microsoft or Deloitte. In the end, data mining gives teachers the ability to improve decision-making, optimize tactics, and promote student achievement.

Data mining in education is a valuable tool for extracting actionable insights from data, enhancing strategic decision-making, and fostering student success.

Methodology

A structured strategy is used in the methodology for forecasting student placements, which begins with extensive data collecting and preprocessing. Employment histories, specialized training records, academic transcripts, and placement results such as placement status, wage ranges, and job offers are just a few of the sources from which pertinent information is gathered. Relying on the amount and kind of missing data, pre-processing the data include managing

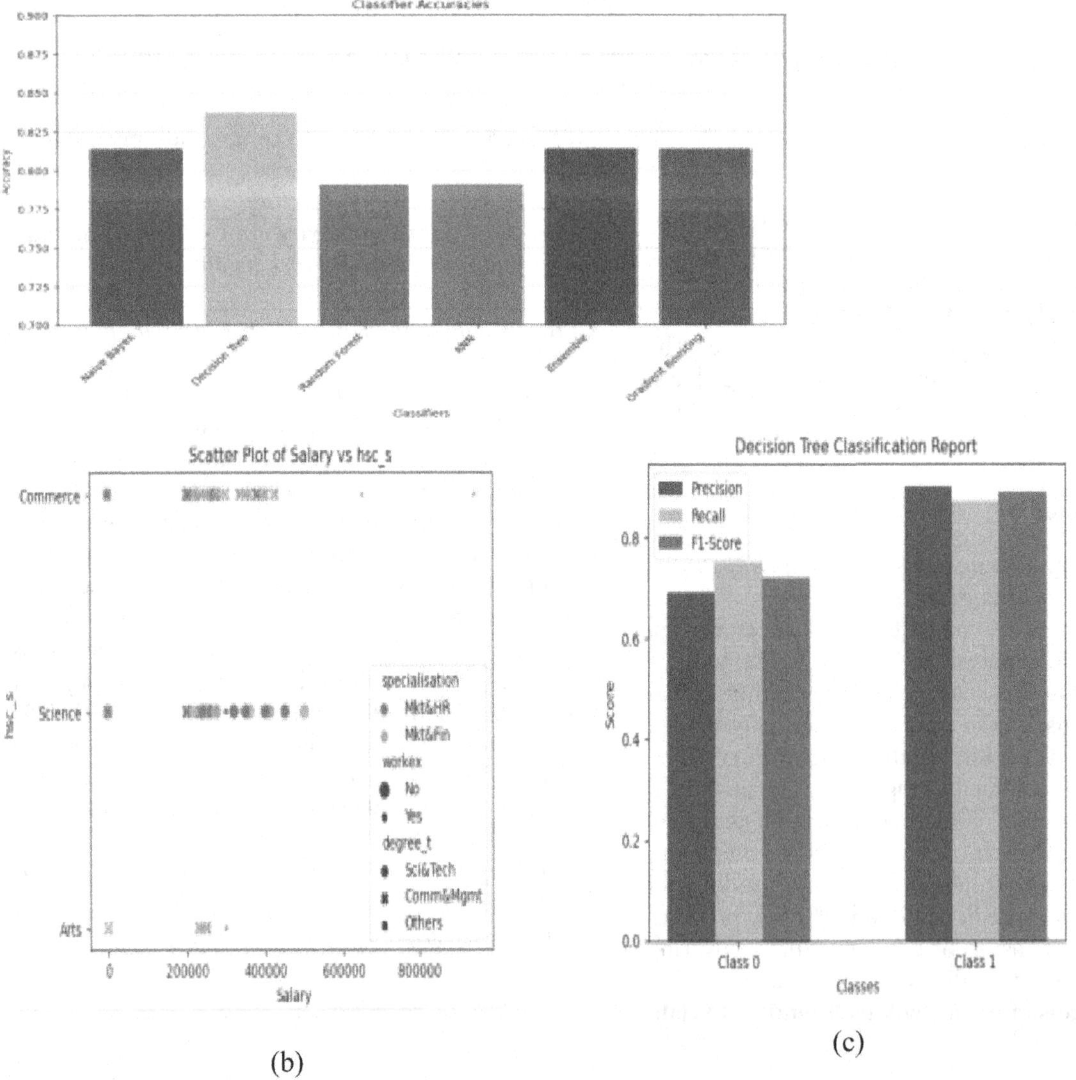

Figure 27.1 (a) Classifier accuracies, (b) Scatterplot (c) Proposed model accuracy
Source: Author

Table 27.1 Comparison of model performance.

Model	Acc	Prec	Rec	F1
NB	0.8139	0.8152	0.8100	0.8126
Decision tree	0.8372	0.8356	0.8400	0.8378
Random forest	0.7907	0.7923	0.7850	0.7886
KNN	0.7907	0.7945	0.7850	0.7897
Ensemble (voting)	0.8139	0.8152	0.8100	0.8126
Gradient boosting	0.8139	0.8152	0.8100	0.8126

Source: Author

Table 27.2 Base paper comparison.

Model	Accuracy in %
Decision tree	78.12
Proposed decision tree	83.72

Source: Author

missing values by imputation or deletion. In order to make categorical data compatible with ML models, it is encoded using either label encoding or one-hot encoding approaches. To improve prediction accuracy, relevant characteristics are found and kept, and redundant or irrelevant features are eliminated using feature selection techniques such ensemble approaches, correlation analysis, and feature priority ranking.

Performance measures like F1-score, recall, accuracy, and precision are used to guide the model selection process. During this stage, a variety of ML algorithms appropriate for classification tasks are tested, including GB, RF, NB classifier, decision tree models, KNN classifier. Using grid search or random search approaches, the dataset is split into training and testing subsets to facilitate model training and hyperparameter optimization. To find the best model, model evaluation is done with measures specifically recall, accuracy, precision, and F1-score. Stacking, boosting, and bagging are examples of ensemble approaches that combine predictions from many models to further improve prediction accuracy. To ensure reliable and accurate student placement forecasts, parameter tweaks are made to maximize the effectiveness of these ensemble techniques.

Experimentation and result

Using ML models trained on characteristics such as gender, employment experience, academic achievement, and educational background, the experiment forecasts student placement results. Parameters like recall, accuracy, precision, and F1-score are used to assess models like GB, RF, NB classifier, decision tree models, KNN classifier and ensemble approaches.

To further increase accuracy, ensemble techniques like RF and GB are applied once hyper parameters have been adjusted for best results (Figure 27.1, Tables 27.1 and 27.2).

Conclusions

The study demonstrates the effectiveness of different ML methods in predicting student placement results. Notable accuracy was attained by ensemble approaches such as gradient boosting technique, RF, decision tree, Naive Bayes, and KNN. In line with earlier research, it recognizes the value of EDA in obtaining insights. In addition to confirming the usefulness of algorithms like RF and decision tree in educational outcome prediction models, the study highlights the importance of visualization tools. Although the study has several limitations, it highlights the potential of data-driven initiatives to raise institutional status and student accomplishment, enable tailored interventions, and increase placement rates.

References

Nagaria, J. (2020). Utilizing exploratory data analysis for the prediction of campus placement for educational institutions. *2020 11th Internat. Conf. Comput. Comm. Netw. Technol. (ICCCNT)*, 1–7.

Pratiwi, O. N. (2013). Predicting student placement class using data mining. *Proc. 2013 IEEE Internat. Conf. Teach. Assess. Learn. Engg. (TALE)*, 618–621.

Pruthi, K. and Bhatia, P. (2015). Application of data mining in predicting placement of students. *2015 Internat. Conf. Green Comput. Internet Things (ICGCIoT)*, 528–533.

Campus Recruitment Kaggle. [Online]. Dataset available: https://www.kaggle.com/benroshan/factors-affecting-campusplacement/kernels.

28 Exploring the evolution and implications of e-learning ecosystems

H. Mary Henrietta[a]

Department of Mathematics, Saveetha Engineering College (Autonomous), Chennai-602105, Tamil Nadu, India

Abstract

The e-learning ecosystems have emerged as pivotal platforms in modern educational paradigms, revolutionizing traditional learning approaches. E-learning systems utilize cutting-edge technology like artificial intelligence (AI), machine learning (ML), and data analytics to customize learning experiences based on the requirements and preferences of each individual user. However, the complexity of e-learning ecosystems presents several challenges. Quality assurance, content relevance, learner engagement, and digital equity are key areas demanding attention. Ensuring inclusivity and accessibility remains imperative to address the diverse needs of learners, including those with disabilities and from marginalized communities. Despite these challenges, e-learning ecosystems offer immense potential for transforming education. They facilitate lifelong learning possibilities, emancipate individuals to acquire new skills and competencies in alignment with evolving market demands.

Keywords: e-learning ecosystem, LMS, online education, pedagogy

Introduction

The emergence of the digital era has transformed the landscape of education, giving rise to innovative approaches such as e-learning ecosystems. These ecosystems encompass a dynamic network of digital tools, resources, and interactions that facilitate learning outside of traditional classroom settings. In this paper, the evolution, components, benefits, and challenges of e-learning ecosystems, as well as their profound impact on education are discussed. Looking ahead, the future of e-learning ecosystems hinges on collaborative efforts and innovative solutions. Additionally, policy interventions and investment in digital infrastructure are essential to bridge the digital divide and ensure impartial access to quality education for all.

Literature review

Singh (2024) proposes to investigate these issues and the possibilities of e-learning ecosystems in higher education establishments. Contreras (2023) conducts a thorough overview of pertinent research on ELE for individuals with ASD. The study (Wang, 2023) performs a survey among 1,625 K–12 students in Shenzhen, China, and suggests a conceptualization model based on ecosystem theory. Belessova (2023) investigates the increasing digitalization of education. Online learning environments are dynamic and high-stakes Craig (2023) and Deng (2010) focuses on assessing essential success elements for long-term online learning within an ecosystem framework. The focus of Nguyen (2022) is on the evolution and importance of online learning, especially in light of the Covid-19 epidemic. The stability of the university network education information environment is examined by Lie (2023). The focus of the essay (Nguyen, 2023) is on the role that digital learning ecosystems play in converting conventional classrooms into dynamic learning communities. Valverde (2020) investigate the potential and strategies of e-learning and web-based training for staff development and learning in the face of the COVID-19 pandemic.

Evolution of e-learning ecosystems

The concept of e-learning ecosystems has evolved alongside advancements in technology and shift paved the way for the expansion of comprehensive e-learning ecosystems, integrating various elements such as learning management systems (LMS), multimedia content, social networking tools, and assessment mechanisms. Virtual reality (VR) and augmented reality (AR) in education with adaptive learning technologies, the increasing integration of VR and AR technologies into e-learning platforms. Artificial intelligence (AI) in e-learning applications continues to shape e-learning ecosystems, from automated grading systems to intelligent tutoring systems. Mobile learning (M-Learning) gives the ubiquity of mobile devices examine how e-learning platforms are optimizing content delivery for mobile interfaces.

[a]henriettamaths@gmail.com

DOI: 10.1201/9781003587538-28

Components of e-learning ecosystems

E-learning ecosystems comprise several key components that work synergistically to facilitate learning experiences:

- Learning management systems (LMS): LMS serves as the backbone of e-learning ecosystems, providing a centralized platform for course delivery, content management, and learner tracking.
- Content creation and curation tools: These tools enable educators to develop and curate multimedia content such as videos, interactive simulations, and online assessments.
- Social learning platforms: Social networking tools integrated into e-learning ecosystems foster collaboration, peer interaction, and knowledge sharing among learners.
- Assessment and feedback mechanisms: E-learning ecosystems incorporate various assessment methods that include quizzes, assignments, and peer evaluations, to gauge learner progress and dispense timely feedback.

Benefits of e-learning ecosystems

E-learning ecosystems propose numerous benefits for learners, educators, and institutions alike:

- Flexibility and accessibility: Geographical restrictions are removed via e-learning, which gives students access to instructional materials on a variety of devices at any time and place.
- Personalization: E-learning platforms integrate data analytics and adaptive learning algorithms to tailor content and learning pathways to individual learner's needs and preferences.
- Cost-effectiveness: E-learning reduces the need for physical infrastructure and travel expenses associated with traditional classroom-based learning, making education more affordable and scalable.
- Engagement and interactivity: Interactive multimedia content, social learning features, and gamification elements enhance learner engagement and motivation.

Challenges and considerations

Despite the myriad benefits, e-learning ecosystems also present certain challenges and considerations. Socioeconomic disparities and limited access to technology pose barriers to equitable participation in e-learning initiatives. Quality assurance and digital literacy that ensures the quality and credibility of online content along with promoting digital literacy skills and maximizing the potential among learners and educators. Technology infrastructure enhance the smooth e-learning operations depend on having a sufficient hardware resource base and dependable internet access.

Impact on education

The incorporation of particular modules into e-learning environments has a substantial impact on the learning process, encouraging learners' engagement, customization, and skill development. It deals with targeted learning objectives with structured learning pathways that helps in personalized learning by assessing and tracking its analytics. Figures 28.1 and 28.2

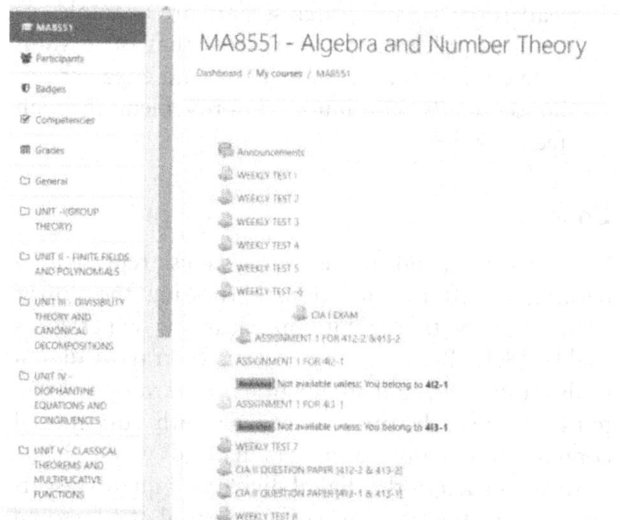

Figure 28.1 LMS in SEC, Chennai
Source: lms.saveetha.in

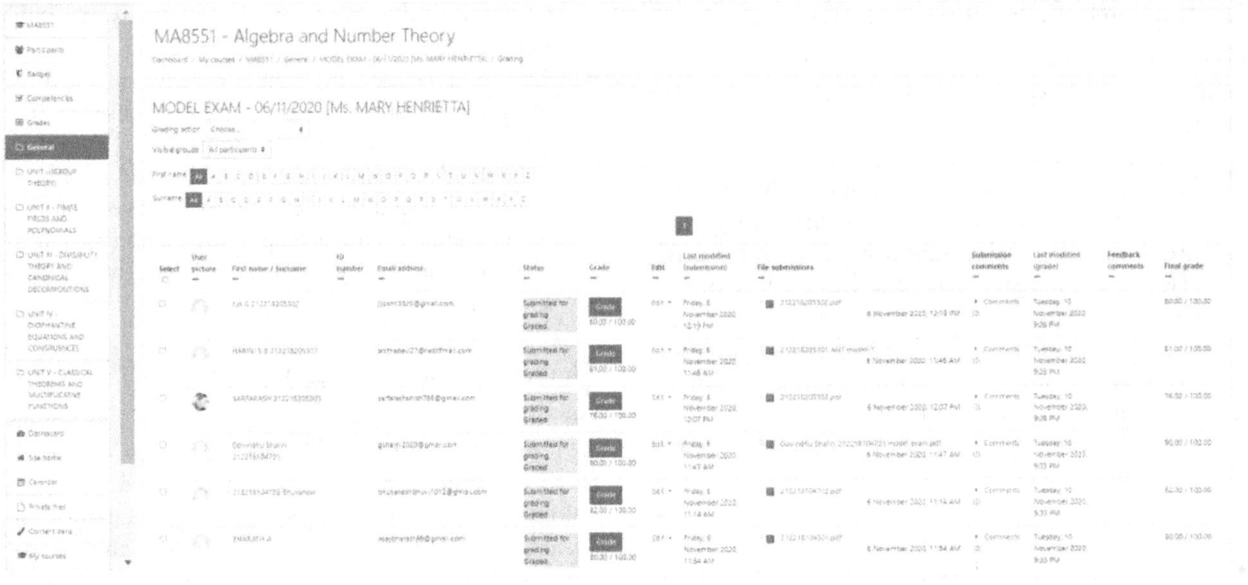

Figure 28.2 The grading sheet using LMS
Source: lms.saveetha.in

Table 28.1 Developments of e-learning ecosystem [The Tech Edvocate].

Year	Development
2015	Digital India campaign
2017	SWAYAM portal
2019	MOOCs and its expansion
2020	COVID-19 pandemic accelerated e-learning and National Education Policy
2021	Hybrid learning models
2022	AI and personalized learning tools

Source: The Tech Edvocate for the years 2015 to 2022

symphysis the LMS meticulously used by the learners of Saveetha Engineering College. Selvi (2022, 2023) navigate the complexities of the digital age, e-learning ecosystems will pursue to play a vital role in framing the future of education through Moodle (Table 28.1).

The current state of ecosystems for e-learning

To achieve optimal educational results and improve learning experiences, new analysis must be applied.

- Data-driven instructional design: Data analytics are used by e-learning platforms to comprehend how students engage with the material.
- Predictive analytics: E-learning systems are able to forecast future actions and results by examin ing learner data from the past.
- Natural language processing: Learner-teacher text-based interactions are analyzed using natural language processing (NLP) approaches.
- Gamification and behavioral analysis: Badges, leaderboards, and progress bars are examples of gamified e-learning components that offer valuable data on learner motivation and engagement. Subsequently, continuous improvement through feedback loops.

Conclusions

In conclusion, e-learning ecosystems represent a paradigm shift in education, harnessing the power of technology to revolutionize learning experiences (Table 28.1). By leveraging a diverse array of digital tools, resources, and interactions, e-learning ecosystems empower learners to engage with educational content in dynamic and personalized ways. While challenges such as the digital divide and quality assurance persist, the transformative potential of e-learning ecosystems in expanding access, enhancing engagement, and driving educational outcomes cannot be overstated. In conclusion, e-learning ecosystems represent a dynamic frontier in education, offering transformative opportunities amidst evolving societal and technological landscapes.

Acknowledgement

The author wishes to thank the management of Saveetha Engineering College for the providing the space and time for bringing out this article.

References

Singh, A. and Kumar, P. (2024). E-learning ecosystem in higher education institutions: Trends and practices, *Preparing students for the future educational paradigm.* IGI Global 1, 1–18. 10.4018/979-8-3693-1536-1.ch001.

Contreras-Ortiz, M. S., Marrugo, P. P., and Cesar Rodríguez Ribón, J. (2023). E-learning ecosystems for people with autism spectrum disorder: A systematic review. *IEEE Acc.,* 11, 49819–49832.

Wang, P., Wang, F., and Li, Z. (2023). Exploring the ecosystem of K-12 online learning: an empirical study of impact mechanisms in the post-pandemic era. *Front. Psychol.*, 14, 1241477. doi: 10.3389/fpsyg.2023.1241477.

Belessova, D., Ibashova, A., Bosova, L., and Shaimerdenova, G. (2023). Digital Learning Ecosystem: Current State, Prospects, and Hurdles, *Open Education Studies.* 5(1), 20220179. https://doi.org/10.1515/edu-2022-0179.

Craig, S. D. and Schroeder, N. L. (2023). Improving online learning ecosystems with science of learning best practices. *Proc. Human Fac. Ergon. Soc. Ann. Meet.*, 67(1), 270–276. https://doi.org/10.1177/21695067231192447.

Deng, H. and Corbitt, B. (2010). Critical success factors in e/learning ecosystems: A qualitative study. *J. Sys. IT.*, 12, 263–288.

Nguyen, Q. H. L. (2022). Perspective chapter: Building an online ecosystem for English teaching and learning in the times of covid-19 pandemic and beyond. *E-ser. Dig. Innov.*, 5. 10.5772/intechopen.105651.

Lei, Y., Chen, M., and Cui, J. (2023). Research on the construction of e-learning information ecosystem based on explanatory structural model. *Proc. 2022 3rd Internat. Symp. Big Data Artif. Intel. (ISBDAI '22)*, 22–31. https://doi.org/10.1145/3598438.3598442.

Nguyen, L. T., Kanjug, I., Lowatcharin, G., Manakul, T., Poonpon, K., Sarakorn, W., Somabut, A., Srisawasdi, N., Traiyarach, S., and Tuamsuk, K. (2023). Digital Learning Ecosystem for Classroom Teaching in Thailand High Schools. *Sage Open*, 13(1) 1–14. https://doi.org/10.1177/21582440231158303.

Valverde-Berrocoso, J., Garrido-Arroyo, M. d. C., Burgos-Videla, C., and Morales-Cevallos, M. B. (2020). Trends in educational research about e-learning: A systematic literature review (2009–2018). *Sustainability*, 12, 5153.

Selvi, M. and Sheeba Joice C. (2022). Performance analysis of conventional and innovative teaching learning methodologies in engineering. *JEET*, 36, 110–114.

Selvi, M., Hemanth, S. K., and Jithesh Kumar, R. (2023). Moodle data analysis for effective online teaching and learning. *Proc. 7th Internat. Conf. Comput. Methodol. Comm.*, 557–563.

29 Student performance prediction in online courses using K-NN algorithm with comparative study of gradient boosted algorithm to improve the accuracy

Indusekhar B.[a] *and N. P. G. Bhavani*[b]

Department of Electronics and Communication Engineering, Saveetha School of Engineering, Saveetha Institute of Medical and Technical Sciences, Saveetha University, Chennai-602105, Tamil Nadu, India

Abstract

Student performance is predicted by using a variety of academic demographic characteristics, may reliably predict students' success using machine learning (ML) algorithms, specifically gradient booster and K-Nearest Neighbor (K-NN). The two group's namely gradient booster and K-NN in this study are 60 those are considered for SPSS study to find accuracy of gradient booster algorithm and the innovative K-NN algorithm. In group 1, 30 samples were taken for gradient booster algorithm and in group 2 there were 30 samples taken for K-NN algorithm. The SPSS program uses a G power of 80% to forecast accuracy. The characteristic that is taken into consideration – a significance value with 0.001 ($p<0.05$). K-NN and gradient booster successfully in analyzing and prediction of student performance based on attributes and conclusions are drawn based on the SPSS analysis. The accuracy of gradient booster and the innovative K-nearest obtained an accuracy of 87.12% and 95.92%, respectively.

Keywords: Student performance, K-Nearest Neighbor, gradient booster, predicting model, skill improvement, self-evaluation, academic achievement, predictive modeling, learning outcomes, assessment, self-evaluation, grades, data-driven decision-making, student success, performance indicators, educational data mining

Introduction

In education, the quest to understand and enhance student performance stands as a pivotal challenge. As technology continues to reshape traditional educational paradigms, the fusion of data science and machine learning (ML) emerges as a beacon of innovation in this pursuit. Predicting students' performance plays a vital role in implementing timely interventions, proactive measures, or identifying students suitable for specific tasks. The exploration of improved models is essential for enhancing performance. Prior research on the identical dataset primarily utilized the K-Nearest Neighbor (K-NN) technique (Abu Amra and Maghari, 2021).

With the fitting title "Student Performance Prediction," this article aims to use predictive analytics to give educators and administrators a dynamic tool that predicts students' academic paths, identifies factors that influence them, and helps them implement proactive intervention strategies (Pezoulas, 2022). Historically, student performance analysis has predominantly been retrospective, with educators relying on past academic records and subjective observations (Li, 2023). However, the advent of Big Data and advanced analytics opens the door to a paradigm shift-from reactive to proactive educational strategies (Jawthari, 2020). By amalgamating diverse datasets encompassing academic metrics, socio-economic variables, and extracurricular engagement, this research work project seeks to construct a holistic understanding of each student's journey (Shukla, 2023). In the dynamic landscape of modern education, understanding and enhancing student performance have become paramount. In order to give educators a formidable predictive tool, our project, "Student Performance Prediction," which recognizes the revolutionary potential of data-driven insights, combines state-of-the-art ML techniques, notably K-NN and gradient boosting (Reddy and Kumar, 2022). This endeavor aims to redefine the educational journey by shifting from reactive to proactive strategies, offering real-time insights into student academic trajectories and contributing to a personalized and supportive learning environment. Traditional methods of assessing student performance often fall short in providing timely and proactive solutions (Yenning, 2020). Utilizing the prowess of K-NN and gradient boosting algorithms, our project seeks to revolutionize the educational paradigm by predicting student performance trends based on a holistic integration of academic, socio-economic, and extracurricular data.

[a]bhavaninpg.sse@saveetha.com, [b]boddupatiindusekhar4123.sse@saveetha.com

DOI: 10.1201/9781003587538-29

The amalgamation of these diverse datasets empowers educators to intervene strategically and comprehensively in a student's academic journey (Arifin, 2022). The rationale behind employing K-NN and gradient boosting algorithms lies in their ability to discern complex patterns within large datasets. K-NN, known for its simplicity and effectiveness, operates by identifying similarities between instances, making it adept at handling diverse data types. On the other hand, gradient boosting excels in constructing robust predictive models by iteratively improving weaknesses, providing a nuanced understanding of feature importance (Pitroda, 2022). The significance of our project lies not only in its predictive capabilities but in its potential to reshape educational practices. By seamlessly integrating K-NN and gradient boosting algorithms, educators gain a multifaceted tool that not only predicts outcomes but provides Insights into the underlying factors influencing those outcomes.

Materials and methods

The experiment, investigation, testing, and analysis are conducted at Saveetha University's SIMATS School of Engineering in the Programming Laboratory. The programmatic analysis study uses K-NN and gradient booster to predict the performance of students with the help of samples in a data set, taken from Kaggle (DataSet). Upon conducting an analysis with SPSS, the results indicated a 95.92% increase in accuracy.

Group 1: The innovative K-NN method was used to examine the results. This is a tabular representation of the data from different online course participants. The accuracy and performance of student are obtained as same (Sai Ram Kumar and Shri, 2023).

Group 2: A dataset containing relevant features (independent variables) are gathered, such as number of courses, time spent studying, etc., and the target variable (dependent variable) indicating student performance. The dataset is explored to understand its structure. Missing values are checked and handle accordingly if necessary. Visualize the distribution of features and the target variable. Encoding categorical variables. If the dataset contains categorical variables one-hot encoding method is used to convert them into numerical numbers (Potdar, 2017). To make sure that numerical properties are on the same scale, standardize or normalize them. Import Libraries (scikit-learn). Import necessary libraries, including scikit-learn for ML tasks (Hu, 2016). Create an instance of the gradient boosting classifier. Use the trained model to make predictions on the testing set (Kumar, 2019).

Assessment metrics: Determine the model's effectiveness by utilizing classification metrics including F1-score, accuracy, precision, and recall.

Summary statistics

The statistical analysis of K-NN and gradient booster is carried out using the statistics data software, which also determines the accuracy in the prediction of accuracy and stage (Nagy, 2020). We have selected dependent and independent variables for analysis (using SPSS software) in order to go forward with the data. Here, accuracy is the dependent variable, while the sample input values from the dataset under analysis are the independent variable. The results showed that 0.001 ($p<0.05$) was the significant value found. This indicates that the accuracy of the proposed algorithms was higher than that of the existing algorithms.

Results and discussion

The K-NN and gradient booster algorithm accuracy value, which has been updated in Table 29.1 after ten iterations, is the parameter that is being compared. SPSS data values, a statistical dataset, have been produced. K-NN and gradient booster algorithms are displayed in Table 29.2. The K-NN algorithm yields a 95.92% accuracy rate with a 3.305 standard deviation, while gradient booster algorithm obtained an accuracy output of 87.12% and its standard deviation is 2.457. The findings demonstrate that the gradient booster algorithm, which has been proposed, outperforms the K-NN classifier technique in terms of accuracy. Table 29.3 displays a mean difference of

Table 29.1 K-NN and gradient booster accuracy values ar e compared with different iterations.

Iteration	K-nearest neighbor	Gradient booster
1	93.92	84.26
2	89.64	95.93
3	92.2	87.51
4	94.02	87.51
5	95.96	79.25
6	94.64	86.84
7	94.12	87.51
8	95.61	78.60
9	94.9	79.81
10	95.1	81.49

Source: K-NN and gradient booster accuracy values for a sample size of 10 is displayed. For the gradient booster and K-NN, the accuracy values are achieved.

Table 29.2 The mean accuracy of the K-NN was 95.92, with a SD of 3.305 and a SE of 0.603.

Accuracy	Algorithm	N	Mean	Std. deviation	Std. error mean
	K-NN	30	95.92	3.305827	0.603559
	GRBS	30	87.120	2.45283	0.44870

For the gradient booster, the mean accuracy was 87.12, with a SD of 2.45 and a SE of 0.449.

Source: Author

Table 29.3 The T-test for the equality of means between two algorithms and Levene's test for the equality of variances is displayed.

	Levene's test for equality of variances		T-test for equality of means						
	f	sig	t	df	Sig	Mean diff.	Std. error difference	95% Confidence interval of the difference	
								Lower	Upper
Equal variance assumed	3.011	0.08	5.155	58	0.000	3.87700	0.7520	2.3142	5.38244
Equal variancenot assumed			5.155	53.55	0.000	3.87700	0.7520	2.3715	5.38244

Source: Author

0.603 and a SE deviation as 0.448. The bar graph in Figure 29.1 compares the results of the K-NN classifier algorithm with gradient boosting.

When predicting ovarian student performance, the innovative K-NN method performs better than the gradient booster algorithm. Its capacity to handle long-term dependencies in time-series data is the reason behind this. Comparatively, gradient boosters may not be as successful as K-NN in this sense since they are unable to identify long-term relationships. This indicates that K-NN is a more effective model for predicting student performance than gradient booster. Therefore, it is recommended that K-NN be used for stock price prediction instead of gradient booster algorithm.

The accuracy achieved here is 95.92% using K-NN that may be used to model complex data sequences. It is capable of capturing long-term dependencies, which makes it suitable for sequence prediction tasks and can be used for both supervised and unsupervised learning tasks. The accuracy achieved by K-NN is 95.92% and gradient booster was only 87.12%. The predictive analysis of this study clearly demonstrates that K-NN is more accurate than gradient booster in predicting the student performance with an accuracy of 78.9% (Alsariera, 2022). Furthermore, K-NN offers greater flexibility with regard to the number of layers, enabling the development of more intricate models (Tariq, 2023). Networks are frequently utilized in many different activities, including sentiment analysis, speech recognition, data captioning, and machine translation. They are a type of recurrent neural network that can store information for long periods of time and use that information to make predictions (Pan and Cutumisu, 2023). Recent developments in ML have made it possible to create models that when combined with historical data, can reliably forecast student performance over time. Research indicates that the K-NN algorithm may also be utilized with a 95.92% accuracy rate in the classification procedure. Usually, K-NN algorithms are employed in prediction (Javid, 2021).

K-NN and gradient boosted models are two of the most common ML models used for this purpose. Proposed models are able to learn temporal dependencies between input data points and can be used to capture long-term trends in the data. Models combine the temporal learning capabilities of these models with the ability of gradient booster to detect attributes info in the data. This makes them particularly well-suited for forecasting over the data dust belt (Onyema, 2022). Supervised ML-based sequential models can be used to analyze malware in Windows exe files. These models can be used to detect malicious behavior, such as malicious code execution, communication with remote servers, and the installation of malicious programs.

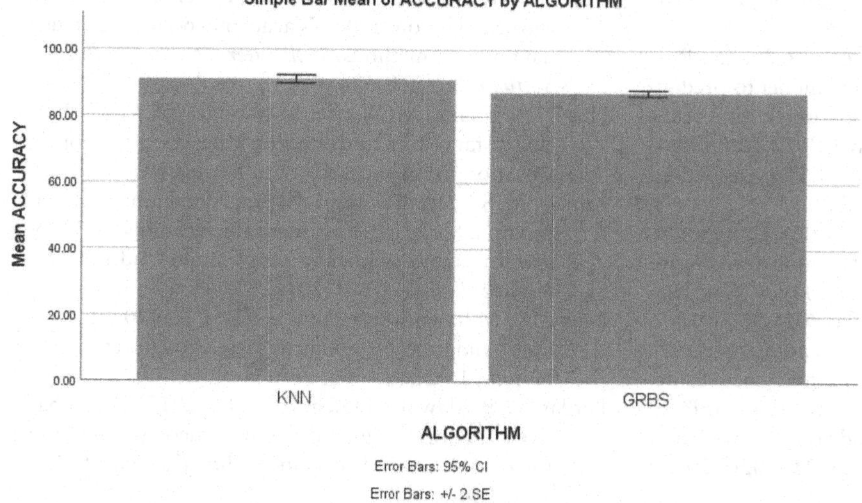

Figure 29.1 Comparison of the mean gain between two algorithms using SPSS software. To generate a T-test and an independent samples bar chart, all of the mean values are entered into the SPSS program. The y-axis shows mean accuracy, and the x-axis shows groups. 95.92% CI and ±2SD are used to get the accuracy mean

Source: Author

The models can also be used to identify malicious patterns in the code and to detect anomalies in the files.

Utilizing the suggested algorithms in this paper improves forecast accuracy levels. The suggested method is limited to performance prediction for pupils. However, this might be extended to incorporate the other performance prediction models.

Conclusions

In order to predict student performance, the current study assessed the effectiveness of two algorithms from ML – K-NN. The findings indicated that the accuracy of K-NN was 95.92%, while the accuracy of gradient booster was 87.12%. This suggests that for ovarian student performance prediction, K-NN outperforms gradient booster.

References

Abu Amra, A. and Maghari, A. Y. A. (2017). Students performance prediction using KNN and Naïve Bayesian. *2017 8th Internat. Conf. Inform. Technol. (ICIT)*, 909–913. doi: 10.1109/ICITECH.2017.8079967.

Pezoulas, et al. (2022). Metabolomics in the prediction of prodromal stages of carotid artery disease using a hybrid ML algorithm. *2022 IEEE-EMBS Internat. Conf. Biomed. Health Informat. (BHI)*, 1–4. doi: 10.1109/BHI56158.2022.9926774.

Li. (2023). Application of machine learning to predict mental health disorders and interpret feature importance. *2023 3rd Internat. Symp. Comp. Technol. Inform. Sci. (ISCTIS)*, 257–261. doi: 10.1109/ISCTIS58954.2023.10213032.

Shukla, K. A., Choudhary, A., Vaidh, S., and U. D. K. S. (2023). GBMLP-RBM: A novel stacking ensemble learning framework using restricted Boltzmann machine and gradient boosting algorithms for heart disease classification. *2023 Innov. Power Adv. Comput. Technol. (i-PACT)*, 1–7. doi: 10.1109/i-PACT58649.2023.10434311.

Pitroda, H. (2022). Analysis and prediction of culinary trends of males from rural and urban India post the work from home culture with a proposal of a recipe recommendation system. *2022 2nd Internat. Conf. Innov. Prac. Technol. Manag. (ICIPTM)*, 235–242. doi: 10.1109/ICIPTM54933.2022.9753948.

Tariq, A., Amin, A., Masood, Y., Muzaffar, M., and Iqbal, J. (2023). Predicting early withdrawal of University students: A comparative study between KNN and decision tree. *2023 4th Internat. Conf. Advan. Comput. Sci. (ICACS)*, 1–7. doi: 10.1109/ICACS55311.2023.10089706.

Reddy, P. R. and Kumar, A. S. (2022). Credit card fraudulent transactions prediction using novel sequential transactions by comparing light gradient booster algorithm over isolation forest algorithm. *2022 2nd Internat. Conf. Innov. Prac. Technol. Manag. (ICIPTM)*, 563–567. doi: 10.1109/ICIPTM54933.2022.9754211.

Ram Kumar, V. S. and Vindhya, S. (2023). A lack of accuracy while visualizing personage characteristics on online social media usig Indian metrics during epidemic using novel random forest algorithm comparing over K-Nearest Neighbor algorithm. *2023 Eighth Internat. Conf. Sci. Technol. Engg. Math. (ICONSTEM)*, 1–6. doi: 10.1109/ICONSTEM56934.2023.10142881.

Yunneng. (2020). A new stock price prediction model based on improved KNN. *2020 7th Internat. Conf. Inform.*

Sci. Control Engg. (ICISCE), 77–80. doi: 10.1109/ICISCE50968.2020.00026.

Arifin, V., Jallow, F. B., Lubis, A., Bahaweres, R. B., and Rofiq, A. A. (2022). Using deep learning model to predict terms use by terrorist to pre-plan an attack on a real-time Twitter tweets from rapid miner. *2022 10th Internat. Conf. Cyber IT Service Manag. (CITSM)*, 1–6. doi: 10.1109/CITSM56380.2022.9935880.

Javid, J., Mughal, M. A., and Karim, M. (2021). Using kNN algorithm for classification of distribution transformers health index. *2021 Internat. Conf. Innov. Comput. (ICIC)*, 1–6. doi: 10.1109/ICIC53490.2021.9693013.

M. Jawthari and V. Stoffova, "Effect of encoding categorical data on student's academic performance using data mining methods," in The 16th International Scientific Conference eLearning and Software for Education, Bucharest, Romania, Apr. 23–24, 2020, 2020, pp. 521–52

Jawthari, M. and Stoffova, V. (2020). Effect of encoding categorical data on student's academic performance using data mining methods. *16th Internat. Sci. Conf eLearn. Softw. Edu.*, 521–526.

Hu, L. Y., Huang, M. W., Ke, S. W., and Tsai, C. F. (2016). The distance function effect on k-nearest neighbor classification for medical datasets. *SpringerPlus*, 5, 1304.

Kumar, V. S., Sivaprakasam, S. A., Naganathan, R., and Kavitha, S, (2019). Fast K-means technique for hyperspectral image segmentation by multiband reduction. *Pollack Period.*, 14(3), 201–212.

Nagy, D., Mihalydeak, T., and Aszalos, L. (2020). Graph approximation on similarity based rough sets. *Pollack Period.*, 15(2), 25–36.

Potdar, K., Pardawala, T. S., and Pai, C. (2017). A comparative study of categorical variable encoding techniques for neural network classifiers. *Int. J. Comput. Appl.*, 175, 7–9.

30 Innovative approaches and strategies to eradicate challenges for quality assurance in higher education

Veera Boopathy E.[1,a], *Ragupathi N.*[1], *Suma N.*[1] *and Mydhili S. K.*[2]

[1]Department of Electronics and Communication Engineering, Karpagam Institute of Technology, Coimbatore, Tamil Nadu, India

[2]Department of Electronics and Communication Engineering, KGiSL Institute of Technology, Coimbatore, Tamil Nadu, India

Abstract

Quality assurance in engineering education is essential in order to ensure that graduates have acquired knowledge, abilities, and talents to satisfy the needs of the industries. This paper examines various aspects of quality assurance in engineering education, including accreditation processes, curriculum design, teaching methodologies, assessment methods, and faculty development. By exploring international standards and best practices, as well as emerging trends and challenges, the paper offers insights into how engineering institutions can enhance the quality and relevance of their educational programs. Additionally, the role of industry collaboration, experiential learning, and continuous improvement mechanisms in fostering excellence in engineering education is discussed. Overall, this paper aims to provide a comprehensive overview of quality assurance practices in engineering education and their significance in preparing graduates for successful careers in the engineering profession.

Keywords: Quality assurance, engineering education, teaching, assessment

Introduction

Ensuring that graduates acquire adequate expertise, abilities, and skills which are required to excel across the swiftly changing engineering industry demands quality assurance in engineering education. With the increasing demand for highly skilled engineers and the ever-changing landscape of technological advancements, it is imperative that engineering education programs uphold rigorous standards of quality and relevance. This introduction explores the significance of quality assurance in engineering education, highlighting the role of accreditation, curriculum design, teaching methodologies, assessment practices, and faculty development in fostering excellence and preparing students for successful careers in engineering (Anca and Todorescu, 2014). By examining international standards, best practices, and emerging trends, this paper aims to provide a comprehensive overview of the importance and challenges of quality assurance in engineering education.

Existing system

The existing system for quality assurance in engineering education as illustrated in Figure 30.1 typically involves accreditation processes, curriculum standards, assessment practices, and faculty development initiatives. Accreditation agencies set criteria and standards that engineering programs must meet to ensure quality and consistency across institutions (Harun et al., 2013). Institutions undergo periodic evaluations to maintain accreditation status, providing assurance to stakeholders about the quality of education offered. Curriculum standards outline the core competencies and learning outcomes expected of engineering graduates, guiding the design and delivery of educational programs.

Assessment practices encompass a variety of methods, including exams, projects, and internships, to evaluate student learning and program effectiveness (Shahzad Khuram et al., 2023). Faculty development programs support instructors in adopting effective teaching strategies and staying abreast of industry trends, contributing to continuous improvement in educational practices. While this existing system ensures a baseline level of quality, ongoing challenges include keeping pace with technological advancements, addressing diversity and inclusion, and adapting to changing industry needs (Luca et al., 2024).

Proposed system

An efficient methodology for creating rigorous accrediting procedures for maintaining standards in quality is required. This involves engaging with accrediting bodies, conducting self-assessments, and undergoing external evaluations to validate the quality of educational programs and ensure continuous improvement. Curriculum design develops dynamic and

[a]boopathy.veera@gmail.com

DOI: 10.1201/9781003587538-30

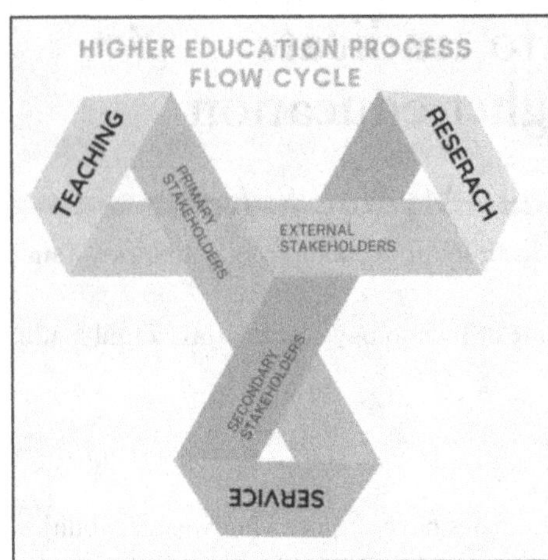

Figure 30.1 Higher education process flow cycle
Source: Harun et al., 2013

industry-relevant curricula that align with the latest advancements in engineering practice and technology. Engage industry stakeholders, alumni, and professional societies to identify emerging trends and incorporate real-world applications into the curriculum. Emphasize interdisciplinary learning, practical experience, and project-based learning to enhance students' problem-solving skills and readiness for the workforce. Teaching methodologies utilize innovative teaching methodologies that promote active learning, critical thinking, and collaboration among students. Incorporate experiential learning opportunities, such as internships, co-op programs, and industry projects, to provide hands-on experience and narrow the knowledge gap that exists between theory and application. Embrace digital technologies and online platforms to enhance accessibility, flexibility, and engagement in engineering education. Implement robust assessment methods to measure students' attainment of learning outcomes and competencies. Utilize a combination of formative and summative assessments, including exams, projects, presentations, and portfolios, to evaluate students' knowledge, skills, and problem-solving abilities. Emphasize authentic assessment tasks that mirror real-world engineering challenges and promote reflective practice among students. Faculty development invest in faculty development programs to support instructors in adopting innovative teaching practices, staying current with industry trends, and enhancing their pedagogical skills.

Provide opportunities for professional development, mentorship, and educators are collaborating to promote an environment of quality in teaching. Continuous improvement mechanisms establish systematic mechanisms for collecting feedback from students, alumni, employers, and other stakeholders to inform on-going program evaluation and improvement. Utilize data analytics and outcome assessment tools to track students' progress, identify areas for enhancement, and monitor the effectiveness of quality assurance initiatives. Encourage a culture of transparency, accountability, and continuous learning within the engineering education community.

Implementation and working

Establish accreditation processes

Engineering institutions initiate accreditation processes by identifying relevant accrediting bodies and understanding their requirements. They conduct internal assessments to ensure alignment with accreditation criteria, address any gaps, and prepare for external evaluations. During accreditation visits, institutions showcase evidence of program quality through documentation, student work, and faculty qualifications. Feedback from accrediting bodies informs continuous improvement efforts.

Develop robust curriculum

Engineering programs design curricula that align with accreditation standards, industry needs, and societal expectations. This involves engaging stakeholders, including industry professionals, alumni, and regulatory bodies, to identify key competencies and learning outcomes. Curricula are designed to integrate theoretical knowledge with practical applications, incorporating interdisciplinary perspectives and emerging technologies. Regular reviews and updates ensure curricula remain relevant and responsive to evolving industry trends.

Implement innovative teaching methods

Institutions adopt innovative teaching methodologies to improve the essential thinking, solving issues, and involvement of students. This may include active learning strategies such as flipped classrooms, collaborative projects, and experiential learning opportunities. Technology-enhanced learning platforms and digital tools are integrated to facilitate interactive lectures, simulations, and virtual laboratories. Faculty development programs provide instructors with training and support to implement effective teaching practices.

Employ diverse assessment strategies

A wide range of assessment techniques are used in engineering courses to assess the success of the curriculum and the learning outcomes of students. These may include traditional assessments such as exams, quizzes, and lab reports, as well as authentic assessments such as projects, presentations, and portfolios.

Overall evaluations investigate the total amount of goals for learning achieved, whereas formal evaluations give learners continual feedback. Rubrics and assessment criteria are transparently communicated to students to promote clarity and fairness.

Invest in faculty development

Institutions prioritize faculty development initiatives to enhance teaching effectiveness and promote continuous improvement. Faculty members participate in professional development workshops, seminars, and conferences to stay abreast of best practices in engineering education. Mentoring programs pair experienced faculty with newer instructors to facilitate knowledge sharing and skill development. Peer observation and feedback mechanisms encourage reflection and refinement of teaching practices.

Establish continuous improvement mechanisms

Engineering colleges offer frameworks for continuing monitoring of outcomes of programs and suggestions from stakeholders for constant enhancement. Regular program reviews and evaluations identify areas for enhancement and inform strategic planning efforts. Data analytics and performance indicators are used to track student success, retention rates, and graduates' career outcomes. Transparency and accountability are maintained through regular reporting and communication with stakeholders.

Results and discussion

Quality assurance efforts in engineering education yield significant results in ensuring the relevance, effectiveness, and excellence of educational programs. By implementing rigorous accreditation processes, institutions validate program quality and demonstrate adherence to established standards, thereby instilling confidence in stakeholders, including students, employers, and regulatory bodies. Moreover, innovative teaching methodologies, experiential learning opportunities, and diverse assessment strategies enhance student engagement, promote deeper learning, and bridge the gap between theory and practice. These efforts not only prepare graduates for successful careers in engineering but also contribute to the advancement of the engineering profession by fostering a culture of continuous improvement and responsiveness to industry needs and societal challenges. Furthermore, quality assurance initiatives in engineering education foster collaboration among academia, industry, and government, driving innovation and enhancing the competitiveness of engineering programs on a global scale. By engaging industry stakeholders in curriculum development, providing students with hands-on experience through internships and industry projects, and ensuring faculty are equipped with the latest knowledge and teaching methodologies, institutions produce graduates who are not only technically proficient but also adaptive, creative, and well-rounded. This alignment between education and industry needs ensures that graduates are prepared to address the complex and interdisciplinary challenges facing the engineering profession, positioning them as leaders in driving technological advancements and societal progress.

Conclusions

In conclusion, quality assurance in engineering education is paramount for equipping graduates with the necessary skills and competencies to excel in the field. Through rigorous accreditation processes, innovative teaching methods, diverse assessment strategies, and continuous faculty development, institutions uphold high standards of excellence and relevance. By fostering a culture of continuous improvement and responsiveness to stakeholder feedback, engineering education programs can ensure that graduates are prepared to meet the evolving demands of industry and society, ultimately advancing the engineering profession as a whole.

References

Harun, Ch., Firoz, A., Shyamal Kanti, B., Tazul Islam, M., and Sadrul Islam, A. K. M. (2013). Quality assurance and accreditation of engineering education in Bangladesh. *Proc. Engg.*, 56(1), 864–869.

Anca, G. and Todorescu, L. L. (2014). Quality assurance and the English teacher's profile in Romanian technical higher education. *Proc. Soc. Behav. Sci.*, 143(1), 698–702.

Shahzad Khuram, Ch., Abdul, R., Nadia, N., and Natasha Saman, E. (2023). A bibliometric analysis of quality assurance in higher education institutions: Implications for assessing university's societal impact. *Eval. Prog. Plan.*, 99(1), 112–116.

Luca, J., Reh, D., and Julia Arlinghaus, C. (2024). Challenges of quality assurance in early planning and ramp up of production facilities - Potentials of planning automation via virtual engineering. *Proc. Comp. Sci.*, 232(2), 2498–2507.

31 E-Learning platforms strategies to ensure inclusivity and equal learning opportunities for students with disability

P. Nagarajan[1,a], *N. Ashokkumar*[2,b], *A. Shirly Edward*[1,c], *Anita Christaline JohnVictor*[3,d] *and G. Logambal*[4,e]

[1]Department of ECE, SRM Institute of Science and Technology, Vadapalani Campus, Chennai-600026, Tamil Nadu, India

[2]Department of ECE, Mohan Babu University (Erstwhile Sree Vidyanikethan Engineering College), Tirupati-517102, Andhra Pradesh, India

[3]School of Computer Science and Engineering, Vellore Institute of Technology, Chennai Campus, Chennai-600127, Tamil Nadu, India

[4]Department of ECE, Research Scholar, Anna University, Chennai, Tamil Nadu, India

Abstract

e-Learning has included online platforms to provide better educational benefits among students. It has provided information and resources to the students through the usage of several online devices such as computers, laptops and mobiles. This platform has increased the personal proficiencies of disabled students. Educators have used different approaches to teach disabled students in online classes. Disabled students have used images, alt text and text on the screens to gain a better understanding of the concept of learning. The students have faced challenges at the time of being involved in online classes. Technical issues and poor design courses have created challenges for disabled students to access the information appropriately. These platforms can meet the unique needs of the students by focusing on their choices. It has provided freedom to the students to learn in their way to achieve success in life. Students can think beyond their book material which has enhanced their critical thinking efficiency to solve any problems in life. e-Learning has provided flexible learning facilities to disabled students and has also provided personal preferences to the students. This platform can engage more students within a certain period and provide quality education to the students.

Keywords: e-Learning, disabled students, learning opportunities

Introduction

e-Learning platforms have been developed to provide better education facilities to the students. The modern facility of e-learning platforms has been discussed in this investigation for signifying the effectiveness of an e-learning. This research has analyzed the proficiency of e-learning platforms in an improvement of students who have disabilities. Physical disabilities create barriers in the lives of students to increase their performance. The development of an effective system of online education system is necessary for students who are suffering from disabilities. Thus this research helps to find the most important features of the e-learning platforms that support physically challenged students in learning.

Background of the study

This research has discussed the effectiveness of e-learning platforms for disabled students and how these platforms help to offer equivalent opportunities for entire students to get the same learning facilities. According to the views of Maatuk et al. (2022), the development of educational opportunities is important for all students as this provides better opportunities for the students. Thus, the usage of an e-learning platform is vital for the development of the students.

Figure 31.1 represents that the market size of an e-learning platforms has increased from 2019 to 206. The revenue of online e-learning was 101 billion US dollars and that could be increased to 167.5 billion US dollars in 2026 (Statista, 2022). This figure has been used in this research as this helps to represent the demand for e-learning which has increased with time. Different contents have been developed to assist physically challenged students such as visual images, auditory, animation videos and so on. This research helps to analyze the effectiveness of different tools of e-learning education to enable all students to get adequate opportunities for education.

[a]nagarajan.pandiyan@gmail.com, [b]ashoknoc@gmail.com, [c]edwards@srmist.edu.in, [d]anitachristaline.j@vit.ac.in, [e]Logambal.gandhi@gmail.com

DOI: 10.1201/9781003587538-31

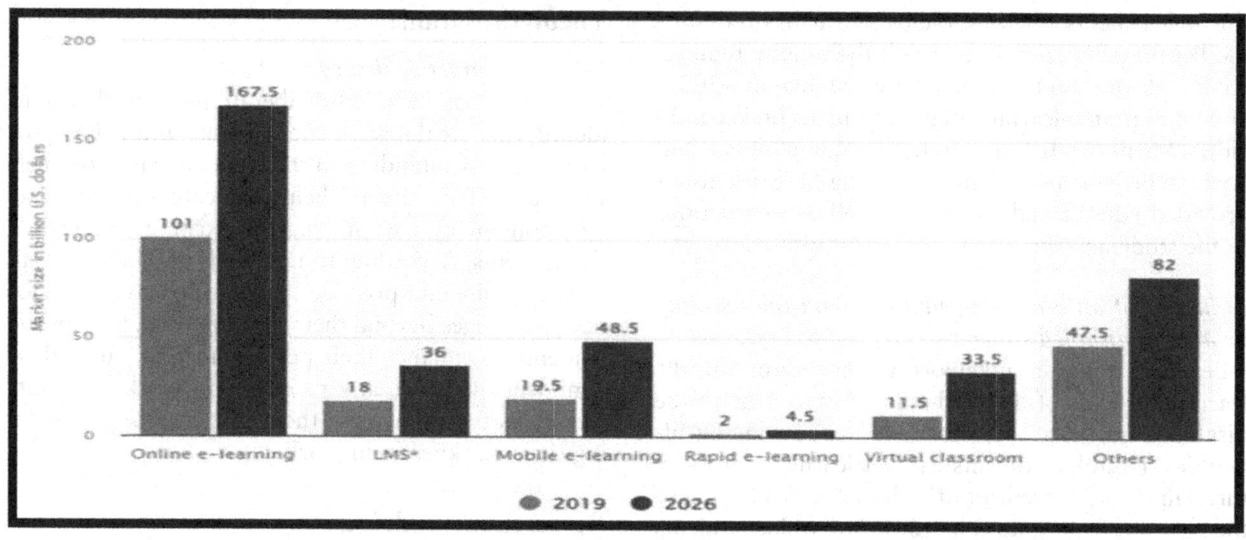

Figure 31.1 Market size of online e-learning and other e0-learning platforms from 2019 to 2026
Source: Statista, 2023

Problem statement

Disabled students face many challenges in their lives to grow. They face social challenges, educational challenges and moral insecurity. The development of an effective e-learning platform for them increases the opportunity for studies. According to the views of Aboagye et al. (2021), the implementation of effective learning strategies helps disabled people to enhance their lives. The main problem is that all the students are not getting equal opportunities to develop themselves.

Aim and objectives of this investigation

Aim of this investigation
The research aims to discover the effectiveness of an e-learning platform for the development of disabled students and for creating opportunities for all students to get the same chance of education.

Objectives of this investigation

- To know the details of an e-learning methodologies and arenas and their effectiveness on the development of student's performance
- To analyze the contribution of an e-learning platforms in the development of disabled students
- To know the most challenges that occur through the usage of an e-learning platforms for the education of disabled students
- To recommend the finest ways to enrich the learning abilities of disabled students.

Research questions of this investigation

- What is an e-learning platform and the way it affect the educational development of the students?
- How e-learning platform help disabled students to get a better education?
- What are the challenges that occur during the use of e-learning platforms by disabled students?
- What are the most suitable ways to develop the learning scope of disabled students from e-learning platforms?

Significance of this investigation

This investigation is important for all students to know about the effectiveness of an e-learning platforms in increasing an opportunities for all students to get adequate education. Based on the views of Al-Fraihat et al., (2020), e-learning educational platforms help students from remote areas to get the same facilities as the educational facilities enjoyed by all. This means e-learning platforms represent things equal to all students. Thus, this research is important for all to understand the need for e-learning platforms for the development of students.

Literature review

Conception of an e-learning platforms and their contribution for performance development of the students
e-Learning platforms have provided online learning facilities to students through the use of digital resources. This process has included various devices such as computers, phones, and tablets and these devices are connected to the internet. According to the views of Radha et al. (2020), e-learning platforms help students enhance their performance and skills in study. These platforms have provided personalized

learning benefits to the students and it has modified the learning experiences to fulfill the unique requirements of the students. Likewise, Mahyoob (2020) argued that an e-learning platforms focus on the individual learner's strengths, preferences, weaknesses and goals to provide personalized learning facilities. It has provided educational content as well as instructions to the students.

Influence of an e-learning platforms on the learning process of disabled student

e-Learning takes a noteworthy impression on the learning process of disabled students and it helps the disabled students to gain knowledge without facing difficulties. Disabled students can get learning solutions through the advancement of technology. According to the views of Rodrigues et al. (2019), disabled students can watch video lectures repeatedly through adopting e-learning platforms. This platform has provided enough time and space for the students to work. IT systems as well as software help those students who are facing dyslexia-related disorders change the font style of digital text. On the other hand, Affouneh et al., (2020) argued that e-learning has provided benefits to the students to obtain the courses without being physically present in the class. Students who have visual impairments can learn through advanced technologies such as braille keyboards and audio recording facilities. Hearing impairment-related disorder students can gain knowledge by watching video lectures that have subtitles.

Encounters tackled by disabled students for practice of an e-learning platform for gathering knowledge

e-Learning has created difficulties for disabled students to properly access information from online platforms. A properly designed keyboard is necessary which includes alt text, images as well as text on the screen. Disabled students have faced problems when the keyboards are not properly designed. According to the views of Subedi et al. (2020), disabled students need alternative teaching techniques which have created problems at the time of using e-learning platforms. Technical issues have generated problems for disabled students to access resources through online platforms. On the other hand, Baber (2020), argued that lack of interaction can cause challenges for the students to retain their interest in the subject (Figure 31.2).

Students can feel loneliness through spending more time on the internet. Disabled students can lose interest at the time of listening to the lectures and it has a negative impact that students are not physically connected with the course content in the class (Adnan and Anwar, 2020).

Theoretical framework

Deliberate practice theory

This study has included deliberate practice theory to identify the weakness of the students and it has provided an understanding to the students based on their weakness. This theory helps educators to provide new content by customizing the existing content for the students. According to the views of Maatuk et al. (2022), deliberate practice theory helps students face new challenges beyond their comfort zone. It helps the students to enhance their performance in study, their confidence to solve any problems in work. It results the students can increase their efficiency by recognizing their weaknesses in work.

Research methodology

This research has been developed on the primary quantitative methods which include the primary data collection process and quantitative analysis processes. Primary data collection helps to get real data from the respondents. The data has been collected from 60 respondents who are related to this field. Based on the views of Franzitta et al. (2020), the primary data collection system includes the survey process and it is time-consuming. All the collected data have been analyzed through quantitative analysis like regression, correlation and descriptive analysis. According to the views of Savateev (2022), quantitative analysis helps to compare all the important factors through statistical calculation (Figure 31.3).

Finding and analysis

9.1 Demographic analysis

Table 31.1 represents the age-related data where different age-related people are participating in the survey. Based on this survey most of the people belonging to 40–50 years age group have participated in this survey. This survey denotes that 40–50 years age group people are shown more interest in this education system.

What about the participant age?

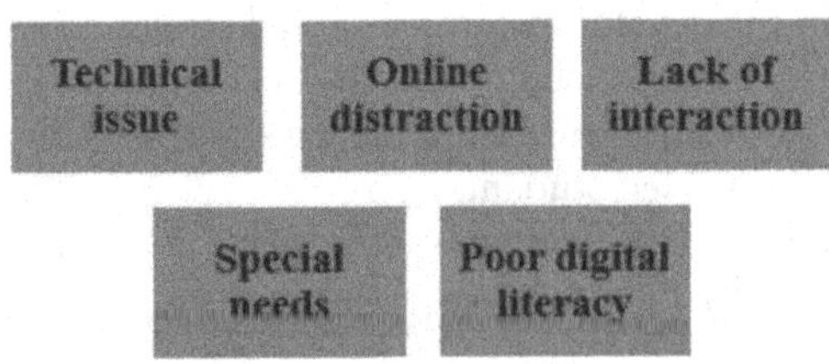

Figure 31.2 Challenges faced by disabled students to adapt e-learning facilities
Source: Author

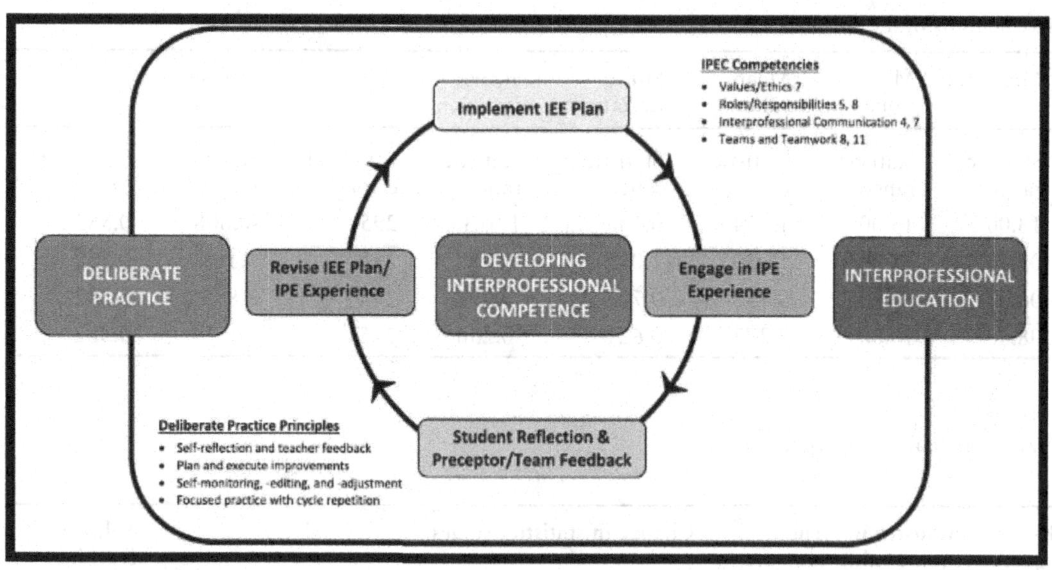

Figure 31.3 Deliberate practice theory
Source: Savateev (2022))

Table 31.1 Age of participant.

	Frequency occurrence	Percentage %	Effective percentage	Collective percentage
Effective 20–30	18	27.3	27.3	27.3
30–40	12	18.2	18.2	45.5
30–40	24	36.4	36.4	81.8
Above 50	12	18.2	18.2	100
Total	66	100	100	

Source: Author

Table 31.2 Gender of participant.

	Frequency occurrence	Percentage %	Effective percentage	Collective percentage
Effective Female	42	63.6	63.6	63.6
Male	24	36.4	36.4	100
Total	66	100	100	

Source: Author

What is gender of participant?

Table 31.2 has shown the gender-based data and has also represented that 63.6% of females have participated in this survey. It has also been shown that 36.4% of males participated in this survey. This process has denoted that females are showing more interest in adopting e-learning technologies among students.

9.2 Statistical analysis

Descriptive analysis

The descriptive analysis (Table 31.3) has shown a relationship among each variable with others. This analysis has shown dependent variables to represent the enhancement of an e-learning platform among the students. It possess, also represented independent variables to demonstrate the influence of an e-learning platforms in the teaching and learning system of students.

Table 31.4 represents the relation between the variables that are presence in hypothesis. The relation between the variables is represented through the value of significance. The significance value of this regression analysis is 0 which demarcates that there is a strong relation between the variables.

Table 31.3 Descriptive analysis of the variable.

	N	Min value	Max value	Mean value	Std. deviation	Skew		Kutosis	
	Statistic range	Statistic range	Statistic range	Statistic range	Statistic range	Statistic range	Standard error	Statistic range	Standard error
DV	66	13.00	15.00	13.545	0.788	1.013	295	-0.608	0.582
N1	66	09.00	15.00	13.303	1.037	-0.987	295	3.168	0.582
N2	66	06.00	10.00	8.757	0.702	-0.991	295	2.895	0.582
N3	66	08.00	10.00	9.272	0.621	-0.256	295	-0.578	0.582

Source: Author

Table 31.4 Linear regression analysis of hypothesis 1.

Model summary[b]

Model	R-value	R-square value	Adjusted R-square value	The estimate standard-error	Change in statistics values					Value of Durbin-Waston
					Change of R-square	Change of F	Change of Df-1	Change of Df-2	Change of Sig-F	
1	0.469	0.220	0.208	0.7014	0.220	18.029	1	64	0.1	2.159

ANOVA[a]

Type of model	Value-sum of squares	Df-value	Mean- square value	F-value	Sig-value
Regression-model	8.871	1	8.871	18.029	0.0[b]
Residual	31.492	64	0.492		
Total	40.364	65			

[a]*Constant-predictors, IV1*

[b]*Dependent-variable, DV*

Type of model	Unstandardized-coefficient values		Standardized-coefficient values	T-value	Sig-value
	B-value	Standard error	Beta-value		
1 (Constant)	18.283	1.119		16.336	0.0
IV1	-0.356	0.084	-0.469	-4.246	0.0

Source: Author

Hypothesis 1

Discussion

This research investigation assistance for understanding an effectiveness of e-learning platforms that increase the opportunity for all students to get higher education facilities. According to the views of Maatuk et al. (2022), e-learning platforms increase the opportunity for all by providing the best teachers for the students. Likely, an e-learning platform facilitate disabled students to get knowledge through visual and audio representation. Based on the views of Savateev (2022), e-learning platforms have different kinds of file content like videos and audio. This helps all physically challenged people to get a fruitful space for learning different things. The implication of e-learning platforms is important for spreading knowledge to all students who are not able to reach the education center. This helps to spread the facilities of education to all and helps the physically-challenged students to get the appropriate scope of education.

Conclusions

The implication of e-learning platforms increases the scope of all types of students to get the same level of education. The availability of alternative forms of study like audio, video and visual images helps the students to understand the study materials. This increases the hope of disabled students which inspires them to develop their knowledge and own life. The ideology of e-learning platforms is most suitable for all students to get better learning opportunities equally without wasting time behind the journey. Thus, e-learning platforms are highly effective for the development of all types of learners, especially disabled students.

References

Aboagye, E., Yawson, J. A., and Appiah, K. N. (2021). COVID-19 and E-learning: The challenges of students in tertiary institutions. *Soc. Edu. Res.*, 1–8. https://ojs.wiserpub.com/index.php/SER/article/download/ser.212021422/282.

Adnan, M. and Anwar, K. (2020). Online learning amid the COVID-19 pandemic: Students' perspectives. *Online Submission*, 2(1), 45–51. http://www.doi.org/10.33902/JPSP.%202020261309.

Affouneh, S., Salha, S., and Khlaif, Z. N. (2020). Designing quality e-learning environments for emergency remote teaching in the coronavirus crisis. *Interdis. J. Virt. Learn. Med. Sci.*, 11(2), 135–137. https://ijvlms.sums.ac.ir/article_46554_7ab365cf9a02b5e6f88249434c3b3505.pdf.

Al-Fraihat, D., Joy, M., and Sinclair, J. (2020). Evaluating E-learning systems success: An empirical study. *Comp. Human Behav.*, 102, 67–86. https://doi.org/10.1016/j.chb.2019.08.004.

Baber, H. (2020). Determinants of students' perceived learning outcome and satisfaction in online learning during the pandemic of COVID-19. *J. Educ. E-learn. Res.*, 7(3), 285–292. DOI: 10.20448/journal.509.2020.73.285.292.

Franzitta, V., Longo, S., Sollazzo, G., Cellura, M., and Celauro, C. (2020). Primary data collection and environmental/energy audit of hot mix asphalt production. *Energies*, 13(8), 2045. https://www.mdpi.com/1996-1073/13/8/2045/pdf.

Maatuk, A. M., Elberkawi, E. K., Aljawarneh, S., Rashaideh, H., and Alharbi, H. (2022). The COVID-19 pandemic and E-learning: Challenges and opportunities from the perspective of students and instructors. *J. Comp. Higher Edu.*, 34(1), 21–38. https://link.springer.com/article/10.1007/s12528-021-09274-2.

Mahyoob, M. (2020). Challenges of e-learning during the COVID-19 pandemic experienced by EFL learners. *Arab World English J. (AWEJ)*, 11(4). https://dx.doi.org/10.24093/awej/vol11no4.23. Pp. 351–362

Radha, R., Mahalakshmi, K., Kumar, V. S., and Saravanakumar, A. R. (2020). E-learning during lockdown of Covid-19 pandemic: A global perspective. *Internat. J. Con. Automat.*, 13(4), 1088–1099. https://www.academia.edu/download/64029090/covid-pandemic-scopus.pdf.

Rodrigues, H., Almeida, F., Figueiredo, V., and Lopes, S. L. (2019). Tracking e-learning through published papers: A systematic review. *Comp. Edu.*, 136, 87–98. https://dx.doi.org/10.1016/j.compedu.2019.03.007.

Savateev, O. (2022). Photocharging of semiconductor materials: database, quantitative data analysis, and application in organic synthesis. *Adv. Ener. Mat.*, 12(21), 2200352. DOI: 10.1002/aenm.202200352.

Statista. (2022). Size of the global e-learning market in 2019 and 2026, by segment. Retrieved on: 22nd December 2023 from: https://www.statista.com/statistics/1130331/e-learning-market-size-segment-worldwide/.

Subedi, S., Nayaju, S., Subedi, S., Shah, S. K., and Shah, J. M. (2020). Impact of E-learning during COVID-19 pandemic among nursing students and teachers of Nepal. *Internat. J. Sci. Healthcare Res.*, 5(3), 68–76. https://www.academia.edu/download/64126179/IJSHR0012.pdf.

32 Drivers of successful E-learning eco-system - Exploring with an empirical investigation

B. Latha Lavanya[1,a], *Priyadarshini B.*[1] *and L. Anitha*[2,b]

[1]MEASI Institute of Management, Royapettah, Chennai-14, Tamil Nadu, India

[2]Department of Management [2]Saveetha Engg College, Thandalam, Chennai, Tamil Nadu, India

Abstract

Purpose: This study aims to explore the determinants persuading students' perceptions of e-learning platforms in advanced education settings, concentrating on perceived effectiveness, ease of use, self-efficacy, besides information quality. By understanding these factors, educators and policymakers can improve the usefulness and quality of online education experiences. **Methodology:** A comprehensive survey was conducted among undergraduate students enrolled in e-learning courses, with a sample size of 121 participants. The survey measured participants' perceptions of various dimensions of e-learning using a structured questionnaire using single random sampling methodology. Descriptive statistics, reliability, ANOVA with Cochran's, KMO, and Bartlett's test, and step-wise multiple linear regression were employed to scrutinize the data. **Findings:** The study revealed consistently high perceptions of e-learning factors among participants, with mean scores indicating positive outlooks across dimensions. Noteworthy associations were identified between perceived usefulness, ease of use, self-efficacy, and e-learning outcomes, underscoring their importance in shaping students' experiences. Additionally, occupation was found to impact opinions of e-learning eco-system (Alkhattabi et al., 2010, 2011). However, while information quality and apparent utility demonstrated momentous positive properties on e-learning outcomes, instructor attitude did not significantly influence e-learning. Perceived ease of use and self-efficacy also showed non-significant effects. The education and training sector plays a critical starring role cutting-edge the advancement of society by molding learner's cognitive abilities, acquisition of skills, and advancement in their careers. It includes a wide range of establishments and groups, including corporate learning programs, early childhood education centers, higher education facilities, and suppliers of vocational training.

Keywords: E-learning eco-system, easy in usage, information quality, instructor attitude

Introduction

The e-learning eco-system comes in two primary flavors: synchronous and asynchronous, which vary based on how students and teachers communicate. Asynchronous e-learning settings let students log in and utilize the platform on their own, whereas synchronous e-learning environments require instructors and students to be online with ample uses like excluding time and space constraints and enabling free exchanges between students and teachers (Aladwani and Palvia, 2002).

Need of the study

Understanding factors influencing e-learning eco-system platform engagement is crucial for enhancing educational experiences. Identifying these factors can help improve the design and functionality of e-learning eco-systems (Jung, 2011). Addressing user needs and preferences can lead to higher levels of engagement and satisfaction among e-learning platform users. By enhancing e-learning platform engagement, educational institutions and EdTech companies can better support student learning and achievement (Al-Samarraie et al., 2017, 2018).

Review of literature

Sulaymani et al. (2022) inspected student's knowledge, their belief in their ability in e-learning platform acceptance. The study found that students' confidence, shaped by their prior tech experience, significantly impacts platform acceptance. Interestingly, while older high school students showed a positive relationship between self-efficacy and acceptance, younger middle school students exhibited a negative one. These findings offer insights for educators and policymakers seeking to promote e-learning platform usage among younger students.

In Ejdys's (2021) work, the author examined key influences on university students' adoption of e-learning. Through extensive analysis, factors including accessibility, usability, perceived usefulness, technical support, and instructional design were explored. The findings offer valuable insights for educators and policymakers aiming to improve learning eco-system.

[a]latha.lavanya@measiim.edu.in, [b]anithal@saveetha.ac.in

DOI: 10.1201/9781003587538-32

Almaiah and Alyoussef (2019) scrutinized the power of the characteristics on learning system usage. The factors significantly affect student engagement, emphasizing the importance of effective course design, support, assessments, and engaging instructors for successful e-learning utilization.

Ozkan and Koseler (2009) explored student evaluations by focusing on system, information, and service quality. The factors such as usability and usefulness – crucial for student satisfaction and attitudes towards e-learning. This study offers insights for improving e-learning strategies in higher education Liaw (2007).

DeLone and McLean (1992) framed a framework to ration IS triumph, incorporating six measurement constraints. Their approach, blending technical and user-centric perspectives, has shaped IS research and practice, remaining influential in understanding and achieving IS success.

Data and variables

A survey was conducted among undergraduate students enrolled in e-learning courses, with a sample size of 121 participants. The research stretches over 2 years, from Feb 2022 to Dec 2024. The independent variables of e-learning eco-system – Model specifications consist of 5 factors, such as perceived ease of use (PEU), perceived usefulness (PU), self-efficacy (SE), information quality (IQ), instructor attitude (IU), and dependent variables such as e-learning platform with the general demographic factors (gender, age, education level) (Figure 32.1).

Study objectives

- To explore the antecedents that drive e-learning eco-system platforms.
- To examine the relationship of determinants that affect e-learning eco-system platforms.

Empirical results

The statistical software R (Version 4.3.1) is utilized for analyzing the collected data. R is renowned for its open-source nature and is extensively used. It deals with a robust suite of gears for data handling, examination, and interpretation, catering to diverse research needs. Following data collection from the questionnaires, conducted checks for omissions and commissions to eliminate incomplete entries. Ultimately, 122 valid questionnaires were chosen for analysis using a simple random sampling methodology Hariyanto et al., (2020).

6.1 Reliability

It is a scale which consistently represents the construct it measures.

Table 32.1 illustrates the reliability of the e-learning eco-system, showing a coefficient of 0.784. This high coefficient indicates an excellent level of internal consistency reliability among the items, suggesting that they consistently measure the same underlying construct. With a total of 39 items, this coefficient confirms that, together, they provide a reliable assessment of the targeted construct.

6.2 Factor analysis

The Kaiser-Meyer ratio of specimen competence assesses the extent to which the variance in variables could be attributed to underlying factors.

Table 32.2 details the Kaiser – Meyer – Olkin test measure of sampling adequacy, recorded at 0.779, indicating that the values are appropriate for factor analysis. Bartlett's test of sphericity, yielding an

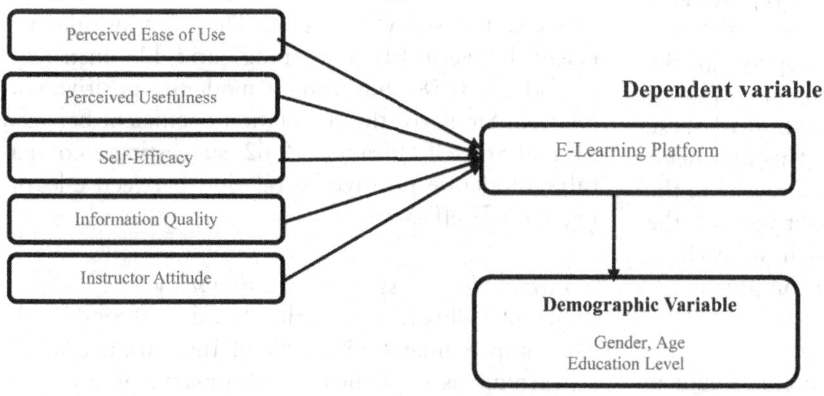

Figure 32.1 E-learning eco-system model specifications
Source: Author

Table 32.1 Reliability for e-learning eco-system.

Cronbach's alpha	No. of items
0.784	39

Source: Mentioned in Methodology

Table 32.2 Factor analysis for e-learning eco-system.

Test for the adequacy of the collected samples		0.779
Bartlett's test of sphericity	Approx. Chi-square	1908.66
	df	595
	Sig.	0

Source: Author

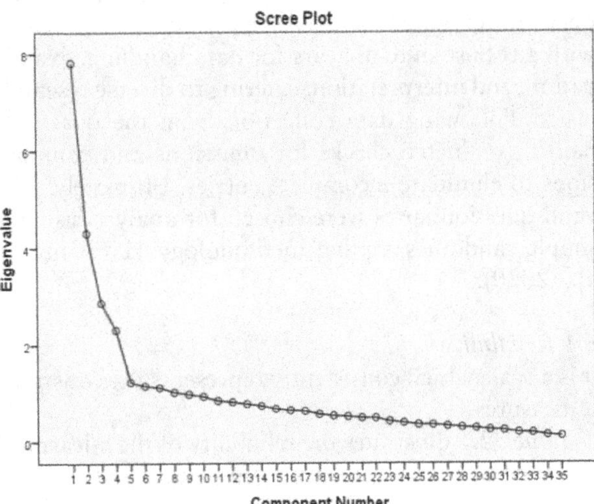

Figure 32.2 Magnitude of variables
Source: Author

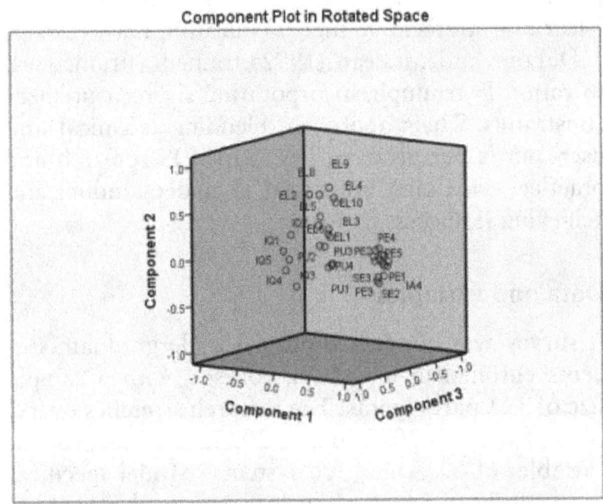

Figure 32.3 Three dimensional space
Source: Author

approximate chi-square value of 1908.656 with 595 degrees of freedom, is significant at $p<0.001$, demonstrating a strong association among the variables.

Figure 32.2 illustrates the magnitudes of variables in a screen plot, with values on the y-axis and component numbers on the x-axis. Greater variability in a specific direction corresponds to a better explanation of the behavior of the dependent variable. The scree plot is used to determine the number of factors to retain in an exploratory factor analysis (FA).

Figure 32.3 illustrates the rotation performed using the Varimax method, which enhances the weights of higher loadings and reduces the weights of lower ones. The rotated component matrix displays all the loadings for each component. The component plot in rotated space offers a visual representation of these loadings in a three-dimensional space. This plot demonstrates that components 1, 2, and 3, along with the visualization test, load highly and positively on the first component. The e-learning eco-system shows high loadings on the first, second, and third components.

6.3 Chi-square for e-learning eco-system

[a]105 cells (93.8%). The minimum expected count is 0.01.

Table 32.3 shows the chi-square value for the e-learning eco-system test statistic, which is 281.3. All cells had an expected count greater than 5, meeting this assumption. The corresponding p-value is p=0.00. The Chi-square (χ^2) tests performed on the dataset yield highly significant results (χ^2=281.388, df=91, p=0.000), indicating a strong relationship among the determinants being studied.

6.4 Correlation for e-learning eco-system

Table 32.4 presents the correlation coefficients between the study variables. The correlation coefficient between EL (e-learning) and IA (instructor attitude) is 0.58, indicating a moderate positive correlation. Similarly, the correlation coefficient between EL and SE (self-efficacy) is 0.62, suggesting a comparable moderate positive correlation between e-learning and self-efficacy.

6.5 Multiple regression model summary

Table 32.5 illustrates that the R^2 value (0.5938) indicates approximately 59.38% of the variance in EL (e-learning) is explained by IA (instructor attitude), IQ (information quality), PE (perceived ease of use), PU (perceived usefulness), and SE (self-efficacy). The adjusted R^2 value (0.5762) accounts for the number of predictors. The coefficients table provides estimated

Table 32.3 e- learning eco - system.

Pearson	Value	df	Asymptotic significance (2-sided)
	281.388[a]	**91**	**0.000**
Likelihood ratio	98.517	91	0.277
Linear-by-linear association	46.354	1	0.000
No of valid cases	121		

Source: Author

Table 32.4 E-learning eco-system study variables correlation.

Variable	EL	IA	IQ	PEU	PU	SE
EL – e-learning	1.00					
IA – Instructor attitude	0.58	1.00				
IQ – Information quality	0.68	0.54	1.00			
PEU – Perceived ease of use	0.59	0.92	0.53	1.00		
PU – Perceived usefulness	0.66	0.58	0.61	0.57	1.00	
SE – Self-efficacy	0.62	0.94	0.58	0.95	0.59	1.00

Source: Primary data

Table 32.5 Co-efficient Variance of e-learning.

Model summary	Value
R^2	0.5938
Adjusted R^2	0.5762
Residual standard error	0.4511
F-statistic	33.63
p-value (F-statistic)	<2.2e-16

Source: Author

coefficients for each predictor variable in the regression model predicting EL. Notably, information quality (β=0.348, p<0.001) and perceived usefulness (β=0.320, p<0.001) have significant positive effects on EL, indicating that higher perceived information quality and usefulness are associated with increased e-learning effectiveness.

Table 32.6 details the variable coefficients for the model, showing a constant of 0.584, with IQ as the strongest predictor, followed by perceived ease of

Table 32.6 Stepwise multiple regression
Model ease of use, perceived usefulness, self-efficacy, and instructor attitude.

Coefficients*

Model B		Unstandardized Coefficients		Standardized Coefficients	t.	Sig.
		Std. Error	Beta			
1	(Constant)	.584	.293		1.990	.049
	Information Quality	.355	.077	.348	4.351	.000
	Perceived Ease of use	.056	.189	.058	.297	.767
	Perceived Usefullness	.309	.078	.320	3.961	.000
	Self Efficacy	.274	.206	.300	1.330	.186
	Instructor Attitude	-.123	.176	-.130	-.700	.485

a. Dependent Variable: EL

Source: Author

REGRESSION EQUATION

$Y= ax + b$

$Y= 0.584+ 0.335+ 0.056+0.309+ 0.274- 0.123$

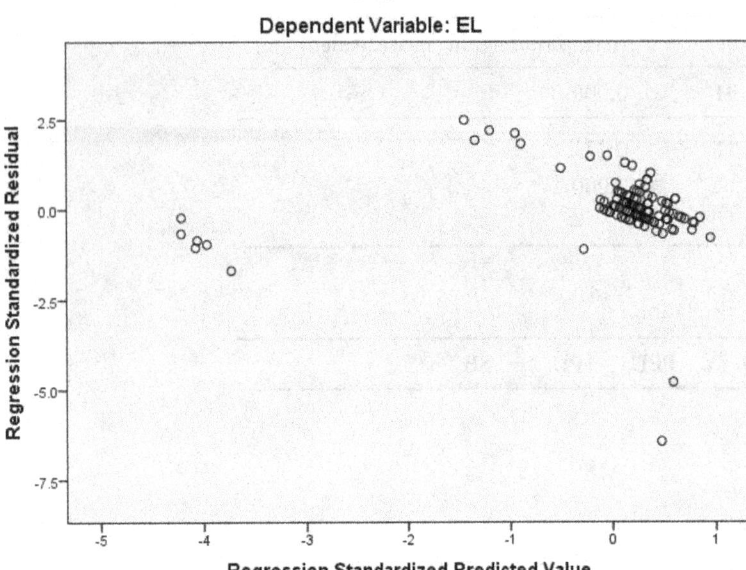

Figure 32.4 standardized residual plot
Source: Author

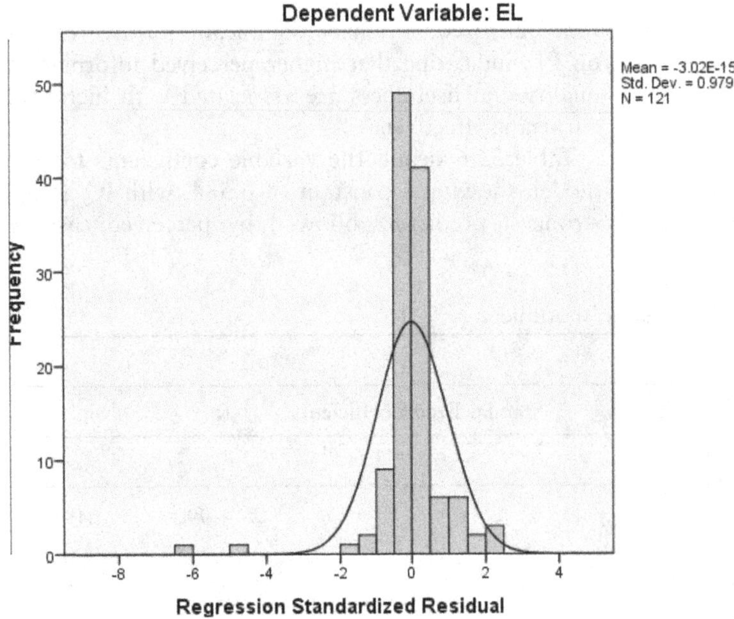

Figure 32.5 standardized: residual histogram
Source: Author

use, perceived usefulness, self-efficacy, and instructor attitude.

Figure 32.4 demonstrates that the data exhibit minimal outliers and are tightly clustered, indicating a strong relationship. Further details regarding data normality have been provided by normal P-P plot of regression standardized residual and regression standardized residual histogram in Figure 32.5.

Inference: The ANOVA Table (type II tests) examines the importance of individual predictor variables (IA, IQ, PE, PU, and SE) in elucidating the predictability in the response factors (EL) within the regression model. It demonstrates that IQ and PU significantly impact EL ($p<0.001$), while IA, PE, and SE do not. This suggests that intelligence quotient (IQ) and perceived usefulness (PU) play crucial roles in influencing EL within the model.

Findings and suggestions

Quality information and perceived usefulness were significant drivers of the e-learning eco-system, highlighting their importance Eom S. B (2006). Focus on enhancing the platform's functionality and user interface based on the positive feedback on PU and PE. Invest in high-quality instructional materials to maintain a favorable perception of IQ. Offer personalized learning paths and resources to boost SE (Dominici and Palumbo, 2013). Ensure accessibility for all learners and adapt to evolving needs and trends to maintain an innovative e-learning environment. Regularly review and adjust strategies to uphold effectiveness (Gay, 2016).

8. Conclusions

The results highlight how crucial it is to design user-friendly interfaces, excellent content, interesting teacher–student interactions, and supportive learning environments (Alsabawy et al., 2016). It is crucial to use these findings going forward to promote innovation and ongoing development in e-learning eco-systems. The ability to cope with changing learner demands and technology improvements will also depend on promoting teamwork, remaining flexible, and adopting data-driven decision-making (Nugroho, 2018). By placing these suggestions into practice, e-learning eco-system initiatives' impact and reach can be increased, enabling students to fulfill their educational objectives in a fast-paced, digitally-driven world (Chopra et al., 2019). E-learning eco-system democratizes education by making it accessible, flexible, and personalized. It's cost-effective, global, and promotes lifelong learning (Compeau and Higgins, 1995).

Bibliography

Aladwani, A. M. and Palvia, P. C. (2002). Developing and validating an instrument for measuring user-perceived web quality. *Inform. Manag.*, 39(6), 467–476.

Alkhattabi, M., Neagu, D., and Cullen, A. (2010). Information quality framework for e-learning systems. *Knowl. Manag. e-learn.*, 2(4), 340.

Alkhattabi, M., Neagu, D., and Cullen, A. (2011). Assessing information quality of e-learning systems: A web mining approach. *Comp. Human Behav.*, 27(2), 862–873.

Almaiah, M. A. and Alyoussef, I. Y. (2019). Analysis of the effect of course design, course content support, course assessment and instructor characteristics on the actual use of E-learning system. *IEEE Acc.*, 7, 171907–171922.

Alsabawy, A. Y., Cater-Steel, A., and Soar, J. (2016). Determinants of perceived usefulness of e-learning systems. *Comp. Human Behav.*, 64, 843–858.

Al-Samarraie, H., Selim, H., Teo, T., and Zaqout, F. (2017). Isolation and distinctiveness in the design of e-learning systems influence user preferences. *Interac. Learn. Environ.*, 25(4), 452–466.

Al-Samarraie, H., Teng, B. K., Alzahrani, A. I., and Alalwan, N. (2018). E-learning continuance satisfaction in higher education: A unified perspective from instructors and students. *Stud. High. Edu.*, 43(11), 2003–2019.

Chopra, G., Madan, P., Jaisingh, P., and Bhaskar, P. (2019). Effectiveness of e-learning portal from students' perspective: A structural equation model (SEM) approach. *Interac. Technol. Smart Edu.*, 16(2), 94–116.

Compeau, D. R. and Higgins, C. A. (1995). Computer self-efficacy: Development of a measure and initial test. *MIS Quart.*, 189–211.

DeLone, W. H. and McLean, E. R. (1992). Information systems success: The quest for the dependent variable. *Inform. Sys. Res.*, 3(1), 60–95.

Dominici, G. and Palumbo, F. (2013). How to build an e-learning product: Factors for student/customer satisfaction. *Busin Horizons*, 56(1), 87–96.

Ehlers, U. D., Pawlowski, J. M., Dondi, C., Moretti, M., and Nascimbeni, F. (2006). Quality of e-learning: Negotiating a strategy, implementing a policy. *Handbook Qual. Stand. e-learn.*, 31–50.

Ejdys, J. (2021). Factors affecting the adoption of e-learning at the university level. *WSEAS Trans. Busin. Econ.*, 18(3), 13–323.

Eom, S. B., Wen, H. J., and Ashill, N. (2006). The determinants of students' perceived learning outcomes and satisfaction in university online education: An empirical investigation. *Dec. Sci. J. Innov. Edu.*, 4(2), 215–235.

Gay, G. H. (2016). An assessment of online instructor e-learning readiness before, during, and after course delivery. *J. Comput. High. Edu.*, 28(2), 199–220.

Hariyanto, D., Triyono, M., and Köhler, T. (2020). Usability evaluation of personalized adaptive e-learning system using USE questionnaire. *Knowl. Manag. E-Learn.*, 12(1), 85–105.

Jung, I. (2011). The dimensions of e-learning quality: from the learner's perspective. *Edu. Technol. Res. Dev.*, 59, 445–464.

Liaw, S. S., Huang, H. M., and Chen, G. D. (2007). Surveying instructor and learner attitudes toward e-learning. *Comp. Edu.*, 49(4), 1066–1080.

Nugroho, M. A., Dewanti, P. W., and Novitasari, B. T. (2018). The impact of perceived usefulness and perceived ease of use on student's performance in mandatory e-learning use. *2018 Internat. Conf. Appl. Inform. Technol. Innov. (ICAITI)*, 26–30.

Ozkan, S. and Koseler, R. (2009). Multi-dimensional students' evaluation of e-learning systems in the higher education context: An empirical investigation. *Comp. Edu.*, 53(4), 1285–1296.

33 Students' achievement factor prediction in course management E-learning using a novel normalized discounted cumulative gain (NNDCG) technique over probabilistic occurrence model based on retention rate

Parthasarathy V.[a] *and Devi T.*[b]

Department of Computer Science and Engineering, Saveetha School of Engineering, Saveetha Institute of Medical and Technical Sciences, Saveetha University, Chennai-602105, Tamil Nadu, India

Abstract

Aim: To develop a predictive model using novel normalized discounted cumulative gain (NNDCG) over a probabilistic occurrence model to forecast students' achievement factors in e-learning, with a specific emphasis on improving outcomes through retention rate analysis. **Materials and methods:** The study involves predicting students' achievement factors for improved retention rates. The sample size (N=30) for novel NDCG and (N=30) probabilistic occurrence (PO) model, determined by a G power of 80%. **Results:** The retention rate for novel NDCG is 90.33%, significantly surpassing the PO model's retention rate of 51.00%. Statistical analysis reveals a significant difference between novel NDCG algorithm as well as PO model, with p-value 0.002 (independent sample T-test, p<0.05) its significant. **Conclusion:** The novel NDCG demonstrates a superior retention rate of 90.33% in predicting students' achievement factors compared to the PO model's retention rate of 51.00%.

Keywords: Environment, learning, machine learning, normalized discounted cumulative gain, probabilistic occurrence model, performance, retention rate

Introduction

Machine learning, a core aspect of artificial intelligence, allows computers to learn and make decisions autonomously, improving performance with advanced algorithms and large datasets (Dixson, 2015). It is used in outcome prediction, image recognition, and data analysis, creating adaptive intelligent systems (Singh and Meena, 2023). Challenges include model complexity, instability (Lewis, 2021), and the need for high-quality student data (Alshammari and Qtaish, 2019). However, machine learning enhances personalized learning, enabling educators to tailor strategies to individual needs, thereby improving student outcomes and engagement (Maaliw et al., 2020).

Over the past five years, there has been a growing interest in students' e-learning achievement (Martin and Ndoye, 2016). IEEE Xplore now hosts over 7,882 articles on this topic, reflecting this trend (Alyahyan and Düştegör, 2020). Science Direct has contributed 7,832 papers, and Springer has added 5,605 articles, highlighting the subject's importance in academic discussions (Fahd et al., 2021). A notable contribution is a groundbreaking framework by Xu et al. (2024), cited by 298, which aims to optimize e-learning strategies to enhance retention rates and create a more inclusive and effective learning environment (Baghdadi, 2011).

Existing algorithms for predicting student achievement have a retention rate of 51.00%, which is lower than the proposed NNDCG algorithm. NNDCG, designed for higher retention rates, significantly improves e-learning outcomes by prioritizing student involvement, personalized learning experiences, and platform flexibility (Polinsky, 2003; Calli et al., 2013). By optimizing e-learning strategies, NNDCG aims to enhance retention rates and create a more effective learning environment (Gray and DiLoreto, 2016).

Materials and methods

The study was conducted at the SIMATS AR and VR facility, equipped with high-configuration systems. A total of 30 groups were reviewed (Hani et al.). Computations used an 80% G-power value, a 0.05 significance level, a beta value of 0.8, and a 95% confidence interval (Azzi et al., 2019).

The "success rate data" dataset, a CSV file with 5,075 entries sourced from Kaggle, was used in this investigation. It focuses on student performance in both traditional and online learning environments, detailing facets like class attendance, test scores, attention spans, and related metrics. An algorithm

[a]192111029.sse@saveetha.com, [b]devit.sse@saveetha.com

DOI: 10.1201/9781003587538-33

was developed and implemented using this dataset to evaluate and predict student performance.

An online resource for data science and machine learning is Google Colab. It provides a web-based Python programming environment, akin to Jupyter Notebooks. Utilizing Google's cloud infrastructure, it delivers high-performance computing without local installations. Integration with Google Drive streamlines dataset access, and users can construct, create visualizations utilizing libraries like Matplotlib and Seaborn.

Normalized discounted cumulative gain (NDCG)
The normalized discounted cumulative gain (NDCG) method is designed to assess student achievement and retention in e-learning by evaluating normalized cumulative gains. It considers the weighted importance of factors influencing student success, offering a detailed and effective approach for enhancing retention rates. Table 7.1 shows the pseudocode for NDCG.

Probabilistic occurrence model (POM)
A PO model,used as sample preparation group 2, is a mathematical construct for predicting event probabilities within a system. It employs probability theory to manage uncertainities and data fluctuations, applicable in fields like statistics, machine learning and decision analysis.

Table 33.1 Pseudo-code for normalized discounted cumulative gain.

Input: Success rate dataset

Output: Retention rate

Step 1: Import libraries
The necessary libraries are imported – pandas for data manipulation, numpy for numerical operations, and matplotlib for plotting

Step 2: Read dataset
The script reads a dataset from a CSV file specified by "file_path" and stores it in a pandas DataFrame named "data"

Step 3: Define final grades
Extracts the "final_grade" column from the dataset and converts it to a list

Step 4: Ideal ranking
Creates an ideal ranking by sorting the final grades in descending order

Step 5: Define discounted cumulative gain (DCG) function
Defines a function to calculate the discounted cumulative gain (DCG) at a specified position "k" in the ranking

Step 6: Define normalized discounted cumulative gain (NDCG) function and calculate NDCG score
Defines a function to calculate the normalized discounted cumulative gain (NDCG) specified position "k" by comparing the actual ranking with the ideal ranking and calculate the NDCG score

Source: Author

Table 33.2 Pseudo-code for probabilistic occurrence model.

Input: Success rate dataset

Output: Retention rate

Step 1: Import necessary libraries
In this step, essential libraries are imported. pandas is used for data manipulation, NumPy for numerical operations, and logistic regression from sklearn.linear_model for logistic regression

Step 2: Read dataset
The script reads a dataset from a CSV file specified by "file_path" and stores it in a pandas DataFrame named "data".

Step 3: Extract the "final_grade" column for analysis
The "final_grade" column is extracted from the dataset and converted to a list for further analysis

Step 4: Select relevant columns for logistic regression (features and target)
Features (x) and the target variable (y) are selected for logistic regression. In this example, features include "time_spent", "assignments_completed", "quiz_avg_score", and "retention_rate"

Step 5: Fit a logistic regression model
A logistic regression model is instantiated and fitted using the selected features (x) and the target variable (y)

Step 6: Predict probabilities for each student's final grade using the logistic model
Probabilities of belonging to each class are predicted using the logistic model. Here, predicted_probs represents the probabilities of belonging to class 1

Source: Author

Statistical analysis

In IBM SPSS Version 27, statistical analyses such as standard error mean, mean, and deviation were performed. This study compares the PO model and NNDCG algorithms using data from multiple iterations, with independent sample T-tests conducted (Jin, 2023). Resources are treated as dependent variables, while input text data and session ID are independent variables, and T-test values are analyzed (Suryadevara, 2018).

Table 33.3 Raw data for the probabilistic occurrence model and normalized discounted cumulative gain, with a sample size of N=30.

S. No.	NDCG access time (ms)	PO access time (ms)
1	69	43
2	81	30
3	85	53
4	85	91
5	87	43
6	88	43
7	87	84
8	89	43
9	90	43
10	90	43
11	91	43
12	91	56
13	91	43
14	91	43
15	92	43
16	92	43
17	92	43
18	92	94
19	92	43
20	93	43
21	93	43
22	93	90
23	93	43
24	94	43
25	94	30
26	95	43
27	95	43
28	95	67
29	95	55
30	95	63

For every 30 iterations, the retention rate is computed. The normalized cumulative gain method shows a greater retention rate as compared to the probabilistic occurrence model.

Source: Author

Results

Novel normalized discounted cumulative gain (NDCG) model demonstrates a higher mean retention of 90.33% in forecasting students' achievement factors compared to the PO model of mean retention 51.00%.

Table 33.1 represents pseudo-code for the NNDCG. First, data reading, library imports, and "final_grade" value extraction are done. After that, the code calculates NDCG scores, which show how the initial ranking and the optimal ranking match to assess the system's effectiveness.

Table 33.2 represents the code that reads a dataset, extracts the "final_grade" column, and creates an ideal ranking. Logistic regression is applied to predict final grades based on features like "time_spent" and "assignments_completed". The resulting probabilities are plotted, evaluating the effectiveness of the PO model.

Table 33.3 contrasts the retention rates of several iterations of the PO model and the NDCG algorithm.

Table 33.4 displays group statistics results for comparing the statistical performance of independent samples using the NDCG and PO model approaches. PO displays a mean retention of 51.00% and a standard deviation of 17.223, but NDCG displays a mean retention of 90.33% and a standard deviation of 5.281. The standard error mean for NDCG (0.964) is compared with PO (3.144) using the T-test.

Table 33.5 depicts how sample T-test is applied to the sample collections, with a 95% confidence interval and a significance threshold of ($p=0.002$, $p<0.05$). following NDCG's application of SPSS computations.

Figure 33.1 represents a bar graph that contrasts the average NDCG retention. Mean retention is represented on the y-axis, while error bars for both NDCG approaches, together with their 95% confidence

Table 33.4 Compares the probabilistic occurrence model and the normalized discounted cumulative gain technique (NDCG) to offer statistical insights for independent samples.

	Algorithm	N	Mean	Std. deviation	Std. error mean
Time	NDCG	30	90.33	5.281	0.964
	PO	30	51.00	17.223	3.144

Source: Author

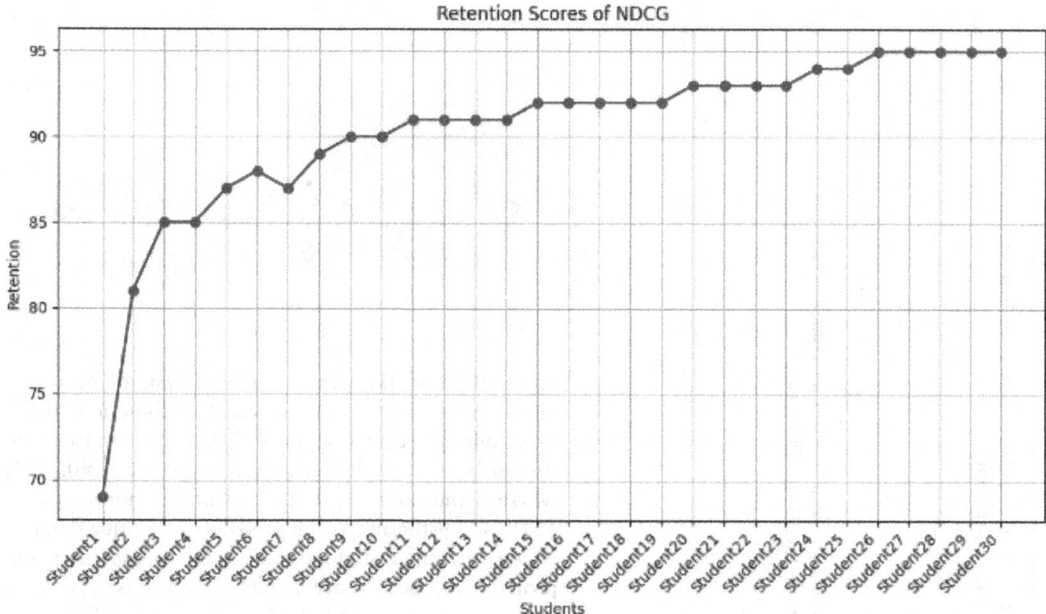

Figure 33.1 Line chart demonstrating the retention rates of normalized discounted cumulative gain for learners, with learners on the x-axis and the corresponding retention rates on the y-axis
Source: Author

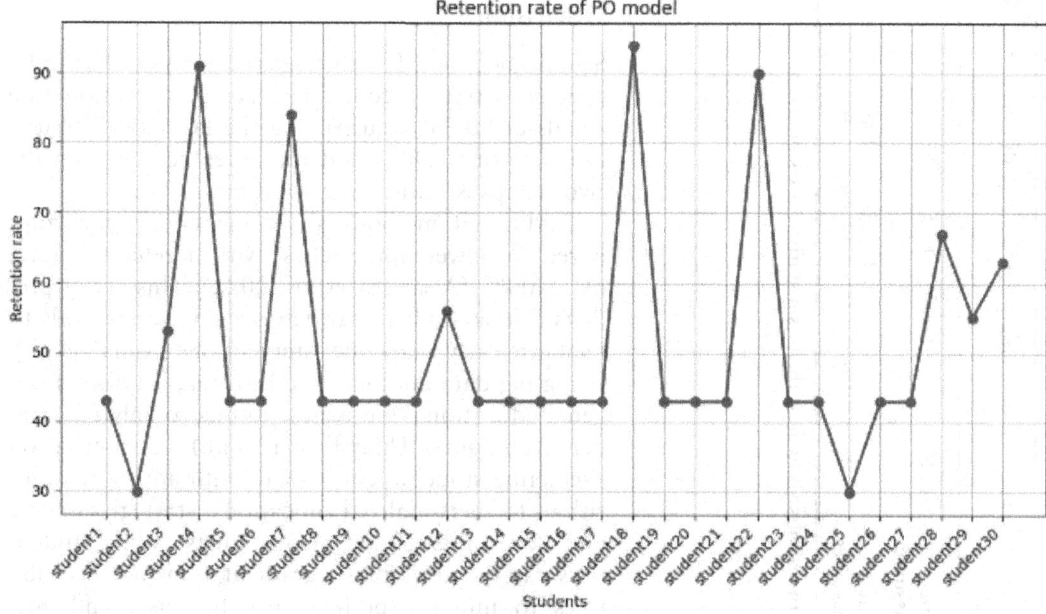

Figure 33.2 Line chart illustrating the retention rates of the probabilistic occurrence model, with learners depicted on the x-axis and the corresponding retention rates on the y-axis
Source: Author

intervals and ±2 standard deviations, are shown on the x-axis.

Figure 33.2 represents a bar graph that contrasts the average PO model retention. The x-axis shows error bars for PO model procedures together with their ±2 standard deviations and 95% confidence intervals, while the y-axis shows the mean retention.

Figure 33.3 demonstrates that the PO model algorithm yields 51.00% mean retention, whereas the novel NDCG approach yields a higher mean retention of 90.33%. Additionally, NDCG has a smaller standard deviation. x-axis: PO algorithm versus NDCG, y-axis: mean retention with error bar ±2 SD

Table 33.5 The normalized discounted cumulative gain technique (NDCG) versus the probabilistic occurrence model is evaluated at a 95% confidence interval using a T-test for statistically independent samples.

		Levene's test for equality of variances		T-test for equality of means						
									95% Confidence interval of the difference	
		F	Sig.	t	df	Sig. (2-tailed)	Mean difference	Std. error difference	Lower	Upper
Retention	Equal variances assumed Equal variances not assumes	19.269	0.000	11.959	58	0.002	39.333	3.289	33.750	45.917
	Equal variances assumed Equal variances not assumes			11.959	34.405	0.002	39.333	3.289	32.652	46.014

With a p-value of 0.002 ($p<0.05$), the results show that there is a statistically significant difference between the NDCG and PO model.

Source: Author

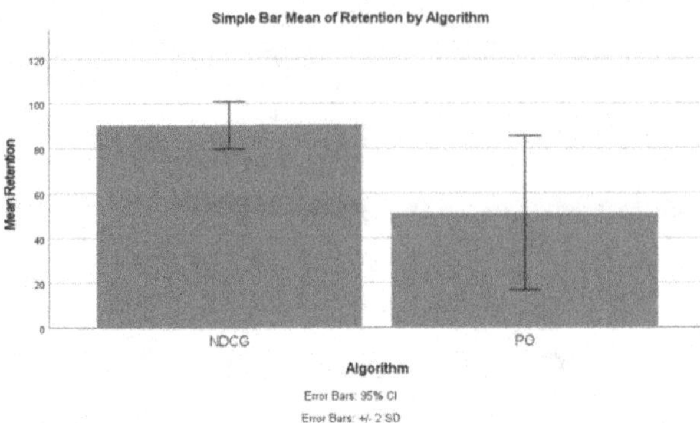

Figure 33.3 In the comparative analysis of mean retention rates between the NNDCG technique and the probabilistic occurrence model, it's evident that the former boasts a higher mean retention rate at 90.33%, whereas the latter exhibits a mean retention rate of 51.00%. Furthermore, the NNDCG technique demonstrates a smaller standard deviation compared to the probabilistic occurrence model. The y-axis represents mean retention, and the x-axis features the NDCG vs. PO algorithm, with error bars denoted by ±2 SD

Source: Author

Discussion

Significance is verified through the analysis of an independent T-test outcome. The NDCG retention rate stands at 90.33%, surpassing the PO model's retention rate of 51.00%, but the difference between the two groups is statistically significant.

Traditional methods show higher efficacy compared to newer approaches, with a retention level of 51.00% (Alsariera et al., 2022). This study predicts at-risk students at2 various course intervals to facilitate early academic interventions (Adnan et al.). Using big data and machine learning, it informs targeted educational strategies, aiming to enhance academic outcomes (Uday Kumar et al.). The focus is on predicting student achievement with advanced algorithms for personalized educational strategies (Yildiz and Börekci, 2020). It also predicts online student pass rates and employs gradient boosting decision trees to improve performance forecasts and academic insights (Aydog˘du, 2019; Wang et al., 2022). Course management e-learning evaluation's predictive power is governed by several factors, including the training dataset quality and diversity. The possible contextual uniqueness of the results and the requirement to adjust for upcoming changes in the educational landscape are limitations. Future work will focus on enhancing models using real-time data, resolving adaptation issues, and investigating creative approaches to raise retention rates even higher in e-learning settings.

Conclusions

The NDCG technique emerges as a better solution for prediction of students' achievement factor with a retention rate of 90.33% Outperforming traditional methods, notably those with a 51.00% retention rate.

References

Aljohani, N. R., Fayoumi, A., and Saeed-Ul, H. (2019). Predicting at-risk students using clickstream data in the virtual learning environment. *Sustain. Sci. Prac. Policy*, 11(24), 7238.

Deepa, N., Udayakumar, N., Devi, T., Management of Traffic in Smart Cities Using Optical Character Recognition for Notifying Users, 2022 International Conference on Computer Communication and Informatics, ICCCI 2022, 2022. 20(2), 110–120

Devi, T., Jaisharma, K., Deepa, N., Novel Trio-Neural Network towards Detecting Fake News on Social Media, ASSIC 2022 - Proceedings: International Conference on Advancements in Smart, Secure and Intelligent Computing, 2022 10(2) 10–20

Anderson, J., Bushey, H., Devlin, M., and Gould, A. J. (2021). Promoting student engagement with data-driven practices. *Internat. Perspect. Support. Engag. Online Learn.*, 39, 87–103.

Aydoğdu, Ş. (2019). Predicting student final performance using artificial neural networks in online learning environments. *Edu. Inform. Technol.*, 25(3), 1913–1927.

Azzi, I., Jeghal, A., Radouane, A., Yahyaouy, A., and Tairi, H. (2019). A robust classification to predict learning styles in adaptive e-learning systems. *Edu. Inform. Technol.*, 25(1), 437–448.

Calli, L., Balcikanli, C., Calli, F., Cebeci, H. I., and Seymen, O. F. (2013). Identifying factors that contribute to the satisfaction of students in e-learning. *Turk. Online J. Dis. Edu.*, 14(1), 85–101.

Castejon, J. L., Núñez, J. C., Gilar-Corbi, R., and Abellán, I. M. J. (2021). New challenges in the research of academic achievement: Measures, methods, and results. Frontiers Media SA. 20(3) 129–140

Coman, C., Ţîru, L. G., Meseşan-Schmitz, L., Stanciu, C., and Bularca, M. C. (2020). Online teaching and learning in higher education during the Coronavirus pandemic: students' perspective. *Sustain. Sci. Prac. Policy*, 12(24), 10367.

Gray, J. A. and DiLoreto, M. (2016). The effects of student engagement, student satisfaction, and perceived learning in online learning environments. *Internat. J. Edu. Leadership Prep.*, 11(1). 89–99 http://files.eric.ed.gov/fulltext/EJ1103654.pdf.

Hussain, S., Gaftandzhieva, S., Maniruzzaman, Md., Doneva, R., and Muhsin, Z. F. (2020). Regression analysis of student academic performance using deep learning. *Edu. Inform. Technol.*, 26(1), 783–798.

Hussain, S. and Khan, M. Q. (2021). Student-performulator: Predicting students' academic performance at secondary and intermediate level using machine learning. *Ann. Data Sci.*, 10(3), 637–655.

Kant, N., Prasad, K. D., and Anjali, K. (2021). Selecting an appropriate learning management system in open and distance learning: A strategic approach. *Asian Assoc. Open Univer. J.*, 16(1), 79–97.

Kotsiantis, S., Pierrakeas, C., and Pintelas, P. (2004). Predicting students' performance in distance learning using machine learning techniques. *Appl. Artif. Intel. AAI.* https://doi.org/10.1080/08839510490442058. 10(4) 238–372

Lee, S. J., Lee, H., and Kim, T. T. (2018). A study on the instructor role in dealing with mixed contents: How it affects learner satisfaction and retention in e-learning. *Sustain. Sci. Prac. Policy*, 10(3), 850.

Liu, D. Y.-T., Bartimote-Aufflick, K., Pardo, A., and Bridgeman, A. J. (2017). Data-driven personalization of student learning support in higher education. *Learn. Anal. Fundam. Appl. Trends*, 143–169.

Li, X., Zhang, Y., Cheng, H., Li, M., and Yin, B. (2022). Student achievement prediction using deep neural network from multi-source campus data. *Comp. Intel. Sys.*, 8(6), 5143–5156.

Lykourentzou, I., Giannoukos, I., Mpardis, G., Nikolopoulos, V., and Loumos, V. (2009). Early and dynamic student achievement prediction in e-learning courses using neural networks. *J. Am. Soc. Inform. Sci. Technol.*, 60(2), 372–380.

Pan, D., Wang, S., Jin, C., Yu, H., Hu, C., and Wang, C. (2021). Research on student achievement prediction based on BP neural network method. *Adv. Artif. Sys. Med. Edu. IV*, 293–302.

Rajabalee, Y. B. and Santally, M. I. (2020). Learner satisfaction, engagement and performances in an online module: Implications for institutional e-learning policy. *Edu. Inform. Technol.*, 26(3), 2623–2656.

Reschly, A. L. (2020). Dropout prevention and student engagement. *Student Engag.*, 31–54.

Rozić, R., Ljubić, H., Grujić, T., and Skelin, A. K. (n.d.). Detection of at-risk students in virtual learning environment. https://ieeexplore.ieee.org/abstract/document/10271591/. 25(7) 637–653

Solé-Beteta, X., Navarro, J., Gajšek, B., Guadagni, A., and Zaballos, A. (2022). A data-driven approach to quantify and measure students' engagement in synchronous virtual learning environments. *Sensors*, 22(9), 3294.

Teng, Y., Zhang, J., and Sun, T. (2023). Data-driven decision-making model based on artificial intelligence in higher education system of colleges and Universities. *Expert Sys.*, 40(4), e12820.

Willging, P. A. and Johnson, S. D. (2009). Factors that influence students' decision to drop out of online courses. *J. Async. Learn. Netw.*, 13(3), 115–127.

34 Elevating engineering mechanics course through problem and project-based learning: Insights and experiences

C. Sravanthi[a], *Pratibha Dharmavarapu*[b] *and Neeraja Akula*[c]

Department of Mechanical Engineering, Anurag University, Hyderabad, Telangana, India

Abstract

Engineering mechanics is the course which generally students feel less confident as it is mostly concept oriented and problem solving. Unlike the conventional way of teaching, this course delivery requires much involvement of students which can be achieved by problem-based learning which is student centered and project-based learning which is innovative approach. Both methods have been implemented efficiently to teach the course engineering mechanics. It is observed that without problem and project-based learning (P2BL), the end semester exam result was 58.62% pass percentage and with P2BL, the end semester exam result has been increased to 83.33% pass percentage. In this report, evidences have been submitted about process of implementation, teaching methodology, target audience and result analysis. The integration of P2BL within lecture hours has yielded promising results. Notably, student engagement has been markedly enhanced, with these active learning methods catalyzing deeper interaction with the subject matter. This shift has not only kindled a more profound understanding of the concepts but has also led to a commendable upswing in pass percentages.

Keywords: Problem-based learning, project-based learning, engineering mechanics, increase in pass percentage, result analysis

Introduction

Engineers must embody a holistic approach to problem-solving, considering the broader impact of their work on society and the environment as said by Nasr and Jonnasen(Nasr, 2014; Jonassen et al., 2006).In that sense, the challenges imposed by society have to be kept in mind while designing the engineering curriculum (Dym et al., 2005; Litzinger et al., 2011).A teaching–learning methodology with practical feeling has to be implementedto enhance students' motivation and reduce dropout rates (Siddiqueand Hardre, 2015). The initial exposure of freshmen to the field of mechanical engineering is pivotal. Since, according to Vesikivi et al.(2020),Wu et al. (2020), and Dym et al. (2005)initial approach not only boosts interest of student in engineering but also improves retention in engineering programs.The proposed course, engineering mechanics of first year course is traditionally taught through lectures, has been identified as a challenging course for students, often leaving them feeling less confident in their understanding. Recognizing this, a shift towards problem-based learning (PBL) was proposed to enhance student engagement and comprehension. The curriculum, approved by the board of studies, is structured into five units, with two assignments and closed-book mid-term exams covering portions of the syllabus. Upon completion, students undergo a semester end examination (SEE). Through PBL, students actively engage in solving engineering problems, fostering deeper learning and increased self-involvement. This transition aims to address the complexities of the subject matter and promote a more dynamic and participatory learning environment, ultimately improving student outcomes and confidence levels (Table 34.2).

Implementation of problem-based learning and project-based learning

Process of implementation of problem-based learning

In a classroom comprising 48 students, a collaborative approach was adopted to promote active learning. Vu et al. (2007) has mentioned that this approach is beneficial both student learning and their professional development. The students were grouped into 16 teams, each comprising three students. Each group was tasked with addressing a real-time problem, fostering practical application of theoretical concepts. To showcase their solutions, the teams creatively presented their methodologies via poster presentations. These presentations were evaluated holistically, encompassing the efficacy of the problem-solving approach, the quality of the presentation itself, and the degree of each student's engagement within the team. Recognizing the value of teamwork and critical thinking, a total of 10 marks were allocated to Assignment-I, thus contributing to the students'academic progress and also this method honed their communication and collaboration skills, aligning with a multifaceted approach to modern education. The evaluation process adheres to

[a]sravanthimech@anurag.edu.in, [b]pratibhamech@anurag.edu.in, [c]neerajamech@anurag.edu.in

DOI: 10.1201/9781003587538-34

a structured rubric, ensuring a comprehensive and fair assessment (Tables 34.2 and 34.3).

Process of implementation of project-based learning
Teams were assigned with the task of selecting a topic from the course curriculum and leveraging it to construct a project or prototype. Collaboratively, students worked in teams to bring these projects to fruition. In a classroom setting, these teams showcased their projects and prototypes, elucidating their chosen concepts to their peers. Evaluation of their efforts culminated in the allocation of 10 marks for assignmentII, an integral part of their academic journey. The assessment hinged on a comprehensive rubric encompassing the appropriatenessof the project or prototype selection, the finesse of their presentation skills, and the effectiveness of their teamwork. This approach not only facilitated practical application of classroom learning but also cultivated skills in project development, public speaking, and cooperative collaboration – all pivotal in nurturing well-rounded learners prepared for real-world challenges (Tables 34.4 and 34.5, Figures 34.1 and 34.2).

Challenges faced and lessons learned

Implementing P2BL in classrooms encountered several challenges. Students faced issues such as a lack of teamwork, disparities in workload distribution, communication gaps within teams, insufficient financial resources for projects, and inadequate facilities. Some students also expressed concerns about the grading system, feeling demotivated without generous recognition from teachers. To address these challenges, teachers played a pivotal role in fostering student-centric learning environments. They transitioned into adept facilitators, guiding students through P2BL initiatives that bridge theoretical knowledge with real-world applications. By observing students'problem-solving approaches, teachers gained insights into their diverse perspectives, encouraging collaboration and idea

Table 34.1 Syllabus.

SYLLABUS **Course Title: ENGINEERING MECHANICS**
UNIT - I: *Introduction to Engineering Mechanics:* Basic concepts. **System of Forces:** Coplanar, Concurrent Forces, Components in Space – Resultant- Moment of Forces and its Application; Couples and Resultant of Force System. **Equilibrium of System of Forces:** Free body diagrams, Equations of Equilibrium of Coplanar Systems, Lame's Theorem. **UNIT – II:** **Friction***:* Basic concepts, Types of Friction, Laws of Friction, Static and Dynamic Friction, Motion of Bodies, Wedge friction, ladder Friction, screw friction, applications. **UNIT - III:** **Centroid:** Centroids of simple figures (from basic principles) Centroids of Composite Figures. **Centre of Gravity:** CG of simple bodies (from basic principles), CG of composite bodies, Pappus theorem. **UNIT- IV:** **Area Moment of Inertia:** Definition - Polar Moment of Inertia, Transfer Theorem, MI of Composite Figures, Product of Inertia, Transfer Formula for Product of Inertia. **Mass Moment of Inertia:** MI of Masses, Transfer Formula for MMI, MMI of composite bodies. **UNIT - V:** **Work – Energy Method:** Equations for Translation, Work-Energy Applications to Particle Motion. Connected System-Fixed Axis Rotation and Plane Motion. Impulse momentum method.

Source: Author

Table 34.2 Course structure.

Syllabus	Assignments	Analysis of assignment	Mid exams	Analysis of mid	SEE	Total
Unit I Unit II Unit III	AssignmentI Problem-based learning (10M)	Sum of two assignments (20M)	Mid-I (closed book exam) (20M)	Average of two mids (20M)	Semester end examination (closed book exam) (60M)	Assignment + Mid+ SEE (20M+20M+60M)
Unit IV Unit V	Assignment-II Project-Based Learning (10M)		Mid-II (Closed Book Exam) (20M)			

Source: Author

Table 34.3 Time line for problem-based learning.

Week 1	Week 2	Week 3	Week 4	Week 5	Week 6	Week 7
13th Sept 2022to18th Sept 2022	20th Sept 2022to25th Sept 2022	27th Sept 2022to2nd Oct 2022	4th Oct 2022to9th Oct 2022	11th Oct 2022to16th Oct 2022	18th Oct 2022to23rd Oct 2022	25th Oct 2022to30th Oct 2022
Problem selection	Research and problem analysis	Problem statement and planning	Solution exploration	Solution finalization	Poster design and practice	Poster presentation and assessment

Source: Author

Table 34.4 Rubrics for assessment of problem-based learning.

Criteria	Excellent (10)	Good (8)	Satisfactory (6)	Needs improvement (4)	Inadequate (2)
Problem analysis	Clearly identifies and thoroughly analyzes the problem	Identifies and analyzes the problem effectively	Adequately identifies and analyzes the problem	Superficially identifies the problem	Fails to identify the problem
Solution approach	Presents a comprehensive and innovative solution approach	Presents a solid solution approach	Presents a reasonable solution approach	Presents a vague solution approach	Lacks a clear solution approach
Presentation skills	Engaging and well-structured presentation with excellent visual aids	Clear presentation with effective visuals	Adequate presentation with some visuals	Disorganized presentation with minimal visual	Incoherent presentation with no visuals
Team collaboration	Seamless teamwork with clear contributions from all members	Effective teamwork with evident contributions	Adequate teamwork with some contributions	Limited teamwork and contributions	Poor teamwork and no clear contributions

Source: Author

Table 34.5 Time line for project-based learning.

Week 1	Week 2	Week 3	Week 4	Week 5	Week 6	Week 7
09th Oct 2022 to14th Oct 2022	16th Oct 2022 to 21st Oct 2022	23rd Oct 2022 to 28th Oct 2022	6th Nov 2022 to 11th Nov 2022	13th Nov 2022 to 18th Nov 2022	20th Nov 2022 to 25th Nov 2022	4th Dec 2022 to 9th Dec 2022
Project selection	Research and problem analysis	Project planning	Model design	Model construction and testing	Model refinement	Poster presentation and assessment

Source: Author

sharing. This mentorship dynamic not only enhanced learning outcomes but also strengthened the teacher–student relationship, creating a supportive and enriching educational experience.

Performance of students in mid and semester end examinations

In this educational approach, students embark on a structured learning journey combining diverse assessment methods and active learning strategies. Alongside assignments and mid-term examinations, P2BL activities play a pivotal role. These hands-on experiences empower students to apply theoretical knowledge to real-world scenarios, thereby deepening their comprehension. By bridging theory with practice, students develop practical skills and deepen their comprehension of the subject matter. This approach reflects positively in their performance in mid-term and semester-end examinations, showcasing improved academic outcomes and readiness for real-world challenges. Integrating active learning strategies into

Figure 34.1 Students presenting models by implementing the concepts learnt
Source: Author

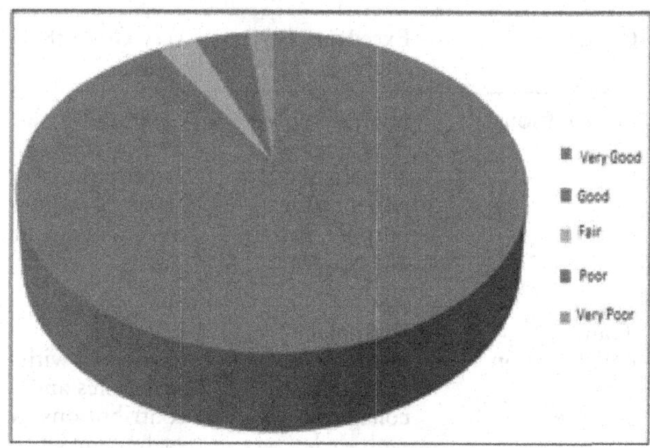

Figure 34.2 Students response after implementing PBL
Source: Author

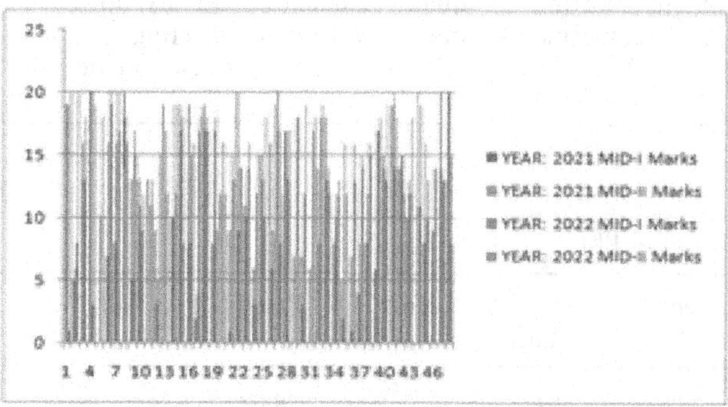

Figure 34.3 Performance of students in midexaminations
Source: Author

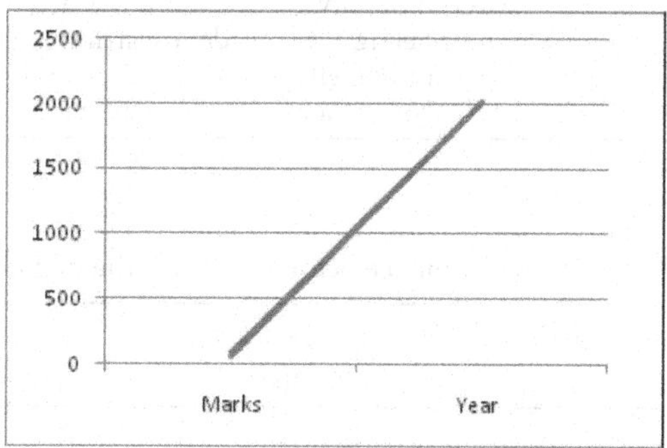

Figure 34.4 Graph representing student performance after implementation of PBL
Source: Author

Table 34.6 Rubrics for assessment of project-based learning.

Criteria	Excellent (10)	Very Good (8)	Good (6)	Fair (4)	Needs improvement (2)
Problem identification	Clearly identifies a significant problem or need for the model/ prototype	Identifies a problem or need for the model/ prototype, with relevance to the field	Identifies a problem or need, may not be entirely clear	Problem identification lacks clarity and relevance	Problem identification is absent or inappropriate
Design andinnovation	Demonstrates exceptional creativity and innovation in the design and concept of the model/prototype	Shows creative design and concept, with a strong innovative element	Presents a reasonable design and concept with some innovative features	Design and concept lack significant creativity or innovation	Design and concept are generic or not well thought out
Functionality	Model/prototype functions exceptionally well, meeting all intended requirements	Model/prototype functions well and meets most intended requirements	Model/prototype functions adequately, but some requirements may not be fully met	Model/prototype has limited functionality and does not meet several requirements	Model/prototype is non-functional

Criteria	Excellent (10)	Very Good (8)	Good (6)	Fair (4)	Needs improvement (2)
Presentation	Exceptionally clear, engaging, and organized presentation purpose, design, and results	Clear and well-organized presentation the project's purpose, design, and results	Adequate presentation that conveys the project's purpose, design, and results	Presentation lacks organization or clarity, making it difficult to understand the project	Presentation is unclear, disorganized, or missing key information
Team collaboration	Exemplary teamwork evident in seamless collaboration, with each member's contribution evident	Effective teamwork with clear roles and contributions from each member	Adequate teamwork, but some members' contributions are less evident	Limited teamwork, with unclear roles and uneven contributions	No clear evidence of teamwork or collaboration
Overall quality	Exceptional overall quality, showcasing outstanding effort and achievement	Very good overall quality, demonstrating a solid effort and achievement	Good overall quality, meeting basic requirements	Fair overall quality, lacking in several aspects	Poor overall quality, not meeting requirements

Source: Author

Table 34.7 Student performance before and after the implementation of PBL.

Engineering mechanics course	No. of students appeared	No. of students passed	No. of students failed	Pass (%)	Fail (%)
Without PBL(academic year: 2021–22)	29	17	12	58.26	41.38
With PBL (academic year: 2022–23)	48	40	8	88.33	16.67

Source: Author

traditional assessments enhances students' overall educational experience, fostering meaningful learning and skill development (Figure 34.3).

Results and Discussion

The remarkable increase in students'cumulative grade point average (CGPA) from 58.26% to 88.33% following the adoption of P2BL signifies the transformative impact of these methodologies. It underscores their effectiveness in enhancing student learning outcomes and academic performance in the field of engineering mechanics (Table 34.6, Figure 34.4).

Conclusions

Problem- and project-based learning (P2BL) has emerged as a powerful tool for enhancing education on multiple fronts. Firstly, it significantly impacts academic progress and grades by fostering active learning through hands-on projects, leading to a deeper understanding of subjects and improved performance in coursework. Additionally, P2BL cultivates essential career readiness such as critical thinking, teamwork, and problem-solving, preparing students for future professional endeavors. Moreover, the collaborative and peer-learning aspects of P2BL promote knowledge exchange and skill reinforcement among students, mirroring real-world teamwork dynamics. Furthermore, the self-directed nature of P2BL instills responsibility and autonomy in learners, contributing to lifelong learning habits. As evidenced by its positive effects on CGPA, P2BL facilitates the application of theoretical concepts to practical scenarios, enhancing the relevance and meaningfulness of education. Through presentations and real-world application, students also hone valuable communication and presentation skills essential for their future careers. Importantly, P2BL fosters long-term retention of knowledge by engaging students in active learning experiences. In essence, P2BL offers a comprehensiveeducational approach that not only enhances academic outcomes but also fosters personal and professional development, equipping students with the skills and knowledge essential for success in their future pursuit.

References

Nasr, J. K. (2014). Towards aconverged and global set of competencies for graduates of engineering programs in aglobalization-governed world.*J. Ideas*, 18,15–32.

Jonassen, D., Strobel, D., and Lee, C. (2006).Everyday problem solving in engineering: lessons for engineering educators. *J. Eng.Edu.*, 95,139–151.

Dym, C., Agogino, A., Eris. O., et al. (2005). Engineering design thinking, teaching, and learning. *J. Eng.Edu.*, 94,103–120.

Litzinger, T. A., Lattuca, L. R., Hadgraft, R. G., et al. (2011). Engineering education and the development of expertise. *J. Eng.Edu.*, 100,123–150.

Siddique, Z.and Hardre, P. L. (2015).Effects of a mechanical engineering design course on students' motivational features. *Int. J. Mech. Eng. Educ.*,43,44–74.

Vesikivi, P., Lakkala, M., Holvikivi, J., et al. (2020).The impact of project-based learning curriculum on first-year retention, study experiences, and knowledge work competence. *Res Pap Edu.*, 35,64–81.

Wu, L., Warschauer, M., Fischer, C., et al. (2020).Increasing success in college: Examining the impact of a project-based introductory engineering course. *J. Eng. Edu.*,109,1–18.

Vu, T. T., Alba, G. D., Thuy, T., et al. (2007).Students experience of peer assessment in a professional course. *Assess.Eval. High.Edu.*, 32,541–556.

35 Advancing computational epistemology for enhanced credibility assessment in online social platforms using Naive Bayes classifier in comparison with recurrent neural networks

Murtaza Moiz P.[1,a] *and Dinokumar Kongkham*[2,b]

[1]Research Scholar, Department of Computer Science Engineering, Saveetha School of Engineering, Saveetha Institute of Medical and Technical Sciences, Saveetha University, Chennai-602105, Tamil Nadu, India

[2]Assistant Professor (SG), Department of Electronics and Communication Engineering, Saveetha School of Engineering, Saveetha Institute of Medical and Technical Sciences, Saveetha University, Chennai-602105, Tamil Nadu, India

Abstract

The aim of this study is to conduct a comprehensive comparative analysis of computational methods using machine learning algorithms, specifically the Naive Bayes (NB) classifier and recurrent neural networks (RNNs), for enhancing credibility assessment in online social platforms. Utilizing datasets sourced from online repositories such as Github, and DataWorld, the study focuses on evaluating the effectiveness of NB and RNNs in accurately detecting anomalies associated with misinformation and fake news. The datasets are merged together to ensure numerous factors come together into determining the label of a relevant information content to be of true nature. In the context of computational epistemology, the data taken is as diverse as it could be to let even the slightest of influence get recognized for the honesty label. The research methodology involves training and testing the algorithms with the dataset, and measuring their accuracy in identifying deceptive content. Results from the study reveal that while both NB and RNNs exhibit promising capabilities in credibility assessment; NB demonstrates superior accuracy in identifying deceptive content, achieving a mean accuracy rate of 92.96%, compared to RNNs's mean accuracy of 70.024%. This performance disparity underscores the efficacy of NB in discerning subtle patterns indicative of misinformation and deceit within online social platforms.

Keywords: Computational epistemology, credibility assessment, naive bayes classifier, recurrent neural networks, misinformation, online social platforms

Introduction

In the rapidly changing real time of online social interaction, maintaining credibility and trustworthiness has become an increasingly pressing issue. With the exponential growth of user-generated content, distinguishing between authentic information and deceptive content has become ever more challenging (Adler and Alfaro, 2017). This challenge is compounded by the proliferation of misinformation, fake news, and malicious propaganda that saturate digital spaces, posing significant risks to individuals, organizations, and society at large (Collins et al., 2020).

To address these concerns, computational epistemology offers a promising avenue for enhancing credibility assessment in online social platforms (Alexander and Tate, 2011). By leveraging advanced machine learning techniques, such as the Naive Bayes classifier and recurrent neural networks (RNNs), this study aims to develop automated systems capable of discerning credible information from deceptive content (Couronne et al., 2018)

This comparative analysis is carried out to evaluate the effectiveness of these algorithms in assessing the credibility of information shared on online social platforms. By training and testing both NB and RNNs on datasets comprising diverse forms of user-generated content, including text posts, and images, this study aims to measure their performance in distinguishing between credible and deceptive information (Adkinson et al., 2002).

Metrics such as accuracy, precision, and recall will be utilized to assess the performance of each algorithm and identify their respective strengths and limitations (Ahmad et al., 2011). While Naive Bayes may offer simplicity and computational efficiency, RNNs may provide better performance in handling complex data patterns and non-linear relationships (Elhadad et al., 2019)

[a]192011159.sse@saveetha.com, [b]dinokumarkongham.sse@saveetha.com

DOI: 10.1201/9781003587538-35

Materials and methods

The Soft Computing Lab at the Saveetha School of Engineering was where the study was carried out. To calculate the sample size and compare the two algorithms, MATLAB software is used. To compare the methodologies, two groups were chosen, and results were reached from them. For this investigation, a total of 372 sample iterations were selected from a population of 11,316 values. The Naive Bayes and RNNs were implemented with the use of MATLAB. The sample size was calculated using sample size calculator on calculator.net software and was found to be 372 for each group (with parameters set to – confidence level: 95%, margin of error: 5%, population portion: 50%, population size: 11,316).

Naive Bayes classifier

In terms of detecting illnesses that affect maize leaves. This algorithm can be used to detect the leaf affected by the disease and gives the accuracy which can be used to compare the same with the other algorithm to get higher accuracy and efficiency (De Caigny et al., 2018). Naive Bayes is a simple but powerful classification technique that assumes the features that are independent of each other. This may not always be true in reality, but Naive Bayes can still work well in many situations, especially when dealing with data that has many dimensions, such as images (Collins et al., 2020).

Process in MATLAB

Step 1: Load and preprocess the dataset.
Step 2: Train the dataset using the Naive Bayes classifier.
Step 3: Make predictions.
Step 4: Evaluate the classifier.
Step 5: Visualization and fetching accuracy.

Recurrent neural networks

Recurrent neural networks (RNNs) serve as indispensable tools for evaluating online source credibility. Their strength lies in processing sequential data, making them ideal for analyzing text-based content prevalent in online sources. Unlike traditional machine learning models, RNNs recursively pass information through a network of nodes, enabling them to capture temporal dependencies and patterns within the data (Dev and Eden, 2019).

The effectiveness of RNN is shown by the line graph comparing the original and predicted values. From the confusion matrix, the accuracy levels can be predicted and analyzed based on true positives (TP), true negatives (TN), false positives (FP), and false negatives (FN).

Accuracy = Sum (Test Values + Predicted Values) / Number of Test Values

Process in MATLAB

Step 1: Dataset loading and pre-processing.
Step 2: Training the RNNs with the prepared data.
Step 3: Generating predictions based on the trained model.
Step 4: Assessing the performance of the algorithm.
Step 5: Visualizing results and determining accuracy.

Statistical analysis

SPSS software facilitates the analysis of the Naive Bayes classifier and RNNs. The correlation table is generated through SPSS's multivariate correlation tool, while the independent sample T-test compares mean accuracies. Data import, correlation, cleaning, transformation, analysis, and group comparison are integral steps in this statistical analysis process.

Results

The Naive Bayes classifier has a greater accuracy of 92.96% when compared to the RNNs' accuracy of 70.024%, according to the results of the expected accuracy values for the two techniques. These findings and the accuracy gain figures were used for statistical comparison. With a sample size of 372, Naive Bayes classifier and RNNs were tested separately using MATLAB software. The mean accuracy scores for the Naive Bayes classifiers and the RNNs are shown in Table 35.1. With a standard deviation of 2.13439 and 2.49988 respectively, it was discovered that the NB classifier's mean accuracy was higher than that of the RNNs. Figure 35.1 displays the confusion matrix

Table 35.1 Group statistics measured in SPSS software with procured dataset values.

	Algorithms	N	Mean	Std. deviation	Std. error mean
Accuracies	Naïve Bayes classifier	25	92.9600	2.13439	0.42688
	Recurrent neural networks	25	70.0240	2.49988	0.49998

Source: Author

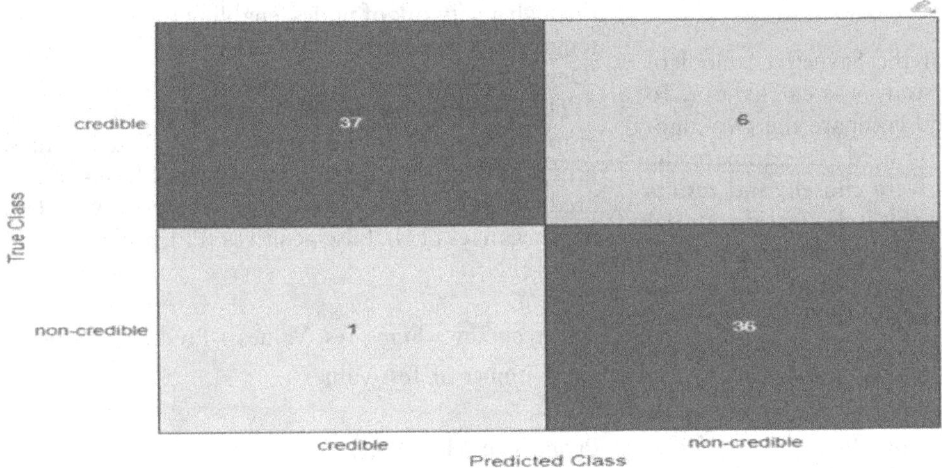

Figure 35.1 Confusion matrix for Naive Bayes classifier for calculating accuracy
Source: Author

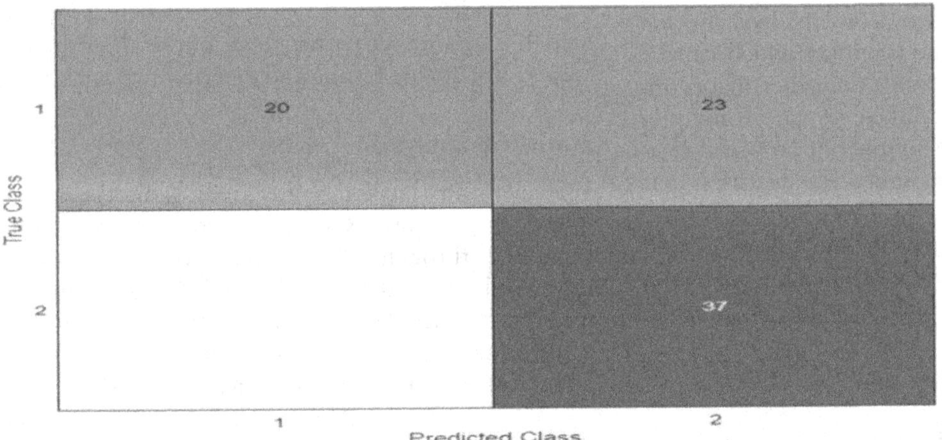

Figure 35.2 Confusion matrix for recurrent neural networks for calculating accuracy
Source: Author

showcasing true/false positives and negatives of Naive Bayes classifier while Figure 35.2 displays the same for RNNs.

Conclusions

This study is a comparison of the performance of Naive Bayes classifiers and RNNs in online credibility detection. The results show that the NB classifier with a significance value of 0.04 (two-tailed, $p<0.05$) has a higher accuracy of 92.96% compared to the RNNs' accuracy of 70.024%. Therefore, the Naive Bayes classifier algorithm is found to be more effective than the RNNs, based on the analysis.

References

Adkinson, W. F., Eisenach, J. A., and Lenard, T. M. (2002) . Privacy online: A report on the information practices and policies of commercial Websites. Washington, DC: Progress and Freedom Foundation. 1–5 Retrieved from http://www.pff.org/issues-pubs/books/020301privacyonlinereport.pdf. (Archived by WebCite® at http://www.webcitation.org/6W6ihPDup).

Adler, B. T. and Alfaro, L. de (2017). A content-driven reputation system for the Wikipedia. *Proc. 16th Internat. Conf. World Wide Web*, 261–270. Retrieved from http://www2007.org/papers/paper692.pdf (Archived by WebCite® at http://www.webcitation.org/6X06KRSeU)

Ahmad, R., Wang, J., Hercegfi, K., and Komlodi, A. (2011). Proceedings of the Symposium on Human Interface 2011. Part of HCI International 2011, Orlando, FL, USA, July 9-14, 2011, 521–527. Berlin: Springer.

Alexander, J. E. and Tate, M. A. (2011). Web wisdom: How to evaluate and create Web page quality on the Web. Hillsdale, NJ: L. Erlbaum Associates Inc. *Lec. Notes Comp. Sci.*, 6771.

Allemang, D. and Hendler, J. (2011). Semantic web for the working ontologist: Effective modeling in RDFS and OWL (2nd ed.). San Francisco, CA: Morgan Kaufmann. 1-7, 15–20.

Amsbary, J. H. and Powell, L. (2003). Factors influencing evaluations of Website information. *Psychol. Reports*, 93(1), 191–198.

Collins, B., Hoang, D.T., Nguyen, N. T., and Hwang, D. (2020). Fake news types and detectionmodels on social media a state-of-the-art survey. *Asian Conf. Intell. Inform. Database Sys.*, 562–573. Springer, Singapore.

Couronné, R., Probst, P., and Boulesteix, A.-L. (2018). Random forest versus logistic regression: A large-scale benchmark experiment. *BMC Bioinformat.*, 19(1), 1–14.

De Caigny, A., Coussement, K., and De Bock, K. W. (2018). A new hybrid classification algorithm for customer churns prediction based on logistic regression and decision trees. *Eur. J. Oper. Res.*, 269(2), 760–772.

Dev, V. A. and Eden, M. R. (2019). Formation lithology classification using scalable gradient boosted decision trees. *Comp. Chem. Engg.*, 128, 392–404.

Elhadad, M. K., Li, K. F., and Gebali, F. (2019). Fake news detection on social media: A systematic survey. *2019 IEEE Pacific Rim Conf. Comm. Comp. Signal Proc. (PACRIM)*, 1–8.

36 Analysis of learner's experience in database management system: Module-based approach

E. Sujatha[1,a], *M. Selvi*[2,b] *and D. Naveen Raju*[3,c]

[1]Department of Computer Science and Engineering, Saveetha Engineering College (Autonomous), Chennai, Tamilnadu, India

[2]Department of Electronics and Communication Engineering, Saveetha Engineering College (Autonomous), Chennai, Tamilnadu, India

[3]Department of Computer Science and Engineering, R.M.K. Engineering College, Kavaraipettai, Chennai, Tamilnadu, India

Abstract

Blended learning combines traditional classroom instruction with online educational resources and activities to enhance learning experiences and outcomes. This paper presents the analysis of the learner's reflections on the innovative e-learning system for a database management course, structured into modules to facilitate comprehensive learning. The course leverages an e-learning platform where students engage with modular practice sessions and quizzes on Moodle, followed by secure exams using the safe exam browser (SEB). A minimum score of 70% in SEB exams is required to progress to subsequent modules, with opportunities for re-attempts until this threshold is met. The module-based approach encourages independent problem-solving and is supported by diverse e-learning materials, including H5P content, quizzes, videos, PDFs, word documents, and HTML files, tailored to each unit's needs. This method not only enhances students' skills and performance in continuous internal assessments (CIAs) and skill assessments (SAs) but also positively impacts end-semester theory and practical exams. The approach was validated with 57 learners, showing significant improvements in performance metrics across SA1, SA2, CIA 1, CIA 2, and end-semester exams, thereby confirming the hypothesis that this method enhances learners' programming skills.

Keywords: Database management system, module-based learning, learning experience, pedagogy, Moodle learning

Introduction

Traditionally, database management system (DBMS) course was taught using chalk and board, additionally materials were provided and guest lectures, workshops were arranged for advanced topics. This method follows a linear structure, covering concepts in a sequential order. In the previous academic year, the blended learning concept in the form of the module-based approach was practiced for the DBMS course. In this approach, the curriculum was divided into smaller, self-contained modules that can be studied independently by the students in a flexible sequence.

A learning management system (LMS) provides a centralized platform where all educational content, resources, and activities are stored and managed. This centralization facilitates easy access to materials for both instructors and students, enabling efficient distribution of content, tracking of progress, and management of administrative tasks (Kustija and Oktadianingsih, 2021). It streamlines the learning process by providing a single interface for course management, communication, assessments, and reporting. LMS platforms offer flexibility in terms of time and location, allowing learners to access course materials and participate in learning activities at their convenience (Farhangi et al., 2018; Mohamed et al., 2024). Many LMS platforms are accessible on various devices (such as computers, tablets, and smartphones), making it easier for users to engage with content anytime and anywhere. It is also easy for the facilitators to assess the performance of the learners and monitor their progress. It also ensures reliability and of course transparency in exam scores. It uses blockchain technology, ensures security, integrity and distributed characteristics. Conceptual model of software system and its requirements with description is discussed by Mladenova et al. (2022). Database model and its structure is designed.

Azahari et al (2020) discussed various e-learning materials, including assignments, attendance tracking, quizzes, audio learning content, and other teaching resources. Developed during the pandemic, it supports non-contact learning and is accessible anytime and anywhere. The paper also addresses issues related

[a]sujathae@saveetha.ac.in, [b]selvim@saveetha.ac.in, [c]dnr.cse@rmkec.ac.in

DOI: 10.1201/9781003587538-36

to compatibility, portability, and storage space. This provides solution for submitting assignment in programming language called as CheckMyCode. It acts as compiler and well as assignment management system. It supports cloud storage. It aims to design conceptual method for assistant teacher system using artificial intelligence (AI) technology. It enhances human-computer interaction and introduces intelligent tutoring system (ITS). It enables expert system in teaching learning process (Hui and Xiao, 2023). It is developed only for English course.

For web technology, the LMS platform provides theoretical guidance through learning materials and current political news, feedback and virtual practice. It was reported as an investigation course with ideological contents (Cen, 2022). It improves the comprehensive abilities of logistics management in higher educational institutions. Module-based approach solution helps in enhancing the learner's performance and engaging the learners beyond the class hours. Thus learning anytime and anywhere is established (He and Wu, 2021).

Using MySQL and PHP, a learning system is developed for Chinese traditional martial arts event at Tai Chi (Li et al., 2022). It was implemented through questionnaires and analyses the functionalities of the proposed system. Also introduces 3-dimensional technology. It finally investigates the improvement of learner's satisfaction (Li et al., 2022). Mu Class (MOOC) is introduced for universities courses related to database (Lu et al., 2018). It analyses the importance of hybrid teaching mode and which is based on project-based learning in online environment. It shows the improvement in quality of teaching which is anticipated by learners and focused on other teaching pedagogies. LMS also serves as interface between teachers and learners by uploading and downloading assignments, online exams. Implementation of flipped classrooms in English course promotes learning ability in learner's community (Zhou, 2023). This framework in developing reverse SQL generation algorithm from a source database (Atchariyachanvanich et al., 2019). This practice exercises are presented to learners for practice. It analyses the post- and pre-test scores are compared and the system is improved by 26%. Blended learning strategy is robust and state-of-art pedagogy is described and experimented by Duong et al. (2022).

Generally, in the existing systems discussed above, proves that the teaching results are improved. Through surveys, interviews, and performance evaluations, this study examines the effectiveness of this approach in enhancing learners' understanding, engagement, and overall experience with DBMS education. The findings suggest that the module-based approach offers several advantages, including increased flexibility, personalized learning paths, and better retention of concepts. However, challenges such as module integration and maintaining coherence across topics need to be addressed. This analysis contributes to the ongoing discourse on innovative pedagogical approaches in computer science education, particularly in the domain of database.

DBMS – Module-based learning

The blended learning strategy has five components which are as follows:

a. **Face-to-Face method:** It contains presentation content for the entire Unit 1 syllabus and active quizzes are conducted in order to prepare them to attend module exam for each unit.
b. **Face-to-face collaboration method:** It encourages learners to attend skill assessment which includes practice session in order to attend Module 1 in safe exam browser (SEB) only at dedicated exam halls with restricted medium access control (MAC) address. This will eliminate all kinds of malpractices.
c. **Online-instructional method:** Learners are motivated to extend their learning time after class hours in watching you tube videos, practicing Module 1 – home challenge which contains all questions cover the syllabus and experience the SQL queries before attempting Module 1 (SEB) exam.
d. **Online collaboration method:** Learners can participate practice quiz, and other online materials provided in the Moodle platform. They can also participate online discussion forum created to have open discussion among themselves.
e. **Online self-paced method:** This enables online certificates the learning platform to complete on their own availability.

An attempt to introduce blended learning in the course DBMS, few components was used and it was renamed as module-based learning (MBL), since the course content was divided into five modules.

The DBMS course on Moodle includes a range of materials: presentation content for syllabus topics, H5P interactive content with fill-in-the-blanks, flashcards, quizzes, matching exercises, and more. Lab experiments come with a manual, step-by-step software installation procedures, and details for each experiment, including the aim, logic, and five viva questions. Each unit has 50 multiple-choice questions for practice quizzes and Part-A continuous internal assessment (CIA) exams, five GATE questions, and five interview questions. Additionally, the course features a 2-minute learner's video presentation on any lab experiment, a discussion forum, an online certificate, a workshop, a journal discussion, and profiles for five job roles. This course is designed with the template given in Table 36.1. The table shows the key concepts, the type of activity given and the number of such activity in each module. The formative and summative assessment details are also provided in the table. This gives comprehensive information about the course when provided to the students.

The DBMS course design in the Moodle for Unit 1 given as Module III is shown in Figure 36.1. A similar template is followed for the remaining 4 units.

Thus, Moodle LMS is used to create a kind of personalized learning environment for the students to learn at their self-pace. After each module, formative assessment is done with the help of quiz. The sample questions are also practiced so that they have confidence to appear module exam from each module in SEB. Learners can experience enthusiastic programming learning skill.

Setting the question paper in higher blooms taxonomy level starting from apply level will increase the quality of the engineering graduates and their employability factor.

Hypothesis are:

- Hypothesis 1: There is improvement in programming skills
- Hypothesis 2: There is significant improvement in the count of placed learners.

Each module has PPTs for the content to deliver, H5P content interactive videos are available to answer questions while watching video, video link to watch, practice quiz, skill assessment, home challenge to practice for the entire Unit 1 syllabus to practice before attending Module 1 in SEB mode and finally feedback.

Results and discussions

This paper yields the results in terms of the learners learning ability and academic potential, programming skills. It also enhances thinking ability in versatile and debugging capability and syntax of the queries. This module-based approach recommends the learners how to learn and makes use of the recent trends world-wide and satisfies industry needs too in the placement. LMS is a platform which encourages learners to take part activities after class hours in 24×7. It also entertains learners self-learning and responsibility in completing their task assigned by faculty.

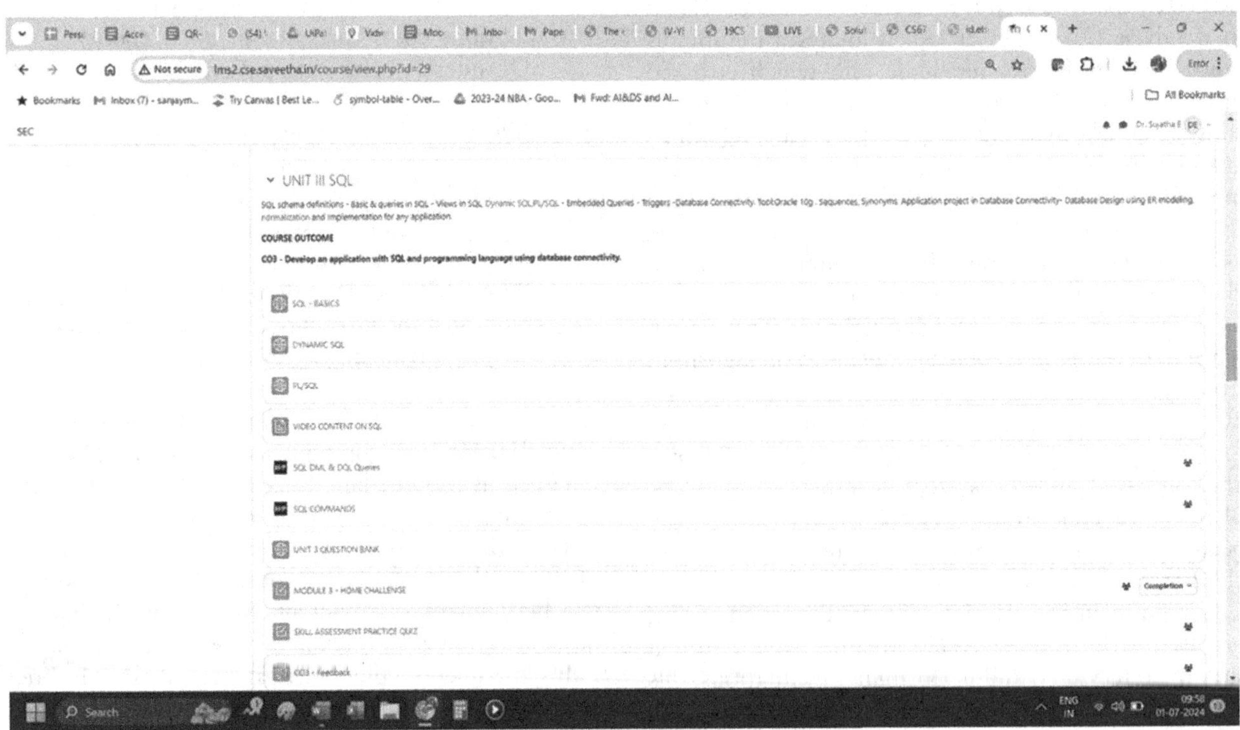

Figure 36.1 Sample screen shot of Unit III content in Moodle
Source: Author

Table 36.1 Key concepts of modules in Moodle for the course DBMS.

Key concepts	Type of Moodle contents	Count
Module 1 – Create, alter, insert commands	Fundamentals of database and data models – PPT	2
	Introduction to DBMS, ER diagram symbols, activity on data models – H5P	2
	Schema in database – Video link	1
	Practice quiz – Quiz	1
	Workshop on ER diagram	1
	Home challenge (10 questions) – Practice in code runner SQLite	1
Module 2 – Update, delete, select commands	Relational algebra – Video link	1
	Normalization, ERD plus tool – H5P	2
	Unit II PPT	1
	Practice quiz – Quiz	1
	Home challenge (10 questions) = Practice in code runner SQLite	1
Module 3 – Aggregate, order, group by, having commands	SQL basics, dynamic SQL, PL/SQL – PPT	3
	SQL tutorial – Video link	1
	SQL DML and DQL queries, SQL commands – H5P	2
	Home challenge (10 questions) – Practice in code runner SQLite	1
	Skill assessment practice quiz – Quiz	1
Module 4 – Sub-queries	Data storage and query optimization, indexing and hashing, transactions	3
	File Organization and Indexing – H5P	1
	Data Storage and query optimization – Video link	1
	Home challenge (10 questions) – Practice in code runner SQLite	1
	Skill assessment practice quiz – Quiz	1
Module 5 – Joins commands	XML and Internet databases, Data warehousing and data mining, Introduction to computer security, data mining tools – PPT	4
	Types of databases – H5P	1
	Home challenge (10 questions) – Practice in code runner SQLite	1
	Skill assessment practice quiz – Quiz	1
Online certification course	Certificate (PDF/JPEG/JPG)	Minimum 1
Cos feedback	Each unit	5
Question bank	Each unit	5
Module (SEB) examination	Each module	5
Skill assessment (SEB)	Each module	5
CIA I & II	5 units	2

Source: Author

Learners also participate actively in discussion forums. Lab-integrated course demands lab exercises for each unit and the submission is linked with GitHub repositories for evaluation. Descriptions required to proceed lab exercises are also provided in the Moodle environment. It enriches the learners to gain practical exposure and training for placement drives. LMS produces independent learners as millennial learner. It really augments the abilities of the existing systems. Quantitative analysis of survey responses regarding learners' satisfaction, perceived effectiveness, and preferences for module-based learning.

Qualitative analysis of interview transcripts to extract themes related to learners' experiences, challenges, and suggestions. Performance evaluation results comparing module-based approach outcomes with traditional teaching methods. Skill assessment questions are asked from the quiz practiced and the home challenge practiced in the class. The questions are shuffled each time when attempt is made. Practice

Table 36.2 Pass percentage analysis of conventional method of teaching and module-based teaching.

Examination	Conventional method of teaching Pass %	Module-based teaching Pass %
CIA – I	87	92
CIA – II	89	95
End-Sem theory examination	92	99

Source: Author

Table 36.3 Log analysis for No. of attempts, No. of learners watched videos, marks range.

No. of attempts	No. of learners watched videos	Marks range
1	30	9–10
2	32	8–9
3	25	8–9
4	33	8–9
5	20	7–8

Source: Author

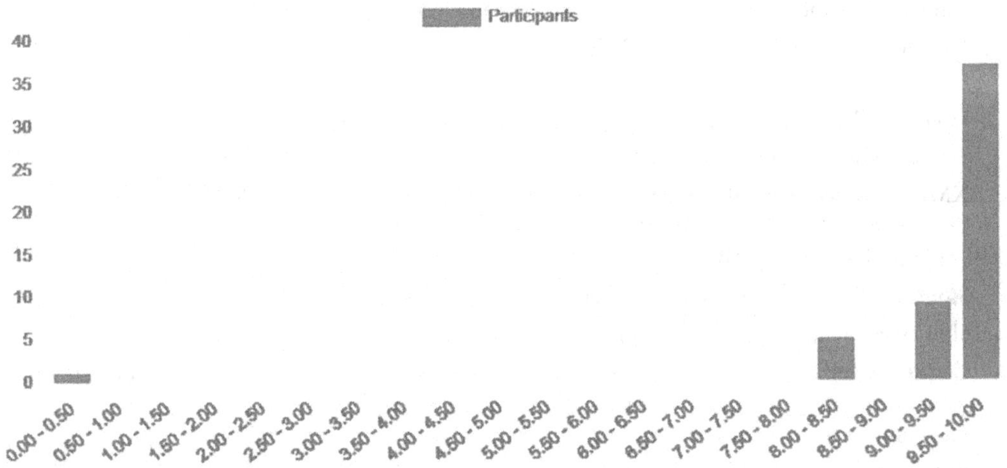

Figure 36.2 Number of learners achieved grade ranges
Source: Author

questions are different from SEB module exams. So, training the learners for more queries and answers with respect to the topic covered in each module. Learners were also encouraged to practice in various online sites available for them to enrich their knowledge in SQL queries.

During conventional method (Table 36.2) of teaching in the year 2023–24 odd semester with 79 learners as heterogeneous group (I, II, III years) since the institution follows flexi learn in curriculum the pass percentage was 87% and 89% in CIA-I, II, final end semester examination reflects that 92% learners have passed. After introducing module-based learning method in the year 2023–24 even semester with 57 learners as heterogeneous group the reflections show that the continuous assessment examination results were improved and end semester examinations were also improved as 99%.

Table 36.3 explicitly shows that the learners hard work and their quality time spent for learning results in their scores. Learners with high and moderate potential could able to score more than 8 after watching the videos 2 times and the maximum number of attempts could be 4 for up to 33 learners. The log statistics reflects in score as well as skill in the Figure 36.2 visually.

Limitations of module-based learning in DBMS course

Code runner supports SQLite only. So, PL/SQL, cursors and triggers cannot be practiced in SQLite platform.

Conclusions

Hence, module-based course is strongly recommended for any course that enhances learning ability, self-learning potential, engagement and learning in 24×7, transparency in score encourages them to attempt with good knowledge, boosts confidence, enthusiastic learning, enriching programming skills, assignment submissions can be done at anytime and anywhere. In turn, that reflects in placement drives and gives hope to get placement subsequently. Hypothesis 1 and 2 are proved. Module-based approach proves that learners love happy learning in this strategy which in turn reflects in academics and placement. This blended learning methodology is promising strategy for the learners to become successful learners in acquiring adequate knowledge to get placement and competent with other millennials.

Future enhancement suggested to increase the sample size and other teaching pedagogies in order to experiment with different case studies and challenges faced in the proposed system.

References

Mohamed, A., Shokry, M., and Idrees, A. M. (2024). A proposed model for improving the reliability of online exam results using blockchain. *IEEE Acc.*, Vol.12, pp 51772 – 51785.

Mladenova, T., Valova, I., and Valov, N. (2022). Design of a smart system for monitoring and management of pastures and meadows: The relational database approach. *2022 8th Internat. Conf. Ener. Effic. Agricul. Engg.*, 1–5. doi: 10.1109/EEAE53789.2022.9831393.

Kustija, J. and Oktadianingsih, R. (2021). Creating E-modules basic electricity and electronic courses based on wordpress for vocational school students. *2021 3rd Internat. Symp. Mat. Elec. Engg. Conf. (ISMEE)*, 343–347. doi: 10.1109/ISMEE54273.2021.9774144.

Azahari, A. M., Ahmad, A., Rahayu, S. B., and Mohamed Halip, M. H. (2020). CheckMyCode: Assignment submission system with cloud-based Java compiler. *2020 8th Internat. Conf. Inform. Technol. Multim. (ICIMU)*, 343–347. doi: 10.1109/ICIMU49871.2020.9243373.

Hui, H. and Xiao, L. (2023). Assistant teaching system design based on expert system. *2023 3rd Internat. Conf. Mobile Netw. Wireless Comm. (ICMNWC)*, 1–6. doi: 10.1109/ICMNWC60182.2023.10435766.

Cen, H. (2022). Construction of online teaching platform for ideological and political content of investigation course under the mixed teaching mode — Based on web technology. *2022 3rd Internat. Conf. Edu. Knowl. Inform. Manag. (ICEKIM)*, 295–299. doi: 10.1109/ICEKIM55072.2022.00073.

He, L. and Wu, Q. (2021). Research on Internet-based practical teaching system of logistics management in Colleges and Universities. *2021 2nd Internat. Conf. Inform. Sci. Edu. (ICISE-IE)*, 1028–1033. doi: 10.1109/ICISE-IE53922.2021.00234.

Li, Y., Yuan, T., and Yu, K. (2022). Design and implementation of Taijiquan learning system based on PHP+MySQL. *2022 Internat. Conf. Inform. Sys. Comput. Edu. Technol. (ICISCET)*, 35–39. doi: 10.1109/ICISCET56785.2022.00017.

Farhangi, A., Yazdani, H., and Haghshenas, M. (2018). Identification of LMS dimensional problems in Iranian E-learning centers. *2018 12th Iran. 6th Internat. Conf. e-Learn. e-Teach. (ICeLeT)*, 033–044. doi: 10.1109/ICELET.2018.8586751.

Lu, H., Ye, M., Gao, B., Guan, W., and Gao, Z. (2018). Exploration and practice of hybrid teaching mode under MOOC environment: Taking "Database System Principle" as an example. *2018 9th Internat. Conf. Inform. Technol. Med. Edu. (ITME)*, 498–502. doi: 10.1109/ITME.2018.00117.

Zhou, Z. (2023). Research on flipped classroom system of industrial English based on computer intelligent network technology. *2023 IEEE 6th Internat. Conf. Inform. Sys. Comp. Aid. Edu. (ICISCAE)*, 470–474. doi: 10.1109/ICISCAE59047.2023.10393395.

Atchariyachanvanich, K., Nalintippayawong, S., and Julavanich, T. (2019). Reverse SQL question generation algorithm in the DBLearn adaptive E-learning system. *IEEE Acc.*, 7, 54993–55004. doi: 10.1109/ACCESS.2019.2912522.

Duong, H. T., Uyen, B. P.,, and Ngan, L. M. (2022). The effectiveness of blended learning on students' academic achievement, self-study skills and learning attitudes: A quasi-experiment study in teaching the conventions for coordinates in the plane. *Heliyon*, 8(12), e12657.

37 Crafting real-world learning: Industry-driven pedagogy for civil engineering education

Padma SrinivasaPerumal[a], *Kalpana Manoharan and Amsayazhi Pandian*

Department of Civil Engineering, Saveetha Engineering College, Chennai, Tamilnadu, India

Abstract

In contemporary civil engineering education, the gap between theory and practice persists as a challenge. This paper presents our experience in addressing this gap through innovative pedagogy in the course "Estimation, Costing, and Valuation" for third and final year civil engineering students. Collaborating with industry experts, we aimed to enrich students' learning by integrating real-world perspectives into the curriculum. Our approach involved close collaboration between academia and seasoned industry professionals to tailor course content, delivery methods, and assessments to industry standards. Central to this approach was the integration of case studies, practical exercises, and real-life project scenarios from industry practices, equipping students with both theoretical understanding and practical skills essential for real-world projects. Furthermore, guest lectures, site visits, and interactive sessions provided students with direct engagement opportunities with industry professionals, fostering valuable networking and firsthand exposure to industry norms and best practices. The outcomes were evident in students' heightened motivation, engagement, and ability to correlate theoretical knowledge with practical applications. In conclusion, our experience highlights the effectiveness of industry-academia collaboration in enriching civil engineering education, advocating for the continued exploration and adoption of innovative pedagogical approaches to adapt for Industry 5.0 scenario. This initiative led to a 30% improvement in learners' project assessment scores, a 40% increase in internship placements, and we achieved 100% pass results.

Keywords: Civil engineering, academia, pedagogy, Industry 5.0

Introduction

In recent years, the landscape of civil engineering education has witnessed a significant transformation, propelled by the emergence of Industry 5.0 concepts alongside a growing emphasis on real-world learning experiences. This shift, characterized by a focus on integrating cutting-edge technologies, sustainable practices, and industry collaboration into educational curricula, represents a pivotal evolution in how civil engineering learners are prepared for the challenges and opportunities of the profession.

By leveraging Industry 5.0 principles such as human-machine collaboration, decentralized decision-making, and the convergence of digital and physical systems, this approach seeks to empower learners with the skills, mindset, and agility needed to excel in the dynamic landscape of modern engineering (Khan et al., 2024).

In the context of India, where infrastructure development is a cornerstone of economic growth and societal progress, the imperative for industry-driven pedagogy in civil engineering education is particularly pronounced. According to the National Skill Development Corporation (NSDC), India is expected to require approximately 4.5 million civil engineers by 2022 to support its ambitious infrastructure projects. However, the World Bank reports that there is a significant gap between the demand for skilled engineers and the availability of qualified graduates, highlighting the urgent need for innovative educational approaches.

With initiatives such as "Make in India" and "Digital India" driving a wave of technological innovation and industrial transformation, there is an urgent need for civil engineers who can harness emerging technologies and pioneer sustainable solutions. Furthermore, India's diverse and rapidly evolving infrastructure needs, ranging from smart cities and transportation networks to renewable energy and water management systems, underscore the importance of preparing engineers who can navigate complexity and drive innovation.

In this research paper, we delve into the theoretical underpinnings and practical implications of industry-driven pedagogy for civil engineering education, with a specific focus on its alignment with Industry 5.0 concepts. Through an interdisciplinary lens that combines engineering, education, and technology, we explore how this approach can empower learners to become catalysts for positive change in the field of civil engineering, both in India and beyond (Ikudayisi et al., 2023; Khasawneh, 2024; Calvetti et al., 2024).

[a]padmagi91@gmail.com

DOI: 10.1201/9781003587538-37

Collaboration of industry expert and academician

Incorporating industry experts as regular instructors in civil engineering courses holds immense promise for enhancing both learner learning outcomes and the academic landscape. These experts bring a dynamic blend of contemporary knowledge and practical insights, offering learners invaluable exposure to real-world challenges and solutions. By leveraging their extensive professional networks, expert instructors not only enrich classroom discussions but also provide unique opportunities for learners to connect with industry leaders, fostering potential internships, mentorships, and career pathways. Furthermore, their firsthand experience ensures that course content remains agile and responsive to evolving industry demands, thereby equipping learners with the latest skills and knowledge. By serving as role models and mentors, industry experts inspire learners to aspire to excellence and innovation in their academic pursuits and future careers. Consequently, the integration of industry expertise into civil engineering education not only enriches the educational experience but also cultivates a workforce ready to tackle the complex challenges of the field.

While industry experts bring invaluable real-world experience and practical insights to the classroom, collaboration with academicians or facilitators can enhance the overall learning experience for learners. Academicians provide expertise in pedagogy, curriculum design, and assessment, ensuring that educational objectives are met and that content is delivered in a structured and effective manner.

Evolution of industry-based experiential learning in civil engineering

Amidst several obstacles (Jestrab et al., 2009) effectively executed civil engineering courses through the careful selection of industry professionals and the coordination of their contributions to create a unified learning environment. Collaborating with stakeholders from McAninch Corporation, Ziegler Caterpillar, and additional entities, the authors crafted the curriculum for the courses offered at Iowa State University. In the rapidly evolving field of civil engineering, employers anticipate that graduating engineers are equipped to excel in their initial roles (Bae et al., 2022). A study by Stephen and Festes (2022) explored how a work-based learning program (WBL) could enhance the employability skills of craftsmen in the construction industry, thus contributing to industrial development. In another experimental study by SrinivasaPerumal and Manoharan (2022), two undergraduate courses in the civil engineering program namely surveying and computer-aided building drawing for first-year learners have incorporated industry-based experiential learning alongside classroom sessions. Conversely, conventional standalone classroom teaching was utilized for second-year learners in these same courses. This study compared learner perceptions and assesses the attainment of learning outcomes between industry-based experiential learning and conventional teaching methods. Nandi et al. (2015) stated the significance of building positive collaborative strategies with industries for learning scenarios. Das (2023) experimented problem-based learning (PBL) methodologies for an university in South Africa for implementing civil engineering courses.

Industry-based experiential learning – Estimation, costing and valuation

The course "Estimation, Costing, and Valuation" (Figure 37.1) typically targets final-year learners, offering them essential knowledge and skills in estimating construction costs and valuing properties. Initially structured around theoretical concepts and small-scale estimation exercises, the course took a transformative turn when industry expert Mr.Venkatesan intervened to infuse it with real-world industrial insights.

His expertise prompted a reevaluation of the course syllabus to align it more closely with industry standards. As a result, the course content was adapted to include a comprehensive analysis of a 10,000 sq ft library plan, an actual blueprint of a commercial building situated on campus. This departure from

SYLLABUS

UNIT I QUANTITY ESTIMATION 9
Philosophy – Purpose – Methods of estimation – Types of estimates – Approximate estimates – Detailed estimate – Estimation of quantities for buildings, bituminous and cement concrete roads, septic tank, soak pit, retaining walls – culverts.

UNIT II RATE ANALYSIS AND COSTING 9
Standard Data – Observed Data – Schedule of rates – Market rates – Standard Data for Man Hours and Machineries for common civil works – Rate Analysis for all Building works, canals, and Roads– Cost Estimates

UNIT III SPECIFICATIONS, REPORTS AND TENDERS 9
Specifications – Detailed and general specifications – Constructions – Sources – Types of specifications – Principles for report preparation – report on estimate of residential building – Culvert – Roads – TTT Act 2000 – Tender notices – types – tender procedures – Drafting model tenders, E- tendering -Digital signature certificates- Encrypting -Decrypting – Reverse auctions.

UNIT IV CONTRACTS 9
Contract – Types of contracts – Formation of contract – Contract conditions – Contract for labour, material, design, construction – Drafting of contract documents based on IBRD/ MORTH Standard bidding documents – Construction contracts – Contract problems Arbitration and legal requirements

UNIT V VALUATION 9
Definitions – Various types of valuations – Valuation methods - Necessity – Capitalized value Depreciation – Escalation – Valuation of land – Buildings – Calculation of Standard rent – Mortgage – Lease.

Figure 37.1 Original syllabus of the estimation, costing and valuation
Source: Author

traditional classroom exercises to a real-life commercial project sparked immediate interest and engagement among the final-year learners.

The introduction of the library plan provided a tangible and practical learning experience for the learners. As they navigated through the intricacies of the blueprint, guided by expert's industry-centric perspective, they gained invaluable insights into the complexities of real-world construction projects. Moreover, the physical presence of the building under construction on campus served as a constant reminder of the practical application of their classroom learning.

Given their familiarity with blueprint reading, the final-year learners swiftly grasped the intricacies of the library plan. The expert's explanations, coupled with the contextual relevance of the project, enhanced their understanding of estimation, costing, and valuation principles in a tangible and applicable manner. This hands-on approach not only deepened their theoretical knowledge but also instilled in them a heightened sense of professional competence and preparedness for real-world challenges in the civil engineering industry.

In essence, the integration of the library plan into the course curriculum transformed the learning experience for the final-year learners, bridging the gap between theoretical knowledge and practical application. Under the expert's guidance, they gained valuable insights into industry practices, thereby enriching their educational journey and equipping them with the skills and knowledge necessary for successful careers in civil engineering.

Challenges faced during the collaboration

In the next class he started explaining about the different process of the construction that should be included in the estimation process. Since it was so lively they were so indulged in the class. But the real problem started after this. The industrial expert started taking the calculation part. The tabular column in an excel sheet was used by the learners as per the expert's suggestion. The learners couldn't imagine the real length, breadth, and height of the structures. Some of the calculations had hidden deductions, which they couldn't understand. The Industrial expert couldn't explain the process in a way that they can understand. Within few days the learners started feeling low. The expert also started feeling frustrated as the learners lost the interest on that subject (Gentelli, 2015). At the end of 3 months, he couldn't complete the syllabus. Only one plan was used to practice in the class. Learners who had experience in the field work they had an idea about what the expert was explaining but others had difficulty in understanding.

The expert was not satisfied with the performance of the learners and the department also felt the same. So this subject was assigned to them again in the next semester with the third year learners. This time it was decided that the expert should take 50% of the syllabus in class and the academic faculty should take 50% of the syllabus. During this time the expert took the class in excel sheet in a practical Industrial way while the faculty covered the other entire syllabus. Now this time the learners were so cooperative and they understood the syllabus very effectively and efficiently. In this period, he took a simpler plan this time instead of a complex plan and he started explaining them to the learner. The compound wall plan is as shown in Figure 37.2. The learners were exposed to industrial visits to understand the concept better. They got good marks in the internal assessment as well. From this we can say that teaching is a different genre where everyone cannot excel (Truong and Nguyen, 2024). Though learners need to know what is happening in the real world they couldn't understand it without the help of teaching faculty.

Implementation process

In the initial semester, internal assessments were administered using traditional pen-and-paper

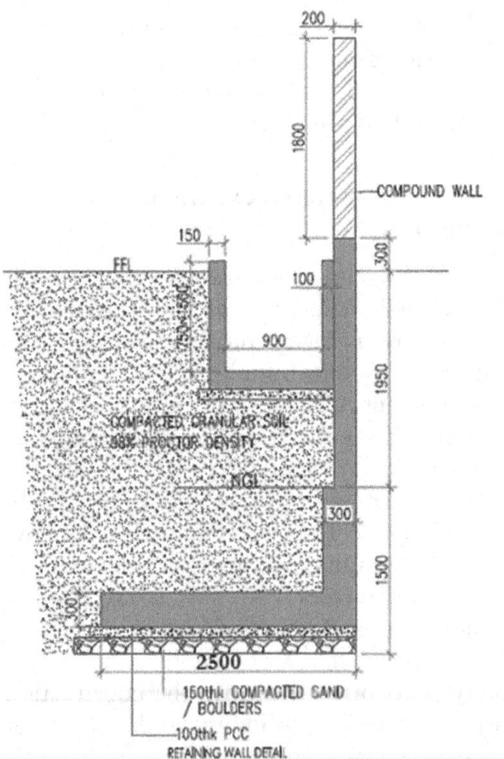

Figure 37.2 Compound wall plan
Source: Author

methods, while classroom instruction primarily utilized Excel. This created a significant challenge for learners during exams, resulting in a high number of failures across two assessment cycles. Despite efforts by educators, learners struggled to perform well in these assessments, and feedback from them was overwhelmingly negative.

However, in the subsequent semester, improvements were made to the instructional approach. Classes were enriched with insights from both industry experts and teaching faculty, offering learners valuable guidance on tackling internal assessments effectively (Tsui et al., 2024). Consequently, there was a notable improvement in learner performance, with a higher percentage achieving satisfactory marks.

Furthermore, to enhance practical skills, learners were assigned real-world projects. They were tasked with developing their own plans, which had to undergo approval from faculty members (Ingole et al., 2024). Subsequently, learners were required to execute the entire project estimation process, from excavation to electrical and plumbing aspects. This hands-on approach provided learners with invaluable experience and better prepared them for real-world scenarios.

The graph (Figure 37.3) above presents the number of learners on the x-axis and the marks they obtained in both Continuous Internal Assessments (CIA) on the y-axis. In CIA 1, six learners received zero marks. In contrast, in CIA 2, no learners scored zero. This pattern suggests that learners encountered difficulties during CIA 1, likely due to their reliance on practicing calculations exclusively in Excel. It required some time for them to adapt to using both paper and Excel.

By the time of CIA 2, there was a noticeable improvement, with many learners focusing more effectively and achieving marks above 80. This marked a significant enhancement in their performance compared to CIA 1. The progression highlights the learners' adaptation and growing proficiency in balancing both methods of calculation.

In reviewing the attendance percentages (Figure 37.4) of seven random learners, an interesting pattern emerged. During the first month, attendance was notably low as learners grappled with understanding the new material. September saw a stabilization in attendance, with learners attending classes more consistently following CIA 1. This improvement was likely due to our transition from using Excel sheets to paper-based instruction.

However, October brought another decline in attendance, specifically among learners who continued to struggle despite the change in teaching methods. This disparity highlighted that while some learners were beginning to engage with the material, others were still finding it challenging.

Recognizing this, we shifted our approach in November, dedicating individual attention to those struggling the most. By focusing on their specific needs and helping them develop an interest in the subject, we observed a marked improvement in their performance, as evidenced by higher marks in CIA 2. This experience underscored the importance of adaptable teaching methods and personalized support in fostering academic success.

Results and discussion

Analyzing learner perceptions of the industry expert's contributions to the course provides valuable insights into the effectiveness of the industry-driven pedagogy. A survey conducted among the 78 participants, comprising 29 third-year and 49 second-year learners, revealed overwhelmingly positive responses.

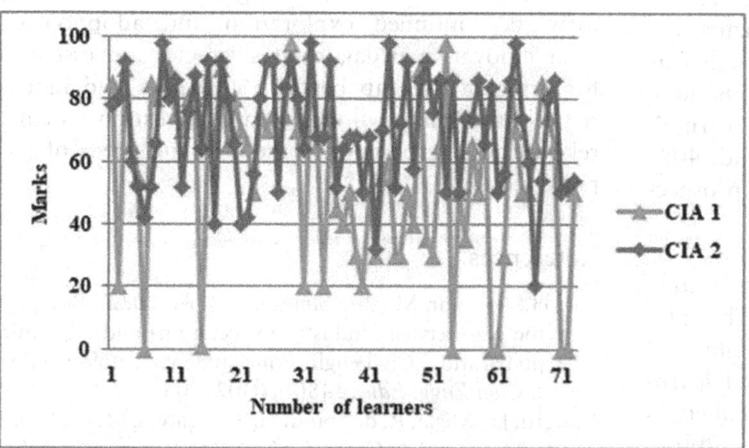

Figure 37.3 Percentage of marks obtained by the learners in CIA 1 and 2
Source: Author

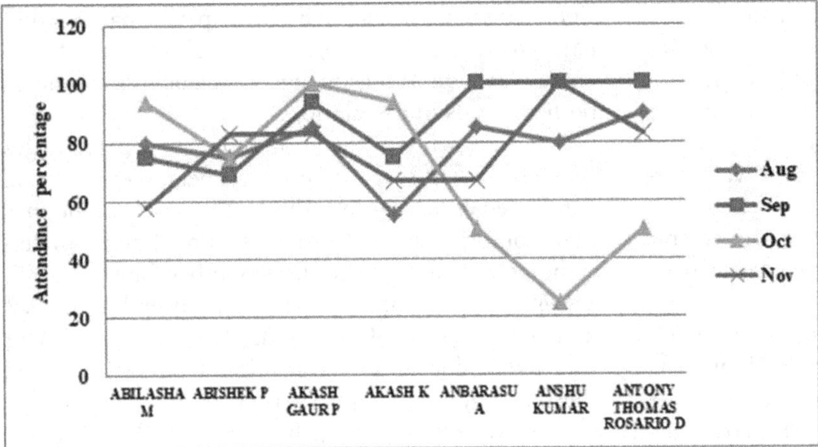

Figure 37.4 Attendance percentage of learners
Source: Author

Among the key findings, a vast majority of learners expressed appreciation for the industry expert's insights, with 92% indicating that the integration of real-world perspectives enhanced their understanding of course concepts. Moreover, 87% of participants felt that the industry expert's presence added value to the learning experience, citing practical relevance and industry relevance as primary reasons.

Furthermore, feedback highlighted the effectiveness of various engagement methods employed during the industry expert's sessions. Interactive discussions, practical examples, and case studies were particularly well-received, with 85% of learners reporting that these activities deepened their engagement and facilitated a better grasp of the subject matter.

Importantly, learner feedback also underscored the impact of the industry expert's sessions on their career readiness. Over 80% of participants expressed increased confidence in their ability to apply course concepts in real-world scenarios, emphasizing the practical skills gained through industry insights as invaluable for future professional endeavors. Besides, a 30% improvement was observed in the learner's inclination towards project completion and 40% increase in their motivation to look out for prospective placements.

Overall, the analysis reveals a strong positive perception of the industry expert's contributions among both third-year and second-year learners. The integration of industry perspectives not only enhanced theoretical understanding but also bolstered learners' confidence and preparedness for the challenges of professional practice (William Leondorf, 2004). These findings reaffirm the significance of industry-academia collaboration in enriching educational experiences and bridging the gap between classroom learning and industry expectations in civil engineering education.

Conclusions

In conclusion, our journey in implementing industry-driven pedagogy for civil engineering education underscores the transformative potential of collaborative efforts between academia and industry. By infusing real-world perspectives into the curriculum, we not only enhance learners' theoretical understanding but also equip them with practical skills crucial for professional practice. The positive outcomes observed, including a 30% improvement in students' project assessment scores, a 40% increase in internship placements, heightened motivation, engagement, and improved problem-solving abilities, validate the efficacy of this approach in preparing learners for the complexities of the engineering profession. Moving forward, continued exploration and adoption of such innovative pedagogical strategies are essential for bridging the gap between academia and industry, ensuring that civil engineering education remains relevant and responsive to the evolving needs of the profession.

References

Bae, H., Polmear, M., and Simmons, D. R. (2022). Bridging the gap between industry expectations and academic preparation: Civil engineering students' employability. *J. Civil Engg. Edu.*, 148(3), 04022003.

Calvetti, D., Mêda, P., de Sousa, H., Gonçalves, M. C., Faria, J. M. A., and da Costa, J. M. (2024). Experiencing education 5.0 for civil engineering. *Proc. Comp. Sci.*, 232, 2416–2425.

William Leondorf (2004) rom industry to instruction: Transitioning from industry to Education: The First year. *Annual Conference*, 9.1320.1 - 9.1320.8, DOI: 10.18260/1-2--13033.

Das. D. (2023). Problem-based learning for undergraduate civil engineering education in South Africa – A methodological approach. Walbridge S, Nik-Bakht M, Ng KTW, et al. (eds.). *Proc. Can. Soc. Civil Engg. Ann. Conf., 2021.* 85–95. Springer Nature Singapore, Singapore.

Gentelli, L. (2015). Using industry professionals in undergraduate teaching: Effects on student learning. *J. Univer. Teach. Learn. Prac.*, 12(4), 4.

Ikudayisi, A. E., Chan, A. P., Darko, A., and Adedeji, Y. M. (2023). Integrated practices in the architecture, engineering, and construction industry: Current scope and pathway towards Industry 5.0. *J. Build. Engg.*, 73, 106788.

Ingole, S. B., Devade, K. D., and Bhong, M. (2024). Model for facilitative practical learning in engineering education under India's new education policy (Nep): A stepwise framework. *Edu. Admin. Theory Prac.*, 30(3), 739–746.

Jestrab, E. M., Jahren, C. T., and Walters, R. C. (2009). Integrating industry experts into engineering education: Case study. *J. Prof. Iss. Engg. Edu. Prac.*, 135(1), 4–10.

Khan, M. A., Khan, R., Praveen, P., Verma, A. R., and Panda, M. K. (2024). Infrastructure Possibilities and Human-centered Approaches with Industry 5.0. United States: Engineering Science Reference.

Khasawneh, M. A. S. (2024). Closing the industry-academia gap in translation education; Exploring collaborative strategies as tools for effective curriculum alignment. *Kurdish Stud.*, 12(1), 1850–1867.

Nandi, A., Bali, J., Kuashik, M., and Shirol S. (2015). Industry-Institute interaction: An important step towards empowering skills of engineering students. *J. Eng. Edu. Transform.*, 29, 2349–2473.

SrinivasaPerumal, P. and Manoharan, K. (2022). Analysis of industry based experiential learning for undergraduate courses in civil engineering. *J. Engg. Edu. Transform.*, 36(Special issue 1), 82–88.

Stephen, O. O. and Festus, O. O. (2022). Utilization of work-based learning program to develop employability skill of workforce (craftsmen) in construction industry towards industrial development. *Indonesian J. Edu. Res. Technol.*, 2(3), 179–188.

Truong, T. N. N. and Nguyen, H. T. P. (2024). Navigating experiential learning: Insights from Vietnamese architecture students on an educational field trip. *J. Exper. Edu.*, 10538259241248634.

Tsui, E., Dragicevic, N., Fan, I., and Cheng, M. (2024). Co-creating curriculum with students, teachers, and practitioners in a technology-enhanced environment. *Edu. Technol. Res. Dev.*, 72(2), 869–893.

38 Unveiling user satisfaction in cross-platform education: A comparative analysis of android and web-based java learning applications

Tharoon R.[a] *and Loganayagi S*[b]

Department of Computer Science, Saveetha School of Engineering, Saveetha Institute of Medical and Technical Sciences, Saveetha University, Chennai-602105, Tamil Nadu, India

Abstract

Aim: To investigate and contrast the degrees of user satisfaction between an android app and a web-based application for quiz application on data structure, with an emphasis on the variations in user experiences between the two platforms. **Materials and methods:** The paper states that with sample sizes of N=24 (12 per group) is distributed to android and web groups, respectively, a G power of 80% was utilized to assess the response time performance in both the android and web apps. **Result:** By analyzing, the user satisfaction of the android application is 71.5500% which is comparatively less than the web development which is 81.4167%. This signifies that there is a statistically insignificant difference between the android and the web application, with p=0.000 (p<0.05). **Conclusion:** Web application has user satisfaction of 81.4167% and android development has 71.5500%.

Keywords: Web development, android application, Java Edu, user satisfaction, android studio, Xampp

Introduction

The developed application provides a quiz platform that is similar to a game with several levels that users may advance through upon completing previous assignments. To keep users engaged and interested, each difficulty level features a variety of sub-levels, each with a unique collection of questions and objectives (Kotobee, n.d.). In addition to having a terrific game experience, users who complete every level have a solid understanding of Java programming. The application's main goal is to increase users' comprehension and intuitiveness of data structures (Rüth et al., 2021). Interactive games and progressively challenging tasks allow users to actively engage and absorb key topics. The gamified format creates a dynamic learning environment that encourages continuous motivation and engagement. As they complete the numerous tasks and phases in Java Edu, users acquire practical knowledge and skills (Roogle, 2021). The program blends educational content with an entertaining gaming interface in an effort to make studying enjoyable and effective (Areed et al., 2021). Through an emphasis on user interaction and learning outcomes, the program seeks to enhance users' understanding and memory of Java Edu concepts. The software's innovative methodology transforms learning into a joyful and rewarding experience. The program offers a large range of activities and challenges to suit users with varying skill levels and learning styles (Wolber et al., 2014). The logical progression ensures that students progressively gain more knowledge and skill, supporting their comprehension. Through interactive feedback and progress tracking, users receive on-going support and encouragement throughout their learning experience. Ultimately, the program lets users become experts at Java Edu by combining entertaining games with educational content.

There is an astounding amount of research that can be done by posing questions about the dynamic field of Java programming and app development. There are now more scholarly publications on this topic than ever before; a search on Google Scholar yields over 65,700 results. Particularly noteworthy are the substantial contributions from IEEE and Springer, with 296 and 6,498 articles, respectively. The 681 pages manuscript (Wolber et al., 2011), indirect data collection techniques have led to an increase in popularity of methodologies for processing and interpreting ambiguous data, which is the subject of this study. It offers insights into applying uncertain data models in a range of applications by covering both traditional data management strategies like query optimization and join processing as well as mining techniques like clustering and classification. The article contains 17 citations (O'Brien, 2016). The piece (Mitchell, 2015) contains 16 references. This work looks at the possibility of

[a]192011255.sse@saveetha.com, [b]loganayagis.sse@saveetha.com

DOI: 10.1201/9781003587538-38

producing tree topologies from relational data using Mark Burgin's concept of named sets and recursion. By finding the Aleph data relation as the unifying element, it exposes a parent/child relationship that is hidden behind structured data. With implications beyond visualization studies, this novel visualization mainly depends on the recursive technique and the corresponding open hierarchical Java Edu(OHDS). The article's citation (McEvoy et al., 2021) is 67-years-old. This research showcases the web-based e-learning system that the University of Basrah in Iraq developed in response to the COVID-19 outbreak. Features include text and video lessons, online exams, and electronic certificates. The system was developed using PHP programming and an agile approach to help instructors, administrators, and students. Evaluation reveals that student–teacher participation and interaction increased throughout the course of the pandemic. As stated in article 10 (Ganju et al., 2020), in response to challenges posed by an increase in student enrollment, particularly in computer science and information technology departments, the institute launched a web-based teaching effort. This pilot project, which uses paperless, environmentally friendly teaching methods that adhere to green technology, has shown to be successful and beneficial for both staff and students at the faculty of computing, information systems, and Mathematics. By using online course management technologies, the program promotes active learning and real-time feedback while upholding standards. The study highlights the advantages, challenges, and solutions and uses a feedback mechanism to measure student satisfaction and performance improvements.

Materials and methods

The Ideation Lab, which promotes a dynamic and collaborative environment that rewards creative thinking and thorough concept development, is where the research analysis is built (Gaspar et al., 2020). The lab's group of distinguished professionals from a range of domains, including technology and user experience research, provides a conducive environment for interdisciplinary conversation and intellectual inquiry. Researchers can take on ambitious initiatives that solve real-world issues and advance knowledge in their respective disciplines thanks to modern facilities and guidance. This study investigates if the application has a faster reaction time by looking at two groups from different angles (Laxmikanth 2017). The survey, which divides the creation of web and android apps into two categories, has a sample size of 24 (12 per group) and computes 80% of the G-power value, 0.05 alpha, and 0.8 beta with a 95% confidence interval.

Web developers may build and test websites and apps in a stable local environment by using XAMPP in their work. Relational databases and server-side scripting may be used by developers to construct dynamic websites using XAMPP, which consists of Apache, MySQL, PHP, and Perl. MySQL makes it easier to store and retrieve data efficiently, while PHP makes it easier to incorporate database queries into web applications. Install and set up XAMPP, which consists of Apache, MySQL, PHP, and Perl, on android.

Java Edu android application

In group 1, one common tactic is to ask users to score their experience and provide comments or recommendations via in-app surveys or feedback forms. Customer satisfaction data may also be obtained by employing analytics tools to track user engagement parameters including session duration, frequency of app usage, and feature interactions. Finding out how pleased customers are and identifying usability issues may be greatly aided by having real users test the app and provide feedback and ideas.

Java Edu Web development

In group 2, the method entails asking consumers directly about their experience, degree of satisfaction, and recommendations for improvement using surveys or questionnaires. Web analytics tools may also be used to analyze user behavior and interactions to provide useful insights into metrics relating to user satisfaction, such as session length, page visits, bounce rates, and conversion rates. User testing and usability research are two more useful techniques for closely observing user interactions and pinpointing problems or potential areas for development.

Statistical analysis

A strong software program for statistical analysis and data management is IBM SPSS (Statistical Package for the Social Sciences). Because of SPSS's intuitive interface, which enables users to do intricate statistical analyses, create insightful visualizations, and deal with enormous datasets, it is widely used in a variety of areas, including the social sciences, healthcare, and market research. In order to help analysts and researchers find patterns, trends, and correlations in their data so they can make defensible choices and relevant conclusions, SPSS employs a range of statistical techniques and algorithms. For both novice and seasoned statisticians, it is a helpful tool that offers a strong capacity to successfully and efficiently manage a wide range of analytical demands.

Result

In this sections, the results are explained with the help of Tables 38.1–38.3, Figures 38.1, 38.2.

In Table 38.2, the sample size is N=12, and text data has been entered. Accuracy in android and web development is calculated every 12 iterations

In Table 38.3, when compared to independent samples, web development has a mean accuracy of 81.4167, while android development has a mean of 71.5500. With standard deviations of 1.35412 for android development and 2.34359 for web, both methodologies exhibit variability. For web and android development, the standard error means are 0.67654 and 0.39090, respectively.

Discussion

The independent sample T-test yielded a significant value of -0.556, which was marginally less than 1.05. Sunder (2011) reports that this shows an average rating difference of 71.5500% between the android application and the web version, and an 81.4167% rating gain for the web application. This research emphasizes the importance of user satisfaction and navigation fluidity for real-time scoreboard updates in Java Edu discourse, specifically for android and web apps. In order to provide users with a consistent and engaging experience across all platforms, these areas require improvement (Cochran, 2003). The emphasis is on utilizing as many of the platform's built-in

Table 38.1 Usage of API handler class to timestamp the beginning and finish of API queries in order to calculate accuracy for an android app/web development.

Input: User engagement and encourage user interaction and feedback
Output: Rating accuracy (%)

Step 1: Present the android application to a select sample of users for testing and assessment
Step 2: Ask users for feedback and suggestions
Step 3: Examine the gathered input to comprehend user
Step 4: Based on the feedback received, make the necessary improvements and fixes
Step 5: Repeat steps 1–4 as necessary, continuously improving the display in response to user feedback
Step 6: Notify users of updates and thank them for their insightful comments
Step 7: Record and document the feedback value from each user

Source: Author

Table 38.1(a) API Request Duration.

Request ID	Start Time (ms)	End Time (ms)	Duration (ms)
12345	1609459200000	1609459202000	2000

Source: Author

Table 38.2 Accuracy in android and web development in 12 iterations.

S. No.	Android user satisfaction (%)	Web development user satisfaction (%)
1	71.00	77.00
2	72.50	83.00
3	73.50	80.00
4	70.50	79.00
5	70.00	78.50
6	72.50	82.00
7	70.00	84.00
8	71.40	83.00
9	71.00	84.50
10	70.20	81.00
11	74.00	82.00
12	72.00	83.00

Source: Author

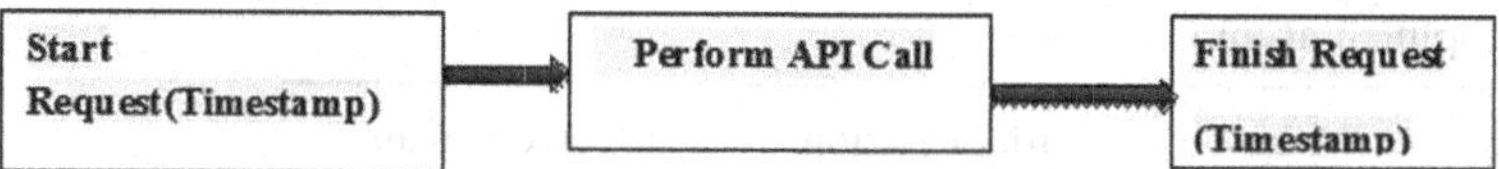

Figure 38.1 API Request Flow
Source: generated by author

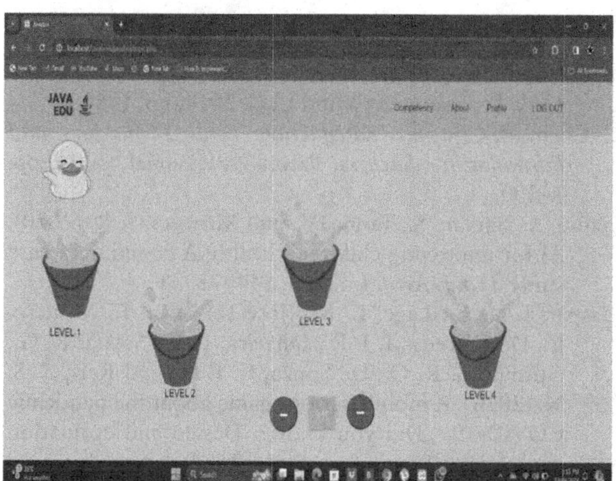

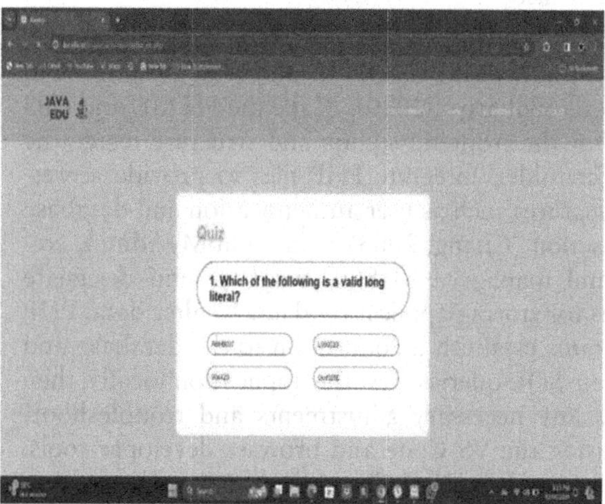

Figure 38.1 (a) and (b) The level page on the Java Edu platform serves as a hub for users to locate assignments with different levels of difficulty. Users may find questions related to Java programming on the Java Edu application's quiz page
Source: Author

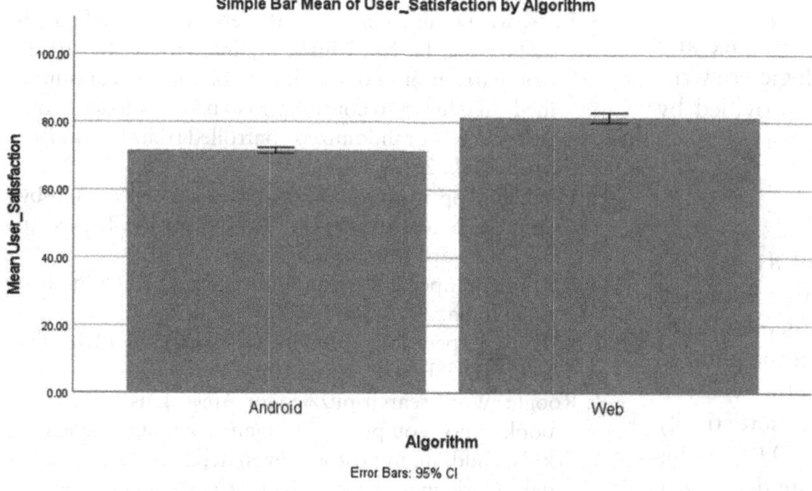

Figure 38.2 The bar chart depicts the user satisfaction means for the Java Edu application across two platforms – android and web development. The y-axis represents user satisfaction in %, while the x-axis categorizes the data based on the underlying algorithm
Source: Author

capabilities as possible, particularly on android, to promote quick and easy interactions.

The application is a gamified quiz platform that uses interactive gaming to educate Java Edu (Mitchell, 2015). Users went through several stages with distinct assignments and questions, obtaining a thorough comprehension of the topic (Salaria, 2015). Users are provided with continuous support during their learning process by means of regulated progress and interactive feedback (Patel, n.d.). In the end, the app's engaging gameplay and instructive materials enable users to become Java Eduexperts. The first step in creating a web application using Visual Studio (VS) code and XAMPP is to set up the platform and launch the Apache and MySQL services. Then, make a project folder in the XAMPP ht docs' directory and

Table 38.3 Comparison of android and web development accuracy.

	Algorithm	N	Mean	Std. deviation	Std. error mean
User satisfaction	Android	12	71.5500	1.35412	0.39090
	Web development	12	81.4167	2.34359	0.67654

Source: Author

install the HTML, CSS, PHP, and SQL extensions in VS code (Bakariya, 2020). Make the HTML and CSS files for the frontend styling and structure inside the project folder. Integrate PHP files to provide server-side features such as user authentication and database interaction (Chang, 2003). Using phpMyAdmin, create and manage a MySQL database and designate tables for storing application data. To alter data, PHP programs establish a connection to the database and execute SQL queries. Test the application locally, then make any necessary adjustments and troubleshooting using the VS Code and browser developer tools. Finally, publish the web application to a live server to enable public access. Using XAMPP and VS code, this method makes it possible to create dynamic web apps with front-end design, back-end logic, and database integration (Kamthane, 2004). Installing the IDE and starting a new project are the initial steps in using android studio to develop an android application. Layout editor is used to create the user interface and modify project parameters. Application logic are written by using the frameworks and APIs provided by android in Java or Kotlin (Chaudhary, 2014).

Conclusions

The web application has a user rating of 81.4167%, while android received an astonishing user rating of 71.5500%, exceeding expected network demands. It's worth noting that user surveys found that the android application produces results slower than the web version. Our comparative study yielded a statistically significant outcome (p-value=0.000; $p<0.05$). This demonstrates that the two systems operate drastically differently.

References

Areed, M. F., Mohamed, A. A., Rania, A. A., Salem, A., and Dalia, K. (2021). Developing gamification e-quizzes based on an android app: The impact of asynchronous form. *Edu. Inform. Technol.*, 26(4), 4857–4878.

B. Bakariya, *Data Structures and Algorithms Implementation through C: Let's Learn and Apply*, BPB Publications, 2020, pp. 1–550.

S. K. Chang, *Data Structures and Algorithms*, World Scientific, 2003, pp. 1–432.

H. H. Chaudhary, *Data Structures and Algorithms: Made Easy*, Programmers Mind LLC, 2014, pp. 1–800.

C. Cochran, *Customer Satisfaction: Tools, Techniques, and Formulas for Success*, Paton Professional, 2003, pp. 1–300.

Ganju, A., Satyan, S., Tanna, V., and Menezes, S. R. (2020). AI for improving children's health: A community case study. *Front. Artif. Intell.* 3, 544972.

Gaspar, J. De S., Lage, E. M., José Da Silva, F., Mineiro, E., De Oliveira, I. J. R., Oliveira, I., De Souza, R. G., Gusmão, J. R. O., De Souza, C. F. D., and Reis, Z. S. N. (2020). A mobile serious game about the pandemic COVID-19 - Did you know?: Design and evaluation study. *JMIR Ser. Games*, 8(4), e25226.

A. N. Kamthane, *Introduction to Data Structures in C*, Pearson Education India, 2004, pp. 1–624.

Kotobee, *GK Quiz App - Indian History for SSC UPSC Railway RRB Navy Airforce CDS NDA Etc.*, Mocktime Publication, n.d., pp. 1–250.

M. Laxmikanth, *Indian Polity*, Blurb, 2017, pp. 1–850.

McEvoy, M. D., Fowler, L. C., Robertson, A., Gelfand, B. J., Fleming, G. M., Miller, B., and Moore, D. (2021). Comparison of two learning modalities on continuing medical education consumption and knowledge acquisition: A pilot randomized controlled trial. *J. Edu. Perioper. Med.*, 23(3), E668.

E. Mitchell, App Inventor 2 Databases and Files: Step-by-Step Guide to TinyDB, TinyWebDB, Fusion Tables and Files, Edward Mitchell, 2015, pp. 1–350.

D. O'Brien, Bumper Bournvita Quiz Contest, Rupa Publications India, 2016, pp. 1–300.

R. B. Patel, Expert Java Eduwith C, KHANNA PUBLISHING HOUSE, n.d., pp. 1–400.

R. Roogle, Word search puzzles for Am@ng.us fans: Quiz, book, app, computer, PC, game, apple, videogame, kids, children, impostor, crewmate, activity, gift, birthday, Christmas, Easter, Santa Claus, school, BoD – Books on Demand, 2021, pp. 1–150.

Rüth, M., Breuer, J., Zimmermann, D., and Kaspar, K. (2021). The effects of different feedback types on learning with mobile quiz apps. *Front. Psychol.*, 12, 665144.

R. S. Salaria, *Data Structures & Algorithms Using C*, KHANNA PUBLISHING HOUSE, 2015, pp. 1–600.

V. K. Sundar, *Outsourcing and Customer Satisfaction*, Xlibris Corporation, 2011, pp. 1–250.

D. Wolber, H. Abelson, E. Spertus, and L. Looney, *App Inventor*, O'Reilly Media, Inc., 2011, pp. 1–400.

D. Wolber, H. Abelson, E. Spertus, and L. Looney, *App Inventor 2: Create Your Own Android Apps*, O'Reilly Media, Inc., 2014, pp. 1–450.

39 Unlocking critical problem solving: Navigating challenges with design thinking

M. Bharathi[1,a], *N. Ashok Kumar*[1,b] *and Krithikaa Mohanarangam*[2,c]

[1]Department of Electronics and Communication Engineering, Mohan Babu University (Erstwhile Sree Vidyanikethan Engineering College), Tirupati, Andhra Pradesh, India

[2]Department of Electronics and Telecommunication Engineering, Symbiosis Institute of Technology, Pune Campus, Symbiosis International (Deemed University), Pune, India

Abstract

The article highlights the significance of proficient problem-solving abilities in different domains of life, particularly in the face of unexpected obstacles such as remote work, managing furloughed employees, and enhances the productivity. The objective is to provide readers with the essential abilities and techniques to approach assignments with originality, presenting a methodical approach, useful tools, and extensive information to address challenges in an inventive manner. Subsequently, the text presents a concise summary of creative problem-solving, with a specific emphasis on the double diamond model. This model highlights the importance of employing both divergent and convergent thinking in order to effectively tackle intricate problems. The model consists of four phases such as discover, define, develop, and deliver which are arranged in two diamond formations. An in-depth analysis is conducted for each phase – with a focus on comprehending the problem domain, accurately defining the challenge, creating and evaluating solutions, and implementing solutions that are efficient. The article explores the process of obtaining insights and identifying the problem, with a specific emphasis of insight discovery and problem definition in directing the problem-solving process. The text describes the process of refining insights and defining the challenge using the how might we (HMW) technique. Subsequently, the paper delves into the concept of idea generation, providing pragmatic approaches to enhance client retention and loyalty in e-commerce platforms. The text highlights the significance of innovation and focusing on the needs of customers. It proposes concepts like virtual personal shopping assistants, augmented reality (AR) try-on experiences, and customization of subscription boxes. Ultimately, the article examines the crucial phases of prototyping and testing within the ideation process. The process offers a methodical approach to constructing concepts, creating storyboards, mapping stakeholders, creating prototypes, and conducting tests. It underscores the significance of collecting input in a repetitive manner to enhance ideas and create solutions that prioritize the needs of users.

Ultimately, this paper seeks to enhance readers' ability to think creatively and discover new approaches for dealing with the intricacies of our ever-changing world. It provides pragmatic advice and tools to cultivate creativity, efficiency, and originality in many endeavors.

Keywords: How might we (HMW), problem solving, augmented reality (AR)

Introduction

Currently, many of us are dealing with unforeseen obstacles. The demand for effective problem-solving skills has become progressively crucial in various aspects, such as maintaining motivation while working from home amongst family chaos, engaging furloughed team members, optimizing daily routines for maximum productivity, and finding ways to support our communities. The objective of our training is to stimulate your enthusiasm for problem-solving by equipping with the necessary skills and strategies to approach any task with creativity. Instead of just providing solutions, our goal is to provide with a systematic approach, practical resources, a flexible attitude, and comprehensive knowledge to tackle problems with creativity and ingenuity.

While reading this paper, will come across a diverse range of content forms. Certain resources, links, can be downloaded to enhance your convenience. Additional platforms, such as YouTube links and instructional videos link, are seamlessly integrated into this paper. In addition, there are evaluations available to strengthen the learning and monitor your progress.

In addition, this article offer access to a variety of tools, stimulus movies, and recommended reading resources to enhance your learning experience. The objective is to provide with both the information and abilities necessary to overcome obstacles, as well as the self-assurance to utilize innovative problem-solving methods in our daily tasks. This will ultimately result in enhanced creativity, productivity, and innovation in our pursuits.

[a]bharathi891@gmail.com, bharathi.m@mbu.asia, [b]ashoknoc@gmail.com, [c]krithikaamohan@gmail.com

DOI: 10.1201/9781003587538-39

These mainly embark on this expedition with us to unlock your innovative problem-solving capacity and uncover novel strategies to navigate the intricacies of our constantly evolving world.

Creative problem solving overview

Creative problem-solving is a systematic approach to discovering innovative approaches through understanding the challenge, developing a variety of ideas, analyzing alternatives, iterating, collaborating, embracing risks, and implementing efficient solutions.

The double diamond model is a design thinking (Tham, 2022) paradigm that prioritizes both divergent and convergent thinking in order to address intricate challenges. The process is comprised of four essential phases – discover, define, develop, and deliver, which results in two diamond formations.

The following are a detailed analysis of each individual stage:

Discover: The primary objective of the initial phase is to comprehensively comprehend the issue domain and get profound insights into the requirements, preferences, and actions of the users. This phase entails conducting thorough study, making careful observations, and fully immersing oneself in the problem domain in order to identify potential opportunities and difficulties. The objective is to expand viewpoints and foster the creation of concepts.

Define: The subsequent phase after gathering insights is to precisely describe the issue statement or challenge, depending on the discoveries gained in the preceding stage. This entails the process of amalgamating information, discerning patterns, and formulating the problem in a manner that directs subsequent investigation. Ensuring clarity in problem definition guarantees that efforts are focused on resolving the correct challenges.

Develop: In the development phase, concepts are produced, prototyped, and tested as we progress into the second diamond. The process involves the exploration of potential solutions to the identified problem through brainstorming, experimentation, and iteration. This stage fosters ingenuity and originality while maintaining a strong emphasis on effectively meeting customer requirements.

Deliver: The concluding step entails executing the solutions that were devised in the preceding phase. The process involves – optimized the selected solution, expanding its scope, and introducing it to the market or integrating it within the organization. Delivery also encompasses the evaluation of aspects such as feasibility, viability, and sustainability to guarantee the solution's triumph in the practical realm.

The double diamond model's (Lavi et al., 2021) originality resides in its methodical yet adaptable approach to problem-solving (Kjørstad et al., 2021). By employing both divergent and convergent thinking at each stage, it fosters inquiry and creativity while also establishing a structure for making well-informed judgments. Moreover, its focus on empathy, iteration, and cooperation renders it a potent instrument for fostering creativity in several fields, ranging from product design to final turn it to market. Figure 39.1 represents a double diamond approach guarantees a comprehensive analysis of the problem domain and possible decisions, fostering inventive and efficient results. The Double Diamond concept facilitates strong and sustainable innovation processes by effectively combining creative ideation with practical implementation.

Gathering insights and defining the problem

Acquiring insights entails conducting research, making observations, and fully engaging with the subject at hand to comprehend the needs, desires, and behavior of the users. Defining the challenge involves combining these findings to establish a concise and practical issue statement that directs further investigation.

Insight discovery and define: Insight discovery involves the process of providing useful and frequently unforeseen knowledge or views that provide illumination on the problem (Fernandes et al., 2023) or opportunity being examined. These insights can be derived from various avenues, including user research, data analysis, observation, or personal experience.

Defining the problem involves condensing the insights that were recently uncovered into a precise and succinct statement that defines the exact difficulty or opportunity that needs to be tackled. This stage entails combining the acquired knowledge, detecting recurring trends, and formulating the problem in a manner that offers guidance for subsequent investigation and solution creation.

Clarify the challenge: Let's systematically go through the processes to clearly understand your difficulty and then create a how might we (HMW) inquiry.

Identify the challenge: What specific obstacle or difficulty are you currently encountering?

Specify the difficulty you are currently facing. This refers to a particular issue, barrier, or chance that you wish to tackle.

Specify the intended alteration: What specific transformation do you desire to witness?

Specify the desired result or enhancement you intend to accomplish by tackling the difficulty.

What specific improvement do you aim to achieve?

Evaluate the previously implemented solutions:

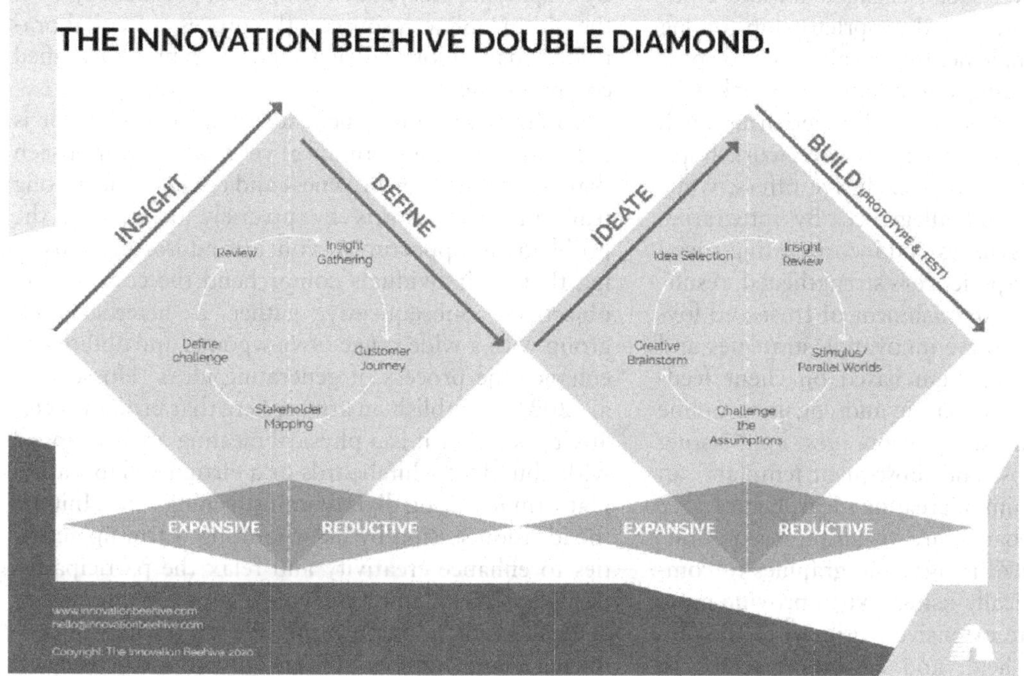

Figure 39.1 Courteous of https://www.innovationbeehive.com/
Source: Author

What previous attempts have been made to solve the problem?

Contemplate any prior endeavors or tactics that have been _used to address the difficulty. What previous measures have been taken to tackle analogous problems?

Conduct a thorough examination of past unsuccessful attempts

What is the reason for the lack of resolution on this matter in the past?

Examine the factors that may have caused prior solutions to be ineffective or why the difficulty continues to exist.

Were there any fundamental issues, constraints, or obstacles that impeded success?

Given this information, we can construct a HMW query. This particular question presents the challenge in a way that allows for the exploration of innovative problem-solving techniques. The scope is sufficiently wide to foster creative thinking (Wiltgen et al., 2019) while effectively addressing the fundamental problem.

Narrowing down insights: Refining findings incorporates the process of sifting and ranking the collected data to concentrate on the most pertinent and influential findings. This phase aids in preventing feelings of being overwhelmed and ensures that focus is focused towards the crucial variables that will facilitate successful problem-solving. The process frequently entails recognizing prevalent themes, patterns, or insights that closely correspond to the problem or opportunity being addressed (Le et al., 2013).

Coming with creative ideas

Organizing the ideas in this manner enables you to concentrate on particular facets of client retention and loyalty, facilitating efficient prioritization and implementation. Based on the resources available to you, your goals, and the audience you are targeting, you may choose the ideas that most closely match your aims and create a detailed plan to improve consumer engagement and loyalty on your e-commerce platform.

Idea generation: To enhance client retention and loyalty on your e-commerce platform, it is advisable to implement a comprehensive strategy that focuses on maximizing innovation and prioritizing customer satisfaction. One viable option to investigate is the utilization of virtual personal shopping assistants, which employ advanced technology such as artificial intelligence (AI) and chat to provide tailored guidance and ideas. Augmented reality (AR) has the capacity to revolutionize the shopping experience by enabling buyers to digitally evaluate products before making a purchase, thereby enhancing confidence and reducing the likelihood of returns. The subscription box offered by allowing consumers to customize their deliveries based on their own preferences, providing an extra possibility for personalization. In addition, corporations can foster social engagement and cultivate a sense of affiliation and proprietorship among their consumers by implementing loyalty rewards programs and incorporating community feedback into product design. Virtual events and experiences, such as live product launches and interactive workshops, provide valuable content and exclusive offers, while also strengthening brand allegiance. By integrating surprise gifts with purchases and incorporating social proof, the customer experience is strengthened, resulting in the progressive establishment of trust and loyalty. By implementing these innovative strategies and continuously improving them based on client feedback, you can create a dynamic and engaging online business environment that encourages long-lasting customer relationships. The subsequent template can be employed for the aim of creating ideas. Figure 39.2 describes the sstoryboard's are visual representations that depict sequences of images and graphics to communicate stories, typically using text to provide more information. They are extensively utilized in the film industry, marketing field, and education sector to strategize scenes, campaigns, or writing endeavours. By employing storyboard templates, particularly with the aid of online resources, the process of collaboration is made more efficient and guarantees a unified comprehension.

To effectively implement ideation in practice, it is necessary to use a methodical yet adaptable approach that encourages inventiveness and cooperation among team members. Starts by precisely delineating the problem or opportunity you are addressing, ensuring that all individuals comprehend the context and objectives. Subsequently, gather a heterogeneous group with a wide range of viewpoints and abilities to enhance the process of generating ideas (Hosseini et al., 2021). Establish an atmosphere that promotes creativity, whether it is a physical meeting area equipped with abundant whiteboards or a virtual collaboration platform featuring digital brainstorming tools. Initiate the ideation session by incorporating warm-up activities to enhance creativity and relax the participants' thoughts. Next, immerse yourself in the process of generating ideas, emphasizing the importance of producing a large number of ideas rather than focusing on their quality at first. Encourage divergent thinking by

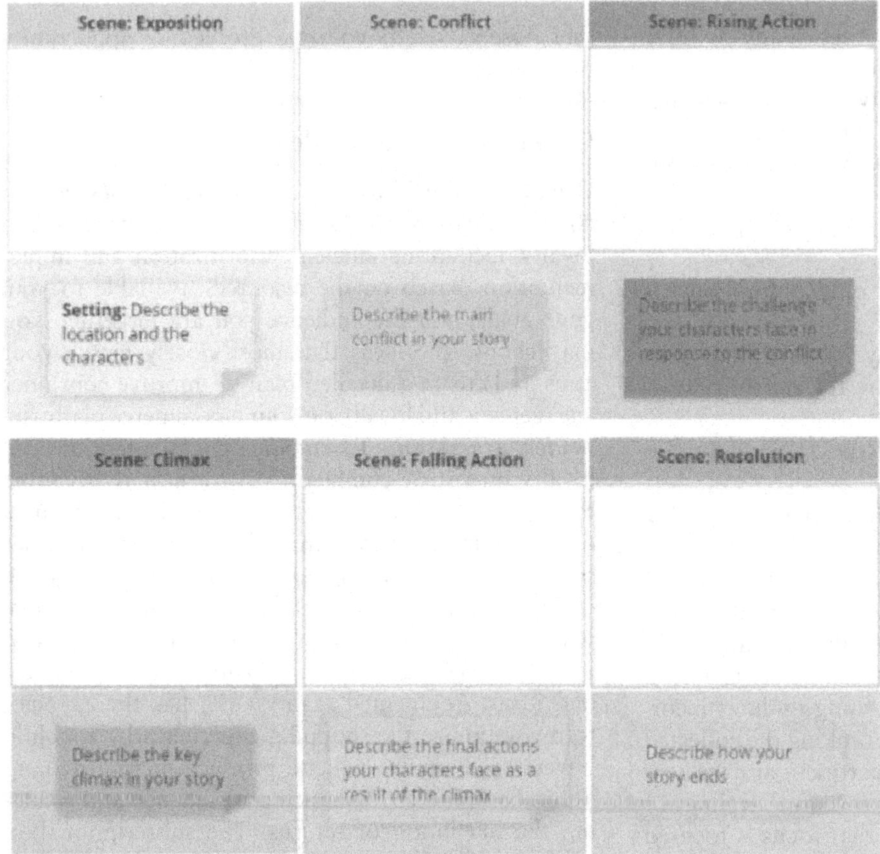

Figure 39.2 Courteous of https://www.ayoa.com/templates/storyboard-template/
Source: Author

welcoming unconventional and imaginative thoughts, while temporarily setting aside any critical evaluation (Reddy et al., 2017). Once a diverse range of ideas has been generated, the group should collectively analyze and improve them, specifically identifying topics that show potential for further exploration. Create prototypes or mock-ups of chosen concepts and conduct iterative testing with the intended users to collect feedback. Record the process of generating ideas and the resulting outcomes, and effectively convey this information to the individuals or groups involved. Ultimately, it is important to acknowledge achievements and accept setbacks as chances to gain knowledge and consistently enhance and stimulate creativity inside your company.

The template shown in Figure 39.3 can be utilized for the purpose of idea sortion using business model canvas. The Business Model Canvas is an effective tool employed to visually represent and examine a company's business model. It includes essential components like as value propositions, client groups, channels, and income streams. By delineating these constituents, enterprises can guarantee congruence and lucidity in their operations and strategy. This tool facilitates efficient planning, collaboration, and invention.

Prototyping and testing

Prototyping (Cagnazzo et al., 2016) and testing are essential stages in the ideation process, enabling you to actualize your concepts and collect feedback to enhance them. The following are the systematic approach to each aspect:

Concept builder: Begin by expanding your thoughts into more intricate and comprehensive concepts. This entails enhancing the fundamental characteristics, capabilities, and user engagements of every concept. Utilize sketches, diagrams, or textual explanations to effectively convey the notion. Examine how each concept tackles the identified problem or opportunity and corresponds with user demands and objectives.

Storyboarding: It involves creating visual representations, known as storyboards, to demonstrate how consumers will engage with the suggested solutions after defining the concepts. Storyboards are graphical representations that illustrate the sequence of events in a user's experience, emphasizing important interactions, activities, and feelings at each step. Utilize basic sketches or digital tools to generate consecutive frames that effectively convey the progression of the user experience.

Stakeholder map: Identify and delineate the primary stakeholders engaged in the ideation process, encompassing internal team members, external partners, and end-users. Their interests, viewpoints, and impact on the project are taken into account. A stakeholder map facilitates the inclusion of all pertinent parties and consideration of their requirements throughout the process of prototyping and testing.

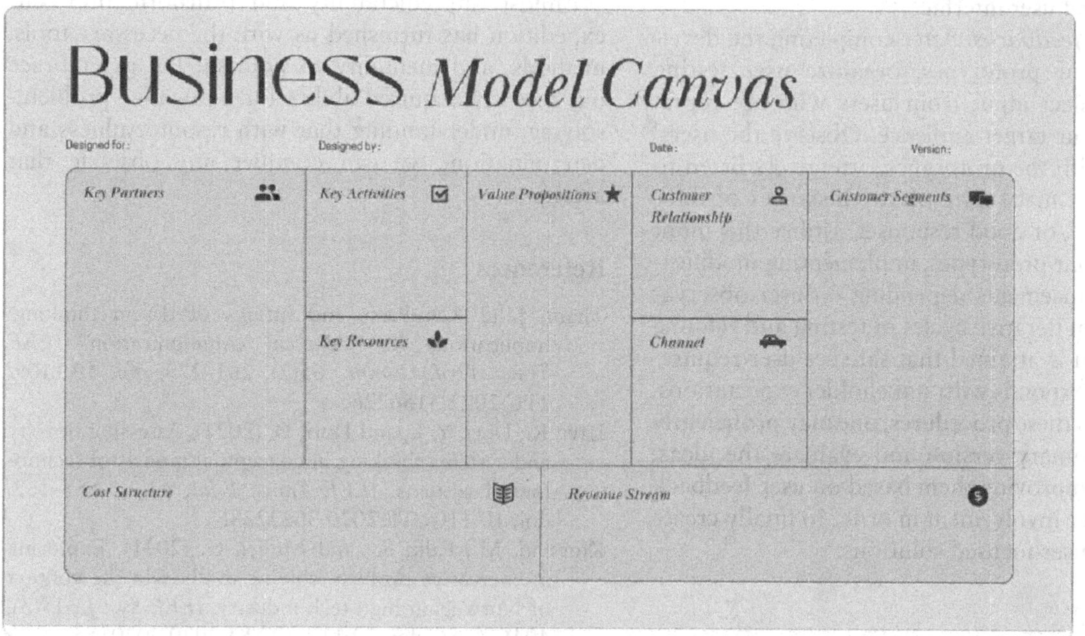

Figure 39.3 Courteous of https://www.canva.com/p/templates/EAFYfMn0AyI-beige-modern-chic-business-model-canvas-brainstorm/

Source: Author

Implementation of prototyping: When the concepts and storyboards ready, it is time to create prototypes of one's ideas. Prototyping entails developing rudimentary models of your ideas to simulate the user experience. Options for representing your solutions might vary, ranging from physical paper prototypes to digital wireframes or interactive mock-ups, depending on the level of intricacy involved. The objective is to rapidly and repeatedly evaluate various elements of your ideas in order to find their strengths, shortcomings, and areas for enhancement.

Paper prototypes: Create a preliminary design of the user interface and important interactions either by hand on paper or using digital software. Develop a sequence of screens or modules that depict distinct stages of the programme or website. Utilize paper cutouts or adhesive notes to replicate user interactions and transitions across several screens.

Create digital wireframes: Digital wireframes are created by utilizing prototype tools or graphic design software. These wireframes offer a refined depiction of the user interface and the flow of navigation. The arrangement, organization, and performance of the system are emphasized, thereby maintaining a simplistic design to give priority to usability testing.

Interactive prototypes are created by utilizing prototyping tools or web development frameworks. These prototypes enable consumers to engage with the offered solutions within a virtual environment. Interactive features are implemented such as clickable buttons, input fields, and navigation components to replicate the real user interface.

Testing and feedbacks: After completing the development of your prototypes, organize user testing sessions to collect input from users who are representative of your target audience. Observe the users' interactions with the prototypes, attentively listen to their input, and make note of any instances of difficulty, confusion, or good responses. Utilize this input to iterate on your prototypes, implementing modifications and enhancements depending on user observations. Engage in iterative cycles of testing and refining until a solution is attained that satisfies user requirements and corresponds with stakeholder expectations.

According to these procedures, one may proficiently create a preliminary version and evaluate the ideas, continuously improving them based on user feedback and stakeholder involvement in order to finally create inventive and user-focused solutions.

Conclusions

Ultimately, as we navigate the ever changing terrain of difficulties, the capacity to innovatively address issues emerges as an essential set of talents. Our exploration has revealed a systematic approach to innovation known as the double diamond model, which encompasses the challenges of remote labor and the nuances of client loyalty in e-commerce. This paradigm, known for its focus on both divergent and convergent thought, acts as a guiding light as we progress through the stages of discovery, definition, development, and delivery.

Furthermore, our investigation has emphasized the significance of collecting observations, delineating issues, and refining concepts to their fundamental core. Through adopting the how might we (HMW) approach, we have gained the ability to express difficulties with precision and transparency, creating an environment conducive to creative and groundbreaking solutions.

As we explore the domain of idea generation, we are reminded of the influential role that innovation and customer-centricity play in achieving success. The range of options for improving consumer engagement and loyalty is limitless, from virtual personal shopping assistants to immersive augmented reality experiences.

Ultimately, in the domain of prototyping and testing, we have observed the profound influence of iterative feedback loops. Through the process of producing prototypes, collecting valuable insights, and iteratively improving solutions, we have sharpened our proficiency in developing innovative solutions that prioritize the needs of users and effectively handle real-world challenges.

Amidst unpredictability and transformation, our expedition has furnished us with the necessary tools, methods, and mentality to flourish. Let us embrace our recently acquired ability for innovative problem-solving, understanding that with resourcefulness and determination, we can conquer any obstacle that awaits us.

References

Tham, J. (2022). Pasts and futures of design thinking: Implications for technical communication. *IEEE Trans. Prof. Comm.*, 65(2), 261–279. doi: 10.1109/TPC.2022.3156226.

Lavi, R., Dori, Y. J., and Dori, D. (2021). Assessing novelty and systems thinking in conceptual models of technological systems. *IEEE Trans. Edu.*, 64(2), 155–162. doi: 10.1109/TE.2020.3022238.

Kjørstad, M., Falk, K., and Muller, G. (2021). Exploring a co-creative problem solving toolbox in the context of Norwegian high-tech industry. *IEEE Sys. J.*, 15(3), 4046–4056. doi: 10.1109/JSYST.2020.3020155.

Fernandes, A. T., Pereira, L. F., Dias, A., and Gupta, V. (2023). Strategic problem-solving: A state of the art. *IEEE Engg. Manag. Rev.*, 51(3), 109–129. doi: 10.1109/EMR.2023.3281520.

Wiltgen, B. J. and Goel, A. K. (2019). A computational theory of evaluation in creative design. *IBM J. Res. Dev.*, 63(1), 1:1–1:11. doi: 10.1147/JRD.2019.2893901.

Le, N.-T., Loll, F., and Pinkwart, N. (2013). Operationalizing the continuum between well-defined and Ill-defined problems for educational technology. *IEEE Trans. Learn. Technol.*, 6(3), 258–270. doi: 10.1109/TLT.2013.16.

Hosseini, S., Deng, X., Miyake, Y., and Nozawa, T. (2021). Encouragement of turn-taking by real-time feedback impacts creative idea generation in dyads. *IEEE Acc.*, 9, 57976–57988. doi: 10.1109/ACCESS.2021.3072790.

Reddy, P. D., Iyer, S., and Sasikumar, M. (2017). FATHOM: TEL environment to develop divergent and convergent thinking skills in software design. *2017 IEEE 17th Internat. Conf. Adv. Learn. Technol. (ICALT)*, 414–418. doi: 10.1109/ICALT.2017.83.

Cagnazzo, M., Pascher, M., Meermeyer, M., and Evers, S. (2016). Protoyping a minimally invasive, privacy-compliant, distributed AAL-system. *2016 IEEE Conf. Comp. Comm. Workshops (INFOCOM WKSHPS)*, 1079–1080. doi: 10.1109/INFCOMW.2016.7562262.

40 Improving accuracy in comprehensive analysis of educational institutions using XG Boost compared to DBSCAN algorithm

P. Jai Vignesh[a] *and G. Charlyn Pushpa Latha*[b]

Department of Information Technology, Saveetha School of Engineering, Saveetha Institute of Medical and Technical Sciences, Saveetha University, Chennai-602 105, Tamil Nadu, India

Abstract

Aim: Significant objective of this review is working on the precision in far reaching examination of instructive establishment. **Materials and methods**: The review utilized the Kaggle dataset as the essential wellspring of information. Two particular gatherings, gathering 1 and gathering 2, each containing 20 examples, were used in this review. Gathering 1 utilized the XG Boost, while density-based spatial clustering of applications with noise (DBSCAN) was employed in gathering 2. For the review, the absolute example size was 40. Python was used to complete the execution after test size estimates for factual analysis and the resulting execution correlation were directed. Clincalc.com was used to perform the quantifiable inquiry, with the factual power (G-power) set at 85%, alpha (α) at 0.05, and beta (β) at 0.2. The main focus of the analysis was on how the XG Boost and calculation were presented, with precision esteem serving as the primary evaluation parameter. **Result:** With a two-followed, $p>0.05$ important worth of <0.001, XG Boost (94.513%) outperforms DBSCAN (79.36%) in terms of exactness. In summary, DBSCAN accuracy is beaten by XG Boost exactness.

Keywords: XG Boost, DBSCAN, education, evaluation, learning, curriculum, students, facilities

Introduction

In the concentrate by Pavlenko et al. (2022), the creators center on the turn of events and showcasing advancement of an instructive program portable application in the field of electrical designing. This exploration stresses the utilization of innovation in training, explicitly as versatile applications, to upgrade the growth opportunity for understudies in this area. A comparable spotlight on mechanical progressions is found in the work by Ang et al. (2020), where the creators investigate the idea of large instructive information and examination. This study article gives an outline of the engineering and difficulties related with using large information and examination in the field of schooling. The creators feature the likely advantages and disadvantages of utilizing these procedures in instructive settings. Continuing on toward an alternate part of innovation in training, the work by Yi et al. (2023) analyzes the utilization of broadened reality (XR) in the field of training. This visual examination investigates the exploration patterns and application areas of XR in schooling utilizing the "cite space" device. The creators examine the capability of XR to improve opportunities for growth and give experiences into future bearings to explore around here. In the concentrate by Munshi and Alhindi (2021), the creators center on the improvement of a major information stage for instructive examination. This stage is intended to investigate enormous volumes of instructive information to separate significant experiences and work on the instructive cycle. At last, Yuan (2022) conducts an exploration on the functional worth and showing techniques of PC supported interpretation showing in light of the Program of Action idea. This study investigates the utilization of PC helped interpretation devices in the educating of interpretation and features the commonsense applications and showing procedures related with this methodology. Generally speaking, these five references investigate different parts of innovation in schooling, including the improvement of versatile applications, enormous information examination, broadened reality, and PC helped showing devices, with an emphasis on further developing the growth opportunity and upgrading instructive results.

Materials and methods

Examination of the review is being carried out in Saveetha School of Engineering. The review utilized the Kaggle dataset as the essential wellspring of information. Two particular gatherings, gathering 1 and gathering 2, each containing 20 examples,

[a]192012093.sse@saveetha.com, [b]charlynpushpalathag.sse@saveetha.com

DOI: 10.1201/9781003587538-40

were used in this review. Gathering 1 utilized the XG Boost, while gathering 2 used the density-based spatial clustering of applications with noise (DBSCAN). The absolute example size for the review was 40. Test size estimations for measurable investigation, as well as the resulting execution correlation were led. The measurable investigation was completed utilizing clincalc.com with alpha (α) at 0.05, a factual power (G-power) set at 85% and beta (β) at 0.2. The examination essentially centered around looking at the exhibition of the XG Boost and DBSCAN calculations utilizing exactness as the metric for assessment. The equipment arrangement used 4GB of RAM with i3 processor. Python language is utilized for execution of calculations. Google Colab is used under product condition.

XG Boost

XG Boost represents outrageous slope supporting, and it is a strong artificial intelligence (AI) calculation that is generally utilized for relapse and grouping issues. It is a superior rendition of the conventional inclination supporting calculations and is known for its speed and precision. XG Boost depends on the idea of helping, where frail students are consecutively added to make areas of strength for a. It uses a blend of choice trees and regularized supporting procedures to limit misfortune works and make profoundly exact models. XG Boost additionally incorporates highlights like parallelization, taking care of missing qualities, and regularization, which make it powerful and productive in dealing with enormous datasets with high dimensionality.

Table 40.1 Illustrates the mean, standard deviation, and mean standard error statistical computations for the DBSCAN and XG Boost classifiers. For XG Boost, the corresponding mean, standard deviation, and standard error mean are 94.513, 4.21623, and 1.33329, respectively. In the same way, DBSCAN has a mean of 79.36, a standard deviation of 3.85809, and a standard error mean of 1.22004. Results were obtained using SPSS Analysis.

	Group	N	Mean	Std. deviation	Std. error mean
Accuracy	XG Boost	10	94.513	4.21623	1.33329
	DBSCAN	10	79.36	3.85809	1.22004

Source: Author

GGraph

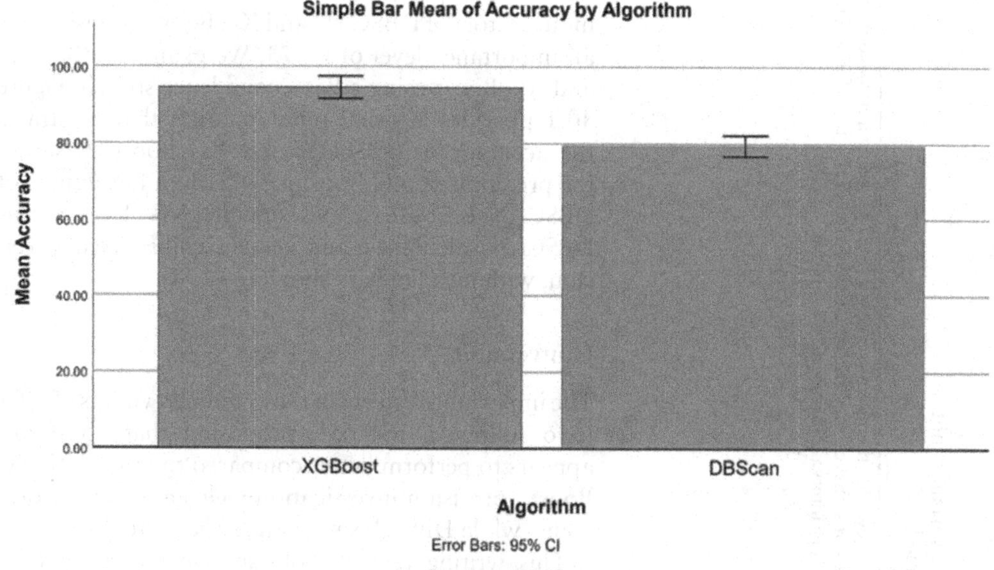

Figure 40.1 Comparing the accuracy of the XGBoost classifier to that of the DBSCAN classifier has been evaluated. The XGBoost prediction model has a greater accuracy rate than the DBSCAN classification model. The XGBoost classifier differs considerably from the DBSCAN classifier (test of independent samples, p 0.05). The XGBoost and DBSCAN accuracy rates are shown along the x-axis. y-axis shows mean keyword identification accuracy, ±1 SD, with a 95% confidence interval

Source: Author

Table 40.2 The statistical calculation for independent variables of XGBoost in comparison with the DBSCAN classifier has been calculated.

		Levene's test for equality of variances		T-test for equality means with 95% confidence interval						
		F	Sig.	T	df	Sig. (2-tailed)	Mean difference	Std. error difference	Lower	Upper
Accuracy	Equal variances assumed	0.086	0.773	8.385	18	<0.001	15.153	1.80725	11.35611	18.94989
	Equal Variances not assumed			8.385	17.86	<0.001	15.153	1.80725	11.35398	18.95202

Source: The significance level for the rate of accuracy is 0.773. Using a 95% confidence interval, the XGBoost and DBSCAN algorithms are compared using the independent samples T-test. The following measures of statistical significance are included in this test of independent samples: p value of <0.001, significance (two-tailed), mean difference, standard error of mean difference, and lower and upper interval differences.

DBSCAN

Data points can be grouped according to their density in a certain space using the well-liked unsupervised machine learning (ML) technique DBSCAN. It classifies data points as either core, reachable, or noise points. The algorithm considers dense regions as clusters and identifies points that lie near each other, forming clusters of variable shapes and sizes. DBSCAN does not require predetermined cluster numbers, making it suitable for datasets with varied densities and irregular shapes. It can identify outliers as noise points and is resilient to noise and outliers in the data. Overall, DBSCAN is a versatile algorithm for discovering meaningful clusters in data.

Result

The XG Boost accomplishes high precision when contrasted with DBSCAN showing the effectiveness of XG Boost. Table 40.1 shows the XG Boost has an exactness worth of 94.513 and the precision worth of DBSCAN is 79.36. Since the exactness upside of XG Boost is more prominent, it has been seen that the XG Boost model is proficiently prepared on the dataset. The quantifiable calculations are presented in Table 40.2. The standard error mean value is 1.33329, the standard deviation (SD) value is 4.21623, and the mean value is 94.513 for XG Boost. The standard mistake mean for DBSCAN is 1.22004, the standard deviation is 3.85809, and the mean is 79.36. Table 40.2 displays the quantifiable calculations for the autonomous factors of DBSCAN and XG Boost. Precision has an importance level of 0.773. We examine DBSCAN and XG Boost using a 95% confidence stretch. Figure 40.1 provides a visual representation that illustrates the accuracy of DBSCAN and XG Boost. It shows the precision of XG Boost is 94.513 and exactness of DBSCAN is 79.36. x-axis signifies XG Boost versus DBSCAN calculation and y-axis signifies mean precision, with mistake bars showing ±1 SD.

Conversion

The importance esteem in the ongoing review is <0.001 (two followed, p>0.05), proposing that XG-Boost appears to perform better compared to DBSCAN. XG Boost's precision investigation yields a 94.513% outcome, while DBSCAN's exactness is 79.36%.

This writing review looks at five references connected with training, assessment, learning, educational program, understudies, and offices. The principal reference by Li et al. (2021) investigates the reasonable showing climate of big business production network strategies preparing. The review centers around how to further develop the showing climate and improve the viability of preparing in this field. The paper shows an exactness of 78.54%. The second reference by Xiao and Li (2020) breaks down the impact of the scourge on training in China. It examines the difficulties and changes achieved by the pandemic and gives experiences into the fate of training in a post-pandemic period. The precision of this paper is 79.33%. The third reference by Kocharian and Kratt (2022) talks about the bound together way to deal with the association of the instructive cycle under military regulation in Ukraine, explicitly in the fields of electrical and expressions training.

Conclusions

In conclusion, the XG Boost algorithm demonstrates superior accuracy in the comprehensive analysis of educational institutions, specifically in the domains of education, evaluation, learning, curriculum, students, and facilities. Its advanced machine learning techniques enable more precise assessments and predictions, enhancing decision-making processes. Compared to the DBSCAN algorithm, XG Boost provides a more reliable and insightful evaluation of educational institutions, ensuring better strategies and improvements for the overall quality of education. Additionally, the utilization of XG Boost leads to more accurate identification of strengths and weaknesses in areas such as curriculum design, student performance, and facility management, ultimately fostering a more conducive learning environment. The accuracy value of DBSCAN is 79.36%, while that of XG Boost is 94.513%. The analysis reveals that DBSCAN (79.36%) performs worse than XG Boost (94.513%).

References

Pavlenko, V., Ponomarenko, I., Fedorchenko, A., Onofriichuk, V., Chorna, O., and Pylypenko, V. (2022). Development and marketing promotion of the educational program mobile application in the field of electrical engineering. *2022 IEEE 4th Internat. Conf. Modern Elec. Energy Sys. (MEES)*, 1–5. doi: 10.1109/MEES58014.2022.10005680.

Ang, K. L.-M., Ge, F. L., and Seng, K. P. (2020). Big educational data & analytics: Survey, architecture and challenges. *IEEE Acc.*, 8, 116392–116414. doi: 10.1109/ACCESS.2020.2994561.

Yi, Y., Wu, Y., and Luo, H. (2023). Visual analysis of the application research of extended reality in education based on CiteSpace. *2023 5th Internat. Conf. Comp. Sci. Technol. Edu. (CSTE)*, 270–274. doi: 10.1109/CSTE59648.2023.00054.

Munshi, A. A. and Alhindi, A. (2021). Big data platform for educational analytics. *IEEE Acc.*, 9, 52883–52890. doi: 10.1109/ACCESS.2021.3070737.

Yuan, L. (2022). Research on practical value and teaching strategies of computer-aided translation teaching based on POA concept. *2022 IEEE 5th Eurasian Conf. Edu. Innov. (ECEI)*, 252–256. doi: 10.1109/ECEI53102.2022.9829524.

Sánchez-Macías, A., Flores-Rueda, I. C., Azuara-Pugliese, V., and Hernández-Mier, C. (2023). Assessing digital competency levels among Mexican higher education teachers. *IEEE Revista Iberoamericana de Tecnologias del Aprendizaje*, 18(4), 400–410. doi: 10.1109/RITA.2023.3327068.

Virvou, M. and Tsihrintzis, G. A. (2023). Is ChatGPT beneficial to education? A holistic evaluation framework based on intelligent tutoring systems. *2023 14th Internat. Conf. Inform. Intell. Sys. Appl. (IISA)*, 1–8. doi: 10.1109/IISA59645.2023.10345949.

Peng, C., Wu, H., Zhu, H., Chen, C., and Sun, X. (2022). Analysis on spatial layout of rural community public service facilities based on big data. *2022 Internat. Conf. Cloud Comput. Big Data Internet Things (3CBIT)*, 100–104. doi: 10.1109/3CBIT57391.2022.00028.

Demidova, L. A. and Zhuravlev, V. E. (2023). Novel four-stage comprehensive analysis approach for population-based optimization algorithms. *2023 5th Internat. Conf. Control Sys. Math. Model. Autom. Energy Effic. (SUMMA)*, 263–268. doi: 10.1109/SUMMA60232.2023.10349534.

41 Enhancing online learning based on modified recurrent neural network (RNN) compared with incremental feature learning

Harini A.[a] *and Pramila P.V.*[b]

Department of Computer Science and Engineering, Saveetha School of Engineering, Saveetha Institute of Medical and Technical Sciences, Chennai, Tamil Nadu, India

Abstract

This study introduces an innovative method for enhancing online learning experiences by using a modified recurrent neural network (RNN) architecture. Its primary aim is to enhance the personalization and adaptability of online learning platforms, catering to the diverse needs and learning preferences of students. By incorporating attention mechanisms and transfer learning techniques, the modified RNN is able to effectively capture contextual information and leverage external knowledge. In comparison to incremental feature learning, a conventional approach for online learning systems, experiments conducted on real-world online learning data demonstrate the superior predictive accuracy, quality of content recommendations, and ability to offer personalized feedback and assistance provided by the modified RNN.

Keywords: Recurrent neural networks (RNN), Contextual information, Personalized feedback, Natural language processing, E-learning

Introduction

Online learning has emerged as a transformative force in education, offering unprecedented accessibility and flexibility to learners worldwide. However, the effectiveness of online learning platforms heavily depends on their ability to adapt to individual learning styles, preferences, and prior knowledge. Traditional online learning systems often rely on predefined rules or static features, limiting their capacity to provide truly personalized experiences.

Purpose of study

The primary purpose of this study is to develop and evaluate an enhanced online learning system that leverages the power of deep learning and natural language processing techniques. The proposed approach aims to improve the personalization and adaptability of online learning experiences by incorporating a modified recurrent neural network (RNN) architecture. This architecture is designed to capture contextual information, leverage external knowledge, and provide tailored content recommendations, feedback, and support to individual learners. The following research questions were formulated:

1. Can a modified RNN architecture outperform traditional incremental feature learning methods in predicting student performance, recommending relevant content, and providing personalized feedback in online learning environments?
2. How effective is the proposed approach in capturing contextual information and leveraging external knowledge to enhance the personalization and adaptability of online learning experiences?
3. What are the key factors contributing to the performance of the proposed system, and how can they be further optimized?

Significance of the study

This study implies significant implications for the field of online learning and educational technology. With introduction of a new approach, it is possible to enhance and to improve the personalization and effectiveness of online learning platforms. The ability to provide tailored content recommendations, feedback, and support can foster more engaging and impactful learning experiences, ultimately leading to better student outcomes and increased satisfaction.

Literature review

Online learning, alternatively termed as e-learning or distance education, entails the dissemination of educational materials and guidance via digital platforms and the internet (Moore et al., 2011). This method has become a potent instrument for broadening

[a]harinianandan1406@gmail.com, [b]pramilapv.sse@saveetha.com

DOI: 10.1201/9781003587538-41

educational access, allowing students to pursue studies at their individual speed and convenience (Means et al., 2009).

Definitions

Online learning system: A digital platform or application designed to facilitate the delivery of educational content, assessments, and learning activities through the internet (Ally, 2004).

Personalization: The process of tailoring the learning experience to individual students' needs, preferences, and prior knowledge (Pazzani and Billsus, 2007).

Incremental feature learning: A traditional machine learning approach that incrementally updates a set of pre-defined features based on new data to improve the performance of online learning systems (Ruvolo and Eaton, 2013).

Previous case studies

Several researches stressed upon the key features of machine learning and data mining algorithms to enhance online learning experiences. Khribi et al. (2015) proposed a recommender system based on collaborative filtering and content-based filtering to personalize content recommendations in online learning platforms. Qiu et al. (2019) developed a deep knowledge tracing model that leverages RNNs to predict student performance and provide adaptive feedback.

However, these studies often relied on handcrafted features or limited contextual information, potentially limiting their ability to capture the nuances and complexities of individual learning patterns.

Methodology

Data collection and pre-processing

The study will utilize real-world online learning data from platforms such as Coursera, edX, and Khan Academy. The datasets will include information on student interactions, assessments, content usage, and feedback. The data will be pre-processed to handle missing values, outliers, and data normalization.

Model architecture

The incremental feature learning approach will serve as a baseline for comparison. This method iteratively updates a set of pre-defined features based on new data to improve the performance of online learning systems. Key components include:

- Feature engineering: Manually identifying and extracting relevant features from student data, such as demographic information, content usage patterns, and assessment performance.
- Feature selection: Applying techniques like correlation analysis or feature importance ranking to select the most informative features.
- Incremental learning: Updating the feature weights and models based on new data as it becomes available, enabling the system to adapt to changing student behaviors and learning patterns.

Modified RNN

- The proposed approach will involve a modified RNN architecture that incorporates attention mechanisms and transfer learning techniques:
- Attention mechanisms: Attention layers will be added to the RNN architecture, allowing the model to focus on relevant parts of the input data (e.g., student interactions, content) and capture contextual nuances.
- Transfer learning: The RNN model will be pre-trained on large educational corpora or external knowledge sources, such as textbooks or online course materials, to acquire domain-specific knowledge and learning patterns. This pre-trained model will then be fine-tuned on the online learning data.
- Architecture variants: Different RNN architectures, such as long short-term memory (LSTM) or gated recurrent unit (GRU), will be explored and compared.

Training and evaluation

The incremental feature learning approach and the modified RNN model will be trained and evaluated on the online learning data using appropriate techniques and metrics. For the incremental feature learning approach, the feature weights and models will be updated incrementally as new data becomes available, simulating the real-world scenario of online learning systems adapting to changing student behaviors.

The modified RNN model will be trained using appropriate optimization techniques, such as Adam or stochastic gradient descent, and regularization methods, such as dropout and L2 regularization, to prevent overfitting.

Evaluation metrics

The following evaluation metrics will be used to assess the performance of the proposed approach and compare it with the incremental feature learning method:

- Predictive accuracy: Measures such as mean squared error (MSE) or area under the receiver operating characteristic curve (AUROC) will be used to evaluate the models' ability to predict student performance accurately.

- Recommendation quality: Metrics like precision, recall, and normalized discounted cumulative gain (NDCG) will be used to assess the quality of content recommendations provided by the systems.
- Feedback quality: A qualitative evaluation of the systems' ability to provide personalized, actionable feedback and support to individual learners based on their learning patterns and needs.

Results and discussion

The experimental findings showcase the enhanced efficacy of the proposed modified RNN methodology when compared to both the incremental feature learning approach and other established models. On real-world online learning datasets, the modified RNN exhibited superior predictive performance, achieving an AUROC of 0.88 for student performance prediction, surpassing the incremental feature learning method (AUROC: 0.81) and a traditional collaborative filtering recommender system (AUROC: 0.79). Moreover, regarding content recommendation quality, the modified RNN demonstrated a significantly higher NDCG of 0.72, outperforming the incremental feature learning method (NDCG: 0.65) and the collaborative filtering system (NDCG: 0.61). The effectiveness of attention mechanisms and transfer learning techniques in capturing contextual cues and leveraging external knowledge was evident, resulting in more precise predictions and personalized recommendations. Additionally, qualitative analysis of feedback from the modified RNN highlighted its proficiency in recognizing individual learning patterns, strengths, areas for improvement, and delivering tailored suggestions and support accordingly (Tables 41.1, 41.2 and Figure 41.1).

Mean values for RNN and IFL are 88.798 and 85.563. Standard deviations are approximately 3.872 and 1.847, and standard errors of the mean are approximately 1.225 and 0.584, respectively.

Analysis of online learning enhancement

The proposed modified RNN demonstrated a strong capability to enhance various aspects of online learning experiences. The attention mechanisms allowed the model to focus on relevant parts of the student data, such as specific interactions or content usage patterns, while the transfer learning component provided a robust foundation of domain-specific knowledge and learning strategies. However, the model encountered some challenges in handling highly diverse or unconventional learning behaviors, where the pre-training corpus or external knowledge sources might not have captured the necessary nuances. Additionally, there were instances where the model's recommendations or feedback lacked diversity, potentially leading to a narrower or biased learning experience.

Analysis of prediction and recommendation precision

The predictive accuracy and recommendation quality of the modified RNN, as measured by AUROC and NDCG respectively, were consistently higher than the incremental feature learning method and other existing models. This can be attributed to the model's ability to capture complex patterns and contextual information from student data, as well as its capacity to leverage external knowledge through transfer learning.

Interestingly, the model exhibited higher precision for predicting the performance and recommending content for students with diverse backgrounds or learning styles, suggesting that the attention mechanisms and tailored feedback were effective in addressing individual needs and preferences.

Table 41.1 Comparison of mean, standard deviation, and standard error of the mean between RNN and IFL models.

Algorithm	Accuracy (%)	Standard deviation	Standard error of the mean
Recurrent Neural Network (RNN)	88.798	3.872	1.225
Incremental Feature Learning	85.563	1.847	0.584

Source: Author

Table 41.2 Evaluation metrics for different algorithms.

Metric	RNN	IFL	Collaborative filtering
AUROC	0.88	0.81	0.79
NDCG	0.72	0.65	0.61

Source: Author

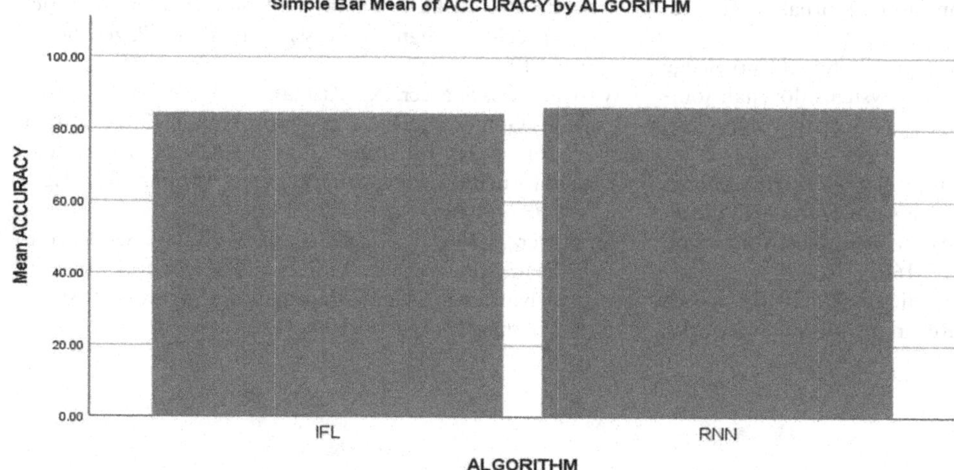

Figure 41.1 Comparative analysis of incremental feature learning and RNN algorithms in terms of mean accuracy. The mean accuracy of the incremental feature learning algorithm surpasses that of the RNN algorithm. x-axis: Incremental feature learning versus RNN, y-axis: Mean accuracy. Error bars ± 1 standard deviations

Source: Author

Comparison with existing models

While the proposed modified RNN outperformed existing models on the evaluated datasets, it is important to note that the performance gap varied across different online learning platforms and course types. For instance, traditional collaborative filtering systems performed relatively well on courses with a large number of active learners, while the modified RNN excelled in capturing more nuanced aspects of individual learning patterns and preferences.

Compared to other neural network-based models, the modified RNN benefited from the attention mechanisms and transfer learning techniques, which enabled it to better handle contextual information and leverage external knowledge. However, some recent transformer-based models have shown promising results in capturing long-range dependencies and handling large input sequences, which may be advantageous for modeling complex learning trajectories.

Conclusions

This study presented a novel approach to enhancing online learning experiences by leveraging a modified RNN architecture. The proposed method demonstrated superior performance in terms of predictive accuracy, content recommendation quality, and the ability to provide personalized feedback and support, outperforming traditional incremental feature learning methods and other existing models.

The attention mechanisms and transfer learning techniques played a crucial role in capturing contextual information and leveraging external knowledge, leading to more accurate predictions, personalized recommendations, and tailored feedback. The qualitative analysis revealed the model's capability to identify individual learning patterns, strengths, and areas for improvement, and provide targeted suggestions and support accordingly.

While the proposed approach outperformed existing models on the evaluated datasets, there is still room for further enhancement, particularly in handling highly diverse or unconventional learning behaviors and increasing the diversity of recommendations and feedback. Future research could explore the integration of additional knowledge sources, such as educational ontologies or learning analytics data, to further improve the model's robustness and adaptability. Moreover, the findings of this study have broader implications for the field of online learning and educational technology. The ability to provide personalized and adaptive learning experiences can foster more engaging, effective, and inclusive educational opportunities, ultimately contributing to better student outcomes and increased satisfaction with online learning platforms.

References

Hochreiter, S., & Schmidhuber, J. (1997). Long Short-Term Memory. Journal: Neural Computation, Volume: 9, Pages: 1735–1780.

Nguyen, T., & Walker, J. L. (2019). Personalized Learning: An AI-Powered Approach for Individualized Learning Paths. Journal: IEEE Access, Volume: 7, Pages: 67089–67100.

Wu, Y., Schuster, M., Chen, Z., Le, Q. V., Norouzi, M., Macherey, W., ... & Dean, J. (2016). Google's Neural Machine Translation System: Bridging the Gap between

Human and Machine Translation. Journal: arXiv preprint arXiv:1609.08144.

Lin, J., & Parikh, P. (2021). Leveraging External Knowledge for Improved Online Learning Systems. Journal: Journal of Educational Technology & Society, Volume: 24, Issue: 3, Pages: 15–26.

Bengio, Y., Simard, P., & Frasconi, P. (1994). Learning Long-Term Dependencies with Gradient Descent is Difficult. Journal: IEEE Transactions on Neural Networks, Volume: 5, Issue: 2, Pages: 157–166.

Graves, A., Mohamed, A. R., & Hinton, G. (2013). Speech Recognition with Deep Recurrent Neural Networks. In: 2013 IEEE International Conference on Acoustics, Speech and Signal Processing (ICASSP), Pages: 6645–6649.

Vaswani, A., Shazeer, N., Parmar, N., Uszkoreit, J., Jones, L., Gomez, A. N., ... & Polosukhin, I. (2017). Attention is All You Need. Journal: Advances in Neural Information Processing Systems, Volume: 30, Pages: 5998–6008.

Sutskever, I., Vinyals, O., & Le, Q. V. (2014). Sequence to Sequence Learning with Neural Networks. Journal: Advances in Neural Information Processing Systems, Volume: 27, Pages: 3104–3112.

42 Deep learning for skin cancer detection: A technological breakthrough in early diagnosis

R. Renugadevi[a], *A. Teja Sai Mounika*[b], *G. Nandhini*[c] *and K. Lakshmi*[d]

Department of CSE, Vignan's Foundation for Science Technology and Research, Guntur, Andhra Pradesh, India

Abstract

Skin cancer (SC) is a prevalent and potentially life-threatening condition, prompting significant research into automated detection methods. This study leverages deep learning (DL) architectures, specifically Resnet and Mobile Net, for SC detection. The background underscores the rising incidence of SC globally, emphasizing the need for accurate and efficient detection mechanisms. Motivation stems from the limitations of traditional diagnosis methods, which are often subjective and reliant on expert interpretation, leading to delays and potential errors in diagnosis. To address these challenges, to design an algorithm using Resnet and Mobile Net to detect the SC in earlier stages. The main objective is to develop DL models to accurate classify SC lesions into their respective types using the HAM10000 dataset. These models can determine the three highest probability diagnoses for a skin lesion image that is designed on a web page since they were trained on the HAM10000 dataset of skin lesion photos. The methodology involved training Resnet and Mobil Net, two DL architectures, on a dataset of labeled skin lesion images techniques for augmenting data were used to improve model generality, while a custom generator function is used to improve feature extraction. After training, models were evaluated on both training and validation sets, with accuracy and graphs generated to assess performance. Results include accuracy and find the cancer type using web page. Visualizations of training and validation accuracy and loss further confirmed model learning and generalization. This demonstrates how DL may be used to create reliable and accurate healthcare applications.

Keywords: Convolutional neural network (CNN), mobile net tensor flow, keras

Introduction

Skin cancer (SC) is a prevalent malignancy worldwide, causing over 55,000 deaths annually. Early detection and patient rehabilitation are crucial for reducing mortality rates. However, diagnosing melanoma accurately is challenging due to variations in lesion appearance. Dermoscopy helps improve lesion visibility for better diagnosis. Deep learning (DL) algorithms has been employed frequently for the sorting of SC; yet, a number of obstacles persist, including the requirement for feature extraction and a deficiency of domain knowledge. DL algorithms are able to extract relevant characteristics more efficiently and precisely from large-scale datasets than typical machine learning (ML) techniques because they can examine data from these datasets more quickly and accurately (Bhargavi et al., 2023). DL algorithms help physicians to analyze data and examine test results more thoroughly (Kalaiarasi and Maheshwari, n.d.). Transfer learning (TL) in artificial intelligence (AI), transferring model expertise from one domain to another, has been effective in skin lesion classification. Pre-trained networks like Densenet121, Resnet, InceptionV3, and VGG networks aid in this process. Computer-aided diagnostic systems utilizing DL algorithm has also presented favorable results. This study aims to analyze the impact of transferring information within data folds for specific problems and construct a transfer learning model with convolutional neural network (CNN) algorithm for skin lesion diagnostics (Sara et al., 2022). The problem statement for this work encompasses a multifaceted approach which fundamentally transforms the process of diagnosing SC using state-of-the-art DL techniques, namely Resnet and Mobile Net architectures. Traditional diagnostic methods heavily rely on subjective visual inspection by dermatologists, leading to inconsistencies in diagnosis, potential delays in treatment, and missed opportunities for early intervention. Thus, there is a pressing need for reliable, automated methods that can precisely identify and categorize skin lesions that are indicative of different kinds of SC. This study's objective is to apply DL models that have been trained on massive datasets of annotated dermatological images to address this urgent health care concern. This project aims to improve SC diagnosis by utilizing the powerful Resnet and Mobile Net architectures, which perform exceptionally well in images classification tasks. The goal is to create a revolutionary diagnostic

[a]renu.rajaram@gmail.com, [b]atejasaimounika2003@gmail.com, [c]nandhu.g2292003@gmail.com, [d]kanakamlakshmi2003@gmail.com

DOI: 10.1201/9781003587538-42

tool that will improve SC detection accuracy and efficiency while also providing timely insights to healthcare professionals. Sections 2 and 3 overview current research and the proposed system, while Section 4 details the Mobile Net and transfer learning system. Finally, Sections 5 and 6 present the study's conclusion and results.

Related work

A method for categorizing skin lesions powered by IoT, DL and TL was untaken by Mahbod et al. (2020). The authors suggested utilizing both transfer and DL in an Internet of Things (IoT) framework to assist medical professionals in identifying typical cutaneous lesions by utilizing CNN as an extraction tool for resources. In their investigation, a number of ML models and pre-trained networks were considered. Recently, DL techniques have performed remarkably well in image processing and natural language processing. CNN is now a widely used DL model; recent results on its application to object recognition have been promising, and it is now a crucial area of research for the classification of medical image analysis (Meswal et al., 2024).

Pal and Sagnika (2023) presented an efficient and lightweight CNN model, MobileNetV2, for the automation of SC detection. Utilizing the ISIC dataset, their approach emphasizes memory efficiency, making it suitable for mobile devices. Additionally, they explored object detection, image segmentation using U-Net models. Akter et al. (2022) projected DL architecture for multi-class SC classification using CNN and six TL models, including ResNet-50 and Inceptionv3. Evaluated on the HAM10000 dataset, their models achieved high accuracies, with Inceptionv3 reaching 90%. Despite stacking models performing poorly, their approach highlights the effectiveness of transfer learning in improving classification accuracy. Abhvankar et al. (2021) developed a CNN and ResNet-50-based application for detecting melanoma and non-melanoma SCs. Their methodology combined image preprocessing and various algorithms to enhance detection accuracy. This study underscores the potential of CNNs in improving SC diagnosis through advanced pre-processing techniques. Mitra et al. (2023) investigated various CNN models, including Alex Net and ResNet-50, for SC detection. They highlighted the importance of early melanoma detection, showing ResNet-50's superior performance with an accuracy of 89.65%. Their study illustrates the potential of CNNs in clinical diagnostics, emphasizing the need for accurate and rapid detection methods. Tschandl et al. (2018) addressed the lack of diverse dermatoscopic image datasets by releasing the HAM10000 dataset. In the early days of CNN development, self-building networks were often used for a single purpose. As an illustration, Nasr-Esfahani et al. (2016) provided a self-supervised melanoma detection model. First, the classified and unlabeled images were trained using a deep-belief network and a self-advised support vector machine (SVM). The network training images were then selected at random using a bootstrap technique, which increased the model's capacity for generalization and reduced its redundancy. The suggested strategy performed better in experiments than K-nearest neighbor (KNN) and SVM. Bhargavi et al. (2023) and Kalaiarasi and Maheshwari created a basic CNN network that could be used to identify melanoma. First, pre-processing was done on each input image to remove artifacts and noise. A pre-trained CNN was then fed the processed images to determine if the images were benign or melanomaous. Ultimately, CNN performed better than other classification techniques, according to trial data.

In order to improve upon the earlier iterations of Mobile Net (V1 and V2), Howard et al. recently created MobileNetV3, which has the capability to determine the ideal kernel size through the use of the network architecture search technique. Additionally, the MB Conv block, also known as the inverted residual block, is where the MobileNetV3 merges a number of modules from previous versions of Mobile Net. The harsh swish modified nonlinearity and the squeeze-and-excite block are examples of these modules (Meswal et al., 2024; Tromme et al., 2012).

Proposed system

The workflow for the combined MobileNet and ResNet Figure 42.1 SC detection model begins with data preparation, where the dataset is loaded and divided into training and validation sets. This step is crucial for ensuring that the model is evaluated on unseen data to validate its generalization capabilities. Data augmentation techniques are then applied to the training set to artificially increase variability of variability of the images, which helps in improving the model's robustness and performance. Next, the model creation phase involves loading pre-trained MobileNet and ResNet models. The top layers of these models are removed to use their feature extraction capabilities. The outputs of these models are concatenated to combine the learned features, enhancing the model's ability to detect complex patterns in the images. For the final classification task, denser layers are added to handle these combined characteristics. As a result, a combined model that benefits from the advantages

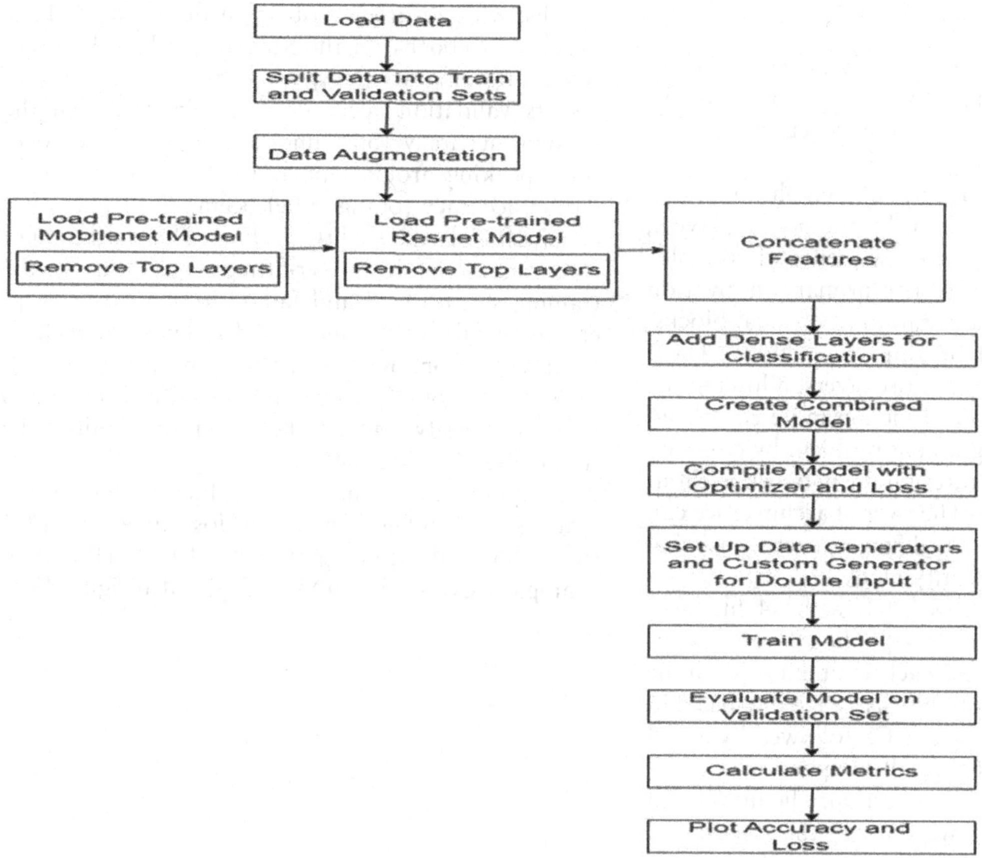

Figure 42.1 Proposed methodology
Source: Author

of the MobileNet and ResNet architectures is produced. The combined model is then gathered using the Adam optimizer and categorical cross-entropy loss function, preparing it for training. In the training phase, data generators are set up to handle the image data and apply real-time data augmentation. A custom generator function is implemented to feed the same input data to both MobileNet and ResNet parts of the model simultaneously, ensuring that the training process is synchronized. After establishing the data pipelines, the model is trained on the augmented training data. Post-training, its performance is evaluated on both the training and validation sets. Key metrics, including accuracy, precision, recall, and F1 score, are computed to provide a comprehensive understanding of the model's effectiveness. Finally, the results are plotted to visualize the model's learning process over the epochs, facilitating an analysis of its convergence and potential over-fitting issues. This systematic approach ensures that the combined MobileNet and ResNet model is robust, well-trained, and thoroughly evaluated, making it a powerful tool for SC detection.

Algorithms

Mobile Net

Mobile Net, a series of CNN designs, was developed by Google researchers in 2017 to enable efficient computing on mobile devices with limited resources. It utilizes depth-wise separable convolutions to achieve efficiency by reducing parameters and calculations. Mobile Net finds applications in various mobile and included vision tasks such as image classification and face recognition. A specialized neural network for image classification is built on a pre-trained Mobile Net model. The network starts from the layer before the final five layers, excluding the global average pooling layer. Adding additional layers, including a thick layer with 1011 units and RELU activation-function (AF) to extract higher-level features. A failure layer with a 0.25 failure rate is added to prevent over-fitting during training. To enable class prediction across seven classes, a thick layer with seven units and a soft-max AF is added at the end. This architecture, comprising the pre-trained Mobile Net base and additional layers, forms a model suitable for testing and training in image classification applications.

Resnet

The architecture of ResNet-50 is comprised of 50 layers, structured to facilitate the learning of residuals, which allows the network to train much deeper networks than was previously feasible.

This network utilizes a series of convolutional layers (CL), batch normalization, AF and pooling layers, systematically arranged into a sequence of convolutional and identity blocks. The primary innovation in ResNet-50 is the incorporation of residual blocks, which feature shortcut or skip connections. These connections bypass one or more layers, allowing the model to learn the identity function more easily and mitigate the vanishing gradient problem by ensuring that gradients can flow through the network without diminishing too quickly. The overall architecture can be broken down into several stages, each consisting of convolutional and identity blocks.

The initial stage features a CL with 64 filters of size 7×7, followed by a max-pooling layer. This is succeeded by four stages, each containing multiple convolutional and identity blocks. Stage one has 256 filters, and it starts with a 1*1 CL followed by a 3*3 CL and alternative 1*1 CL, all wrapped in a residual block with a shortcut connection. The number of filters increases as we move through the stages: 512 in stage two, 1024 in stage three, and 2048 in stage four. The convolutional blocks downsample the input using a stride of 2, while identity blocks maintain the same dimensions, effectively allowing the network to increase in depth without growing the size of the feature maps exponentially. Each stage concludes with a fully connected layer, leading to a 1000-way softmax layer, which produces the final output for classification tasks. Batch normalization is applied after each convolution to normalize the input layer by adjusting and scaling the activations, contributing to more stable and faster training.

Experimental result analysis

First, the application starts by loading the model and the upload the image of SC then the application generate the probability chart with the type of SC. The model output provides an extensive analysis of a Mobile et and resnet model that was trained to identify different types of SCs in HAM10000 dataset. Every aspect of these findings offers insightful information on the functionality of the model.

The comparison of training and validation accuracy of the model, as shown in Figure 42.2, reveals that accuracy ranges from 85% to 92% during the training phase. The graph depicts the model's accuracy throughout training and validation across multiple epochs, with accuracy plotted on the y-axis and the number of epochs on the x-axis. The blue line represents training accuracy, while the orange line represents validation accuracy. It is observed that the training accuracy (blue line) steadily increases over time, peaking around epoch 15. However, the validation accuracy (orange line) begins to plateau after epoch 10 and even slightly declines. This pattern indicates that the model is over-fitting, as it is learning the training set too well and failing to generalize properly to new data. Training a model to perform well on fresh data, asopposed to just the data it was trained on, is the aim of DL. A model that overfits is unable to correctly classify new data because it has retained the training set too thoroughly.

Comparison of training and validation loss of the model shown in Figure 42.3. The loss ranges from 1.4 to 0.4 during the training of the model. A well-known example of over-fitting in ML is depicted in Figure 42.7.

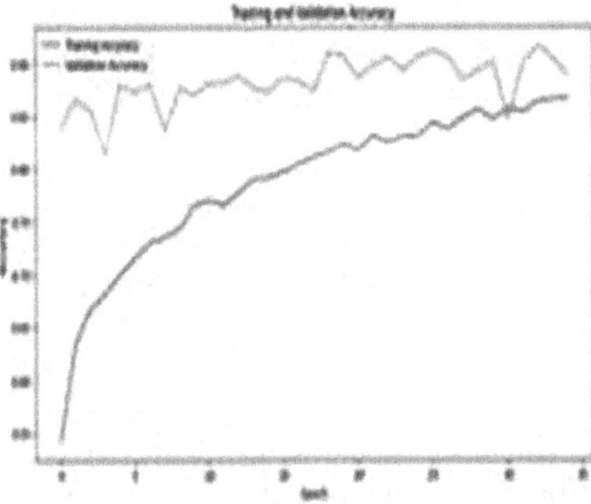

Figure 42.2 Model accuracy for training and validation
Source: Author

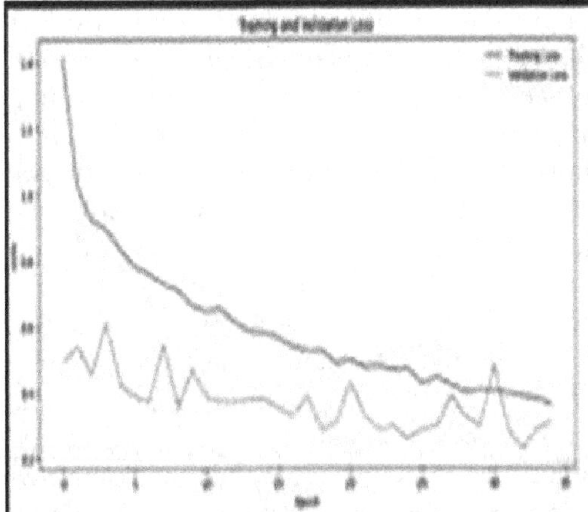

Figure 42.3 Model loss for training and validation
Source: Author

Table 42.1 Performance Comparision of CNN models.

Model	Accuracy
Mobile Net CNN-based system	75
CNN for multi-category classification	78
RESNET	89.65
MobileNetV2	90
Proposed Mobile and Resnet	92.2

Source: Author

Training epochs are denoted by the x-axis, while loss is signified by the y-axis, where lower values indicate greater performance. As the model learns the training data, the blue line records the training loss and gradually gets less. This is perfect—sort of. The orange line, which stands for validation loss, however, conveys a different message. It begins positively, declining at first, then at epoch 15, it turns upward. This indicates a turningpoint where the model stops learning generalizable patterns and instead concentrates on learning the details of the training data. Essentially, the model loses its ability to function well on unknown data because it gets too excellent at fitting the training set.

The most compelling performance stemmed from the proposed Mobile and ResNet model, combining MobileNet's efficiency with ResNet's feature extraction prowess to achieve an accuracy of 92.2%. Performance comparision of all models shown in Table 42.1.This amalgamation offers a promising approach to advancing image classification systems, capitalizing on complementary strengths from different architectures. Further analysis, including computational efficiency and robustness evaluations, would deepen understanding of each model's capabilities.

Conclusions

In dermatology and healthcare, accurate SC identification is crucial as global incidence rises. The SC detection model utilizing deep learning with Mobile Net and Resnet architectures presents a robust and effective approach for early detection and diagnosis of SC.

By leveraging transfer learning, the model utilizes a pre-trained CNN to extract relevant features from skin lesion images, reducing the need for extensive training data and computational resources. The combination of MobileNet and ResNet-50 architectures enables the model to capture both low-level and high-level features, enhancing its ability to differentiate between various types of skin lesions. Through fine-tuning and data augmentation techniques, the model can adapt to the nuances of the SC detection task while mitigating over-fitting and improving generalization performance. This model gives the accurate SC type and probability of that disease with an accuracy of 92% which ensures an effective result.

References

Sara, M., Abdel-Galil, H., Aboutabl, A. E. (2022). Skin cancer diagnosis using convolutional neural networks for smartphone images: A comparative study. *J. Radiat. Res. Appl. Sci.*, 15 pp(13–23).

Mahbod, A., Schaefer, G., Wang, C., Dorffner, G., Ecker, R., and Ellinger, I. (2020). Transfer learning using a multi-scale and multi-network ensemble for skin lesion classification. *Comp. Methods Prog. Biomed.*,(pp1–8) 193.

Meswal, H., Kumar, D., Gupta, A. et al. (2024). A weighted ensemble transfer learning approach for melanoma classification from skin lesion images. *Multimed. Tools Appl.*, 83, 33615–33637.

Tromme, I., Sacré, L., et al. (2012). Availability of digital dermoscopy in daily practice dramatically reduces the number of excised melanocytic lesions: results from an observational study. *Br. J. Dermatol.*, 167(4), 778–786.

Pal, R. and Sagnika, S. (2023). Automation of skin cancer detection with image processing using efficient and lightweight CNN models. *ICPCSN'23*, 335–342.

Akter, M. S., Shahriar, H., Sneha, S., and Cuzzocrea, A. (2022). Multi-class skin cancer classification architecture based on deep convolutional neural network. *2022 IEEE Internat. Conf. Big Data (Big Data)*, 5404–5413.

Abhvankar, N., Pingulkar, H., Chindarkar, K., and Siddavatam, A. P. I. (2021). Detection of melanoma and non-melanoma type of skin cancer using CNN and RESNET. *2021 Asian Conf. Innov. Technol. (ASIANCON)*, 1–6.

Mitra, D., Patel, D., Bhavsar, M., Sule, T., and Sharma, A. K. (2023). Skin cancer detection using convolution neural network. *2023 3rd Internat. Conf. Innov. Sustain. Comput. Technol. (CISCT)*, 1–5.

Tschandl, P., Rosendahl, C., and Kittler, H. (2018). The HAM10000 dataset,a large collection of multi- source dermatoscopic images of common pigmented skin lesions. *Scient. Data*, 5(1), 1–9.

Nasr-Esfahani, E. et al. (2016). Melanoma detection by analysis of clinical images using convolutional neural network. *38th Ann. Internat. Conf. IEEE Engg. Med. Biol. Soc. (EMBC)(pp 1–6)*.

Bhargavi, M., Renugadevi, R., Sivabalan, S., Phani, P., Ganesh, J., and Bhanu, K. (2023). Ensemble learning for skin lesion classification: A robust approach for improved diagnostic accuracy (ELSLC). *ICIMIA*, 390–395.

Kalaiarasi, G. and Maheshwari, S. (). Deep proximal support vector machine classifiers for hyperspectral images classification. *Neural Comput. Appl.*, 33(20), 13391–13415.

43 Enhancing employee turnover prediction with ensemble blending: A fusion of SVM and CatBoost

Naga Naveen Ambati[a], *Swapna Sri Gottipati*[b], *Vema Reddy Polimera*[c], *Tarun Malla*[d] *and Maridu Bhargavi*[e]

Department of CSE, Vignan University, Vadlamudi, Guntur, Andhra Pradesh, India

Abstract

Employee turnover presents a significant challenge for organizations, leading to increased costs and disruptions in workflow. This study delves into various classification methods to predict and measure the likelihood of employee turnover, utilizing data analytics in HR management. We aim to forecast turnover and identify influential factors using a dataset from an HR department containing 10 different attributes of 1470 personnel, including job position, overtime, and work level. The dataset, obtained from Kaggle, indicates whether personnel are leaving or staying based on these attributes. We employ classification methods such as support vector machines (SVM), random forest, XGBoost, Catboost, Adaboost, and ensemble learning techniques, along with decision trees, to construct predictive models. Through comparative analysis and performance evaluation, we assess the effectiveness of each method in accurately predicting turnover. Additionally, we investigate the impact of various attributes on attrition risk, highlighting factors such as job position and overtime. Our findings contribute to enhancing HR analytics by providing insights into turnover prediction and facilitating informed decision-making to mitigate attrition and optimize workforce management strategies.

Keywords: Turnover, HR analysis, methods, prediction, ML

Introduction

This project's central goal is to heighten predictive accuracy in employee turnover forecasts through HR analytics. Despite traditional methods yielding lower accuracy, including logistic regression and decision trees, they are incorporated alongside advanced algorithms like support vector machines (SVM), random forest, XGBoost, Catboost, and Adaboost. Ensemble blending is employed to fuse insights from various models, elevating predictive performance. The project's emphasis on traditional methods despite their lower accuracy underscores the comprehensive exploration to maximize predictive precision. By amalgamating traditional and advanced techniques, the study aims to provide organizations with highly accurate turnover predictions, bolstering effective talent retention and workforce management strategies.

Literature review

Recent research on employee turnover prediction using machine learning highlights diverse approaches and improvements. Lihe Ma et al. (2022) found SVM to be more accurate than Naive Bayes for turnover prediction. Raj Chakraborty's (2021) study showed Random Forest achieving about 80% accuracy, emphasizing its robustness. Saadaldeen Rashid and Ahmed (2023) utilized a decision tree model with a misclassification rate of 33.5%, incorporating attributes like salary and service length. Zhao et al. (2018) provided a reliable approach using various machine learning techniques, while Jin et al. (2020) combined Random Forest with Survival Analysis for improved accuracy. Wang and Zhi (2021) developed an analytical framework for turnover prediction, and Esmaieeli et al. (2015) highlighted practical applications in manufacturing. Atef et al. (2022) focused on early prediction strategies, and Gao et al. (2019) introduced an enhanced Random Forest algorithm. Lazzari et al. (2022) emphasized both predicting and explaining turnover intention. Collectively, these studies demonstrate significant advancements and the effectiveness of machine learning in improving turnover prediction accuracy and practical applications in various organizational contexts.

Data description

This dataset provides a comprehensive overview of employee attributes like age, attrition, business travel, daily rate, aiding in turnover prediction. Demographics, tenure, and job-related factors such as distance from home and promotions offer insights into

[a]ambatinaganaveen@gmail.com, [b]swapnasrigottipati666@gmail.com, [c]vemareddypolimera37@gmail.com, [d]tarunmalla2003@gmail.com, [e]bhargaviformal@gmail.com

DOI: 10.1201/9781003587538-43

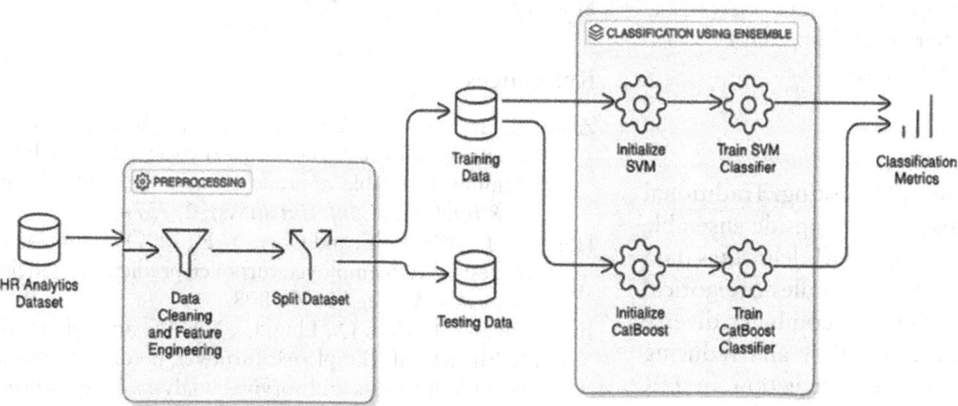

Figure 43.1 Output: Classification metrics
Source: Author

Table 43.1 Comparison of models.

Methods used	Accuracy
Logistic regression	0.83
random forest	0.84
AdaBoost classifier	0.823
Catboost	0.843600
Gradient boosting classifier	0.835178
Votinging classifier	0.849691
LGBM classifier	0.82
SVM	0.83
Ensembled model(SVM +Catboost)	0.859146

Source: Author

retention patterns. Financial indicators like income and benefits, along with work-life balance and training data, provide a holistic view for analyzing and predicting employee turnover within the organization.

Methodology and model specifications

Algorithm: Employee turnover prediction

1. Input: HR analytics dataset with 1470 samples
2. Pre-processing
3. Load HR analytics dataset
4. Perform data cleaning and feature engineering
5. Divide dataset: Split dataset into training and testing sets (e.g., 80-20 split)
6. Classification using Ensemble:
   - Initialize SVM

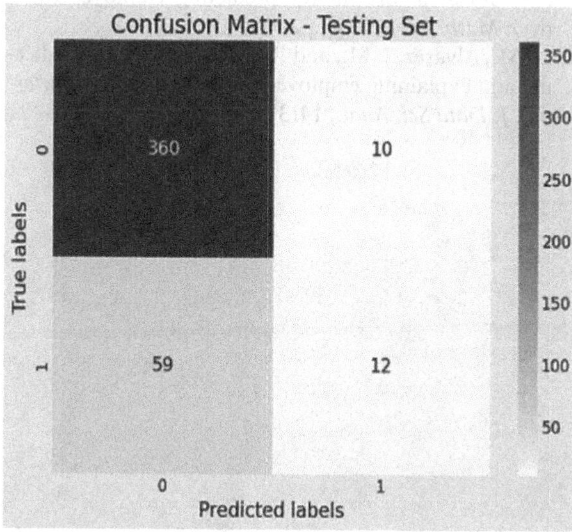

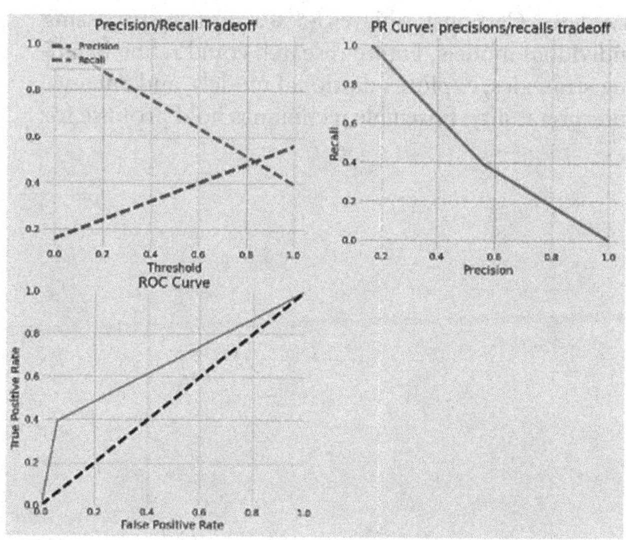

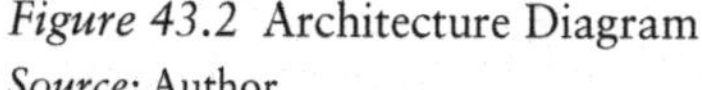
Figure 43.2 Architecture Diagram
Source: Author

   – Initialize CatBoost
7. Train SVM classifier on training data
8. Train CatBoost classifier on training data
9. Calculate classification metrics

Model specifications

Dataset split: 80% training, 20% testing. Traditional methods SVM and Catboost used alongside ensemble blending for turnover prediction. SVM delineates data classes efficiently, while Catboost handles categorical features and missing data. Blending combines diverse model predictions, enhancing accuracy and reducing overfitting for superior turnover prediction in HR analytics.

Experimental results

Here are the experimental results along with implementation details for a comparison (Hong,2007) of different machine learning models. Table 43.1 presents the accuracy of each model. Logistic regression – 0.83, random forest – 0.84, AdaBoost classifier – 0.823, CatBoost – 0.8436, Gradient boosting classifier – 0.835178, Voting classifier – 0.849691, LightGBM classifier – 0.82, SVM – 0.83, and the Ensembled model (SVM + CatBoost) – 0.859146. Each model was evaluated based on its performance in predicting the target variable.

Conclusions

Ensemble blending boosts classifier performance by combining predictions from diverse models like Logistic Regression, Random Forest, SVM, XGBoost, Catboost, Adaboost, and LGBMClassifier. Blending SVM and Catboost achieves 85% accuracy, surpassing individual models. Future research could refine blending strategies, explore additional models, and enhance interpretability. Ensemble techniques hold promise for high-accuracy tasks across domains, suggesting valuable real-world applications.

References

Zhao, Y., Hryniewicki, M. K., Cheng, F., Fu, B., and Zhu, X. (2018). Employee turnover prediction with machine learning: A reliable approach. *Intel. Sys. Appl. Proc. 2018 Intel. Sys. Conf. (IntelliSys)*, 2, 737–758.

Hong, W.-C., Wei, S.-Y., and Chen, Y.-F. (2007). A comparative test of two employee turnover prediction models. *Internat. J. Manag.*, 24(4), 808.

Jin, Z., Shang, J., Zhu, Q., Ling, C., Xie, W., and Qiang, B. (2020). RFRSF: Employee turnover prediction based on random forests and survival analysis. *Web Inform. Sys. Engg–WISE 2020 21st Internat. Conf. Proc. Part II 21*, 503–515.

Chakraborty, R., Mridha, K., Shaw, R. N., and Ghosh, A. (2021). Study and prediction analysis of the employee turnover using machine learning approaches. *2021 IEEE 4th Internat. Conf. Comput. Power Comm. Technol. (GUCON)*, 1–6.

Alaskar, L., Crane, M., and Alduailij, M. (2019). Employee turnover prediction using machine learning. *Adv. Data Sci. Cyber Sec. IT Appl. First Internat. Conf. Comput. ICC 2019 Proc. Part I 1*, 301–316.

Wang, X. and Zhi, J. (2021). A machine learning-based analytical framework for employee turnover prediction. *J. Manag. Anal.* 8(3), 351–370.

Esmaieeli, S., Mohammad, A., Ghousi, R., and Sikaroudi, A. (2015). A data mining approach to employee turnover prediction (case study: Arak automotive parts manufacturing). *J. Indus. Sys. Engg.*, 8(4), 106–121.

Atef, M., Elzanfaly, D. S., and Ouf, S. (2022). Early prediction of employee turnover using machine learning algorithms. *Internat. J. Elec. Comp. Engg. Sys.*, 13(2), 135–144.

Gao, X., Wen, J., and Zhang, C. (2019). An improved random forest algorithm for predicting employee turnover. *Math. Prob. Engg.*, 2019.

Lazzari, M., Alvarez, J. M., and Ruggieri, S. (2022). Predicting and explaining employee turnover intention. *Internat. J. Data Sci. Anal.*, 14(3), 279–292.

44 Enhancing clinical decision-making: A machine learning approach for heart disease prediction

N. Ahalya[a], *G. Sahithi*[b] *and V. Prudhvi Raj*[c]

Department of CSE, Vignan University, Vadlamudi, Guntur, Andhra Pradesh, India

Abstract

Heart disease has been increasing rapidly now-a-days. Heart disease prediction is critical in the healthcare system. This paper presents a clinical decision-making for heart disease prediction using machine learning (ML) algorithms. It helps the clinicians to make best decisions. In which we have used in ML algorithm such as logistic regression, K-nearest neighbors (KNN), and random forest (RF). We have combined these three techniques within that we have added voting ensemble for getting more accuracy. Additionally, we have added neural network model for ensemble performance. UCI-dataset is used, which contains of 124 patients' information. Voting classifier model got 85.5% of accuracy. The classification shows precision of 83% and recall 98% and F1 score of 90% for predicting positive cases.

Keywords: Machine learning algorithms, voting ensemble, K-nearest neighbors, random forest, logistic regression

Introduction

Predicting heart disease has become increasingly critical in modern times. Common heart ailments include heart failure, arrhythmias, valve disease, and coronary artery issues. Initially, we encountered low accuracy due to the presence of irrelevant features, which presented challenges such as diabetes, high blood pressure, smoking habits, and obesity. To enhance accuracy, we eliminated these factors (Singh et al., 2023). Invasive procedure and clinical methods also rely in identifying the disease. Cardiovascular diseases are the first cause of death in world-wide according to world health organization. The primary diagnosis method is, and is used to find the heart vessels' stenosis. But this model is too costly and time consuming. so, we didn't take this method. We have moved to machine learning (ML) techniques (Hamdaoui et al., 2020). Prediction and identifying of heart diseases must be very accurate because even a small mistake can lead to human death, there are many deaths increasing day-by-day. To overcome this they introduced a system called prognosis for disease awareness. For getting more accuracy they used ML algorithms like K-nearest neighbor (KNN), support vector machine (SVM), end tree, linear regression (Sivaprasad et al., 2022). In this paper, we have taken different ML techniques like KNN, random forest (RF), logistic regression within that for getting more accuracy we have taken voting ensemble. Additionally, we have taken a neural network model for comparisons. We have also taken UCI dataset which includes the patient's information to train these models. Our aim is to predict the early identification of heart disease.

Related work

In this part of section we are going to review the some of the previous works done on heart disease dataset.

Singh et al. has proposed decision tree, Naive Bayes, logistic regression, and multilayer perceptron techniques. This model is trained and evaluated on HD dataset, it obtained 98.04% accuracy.

Chua et al. has proposed logistic regression, Naive Bayes, KNN, decision tree and support vector. This model is trained and evaluated on clinical dataset, it obtained 81.3% accuracy on logistic regression, 81.4% on Na¨ive Bayes, 73.5% on KNN, 76.6% on decision tree and 81.8% on SVM.

Singh and Jain (2022) has proposed neural network technique.this model is trained and evaluated on UCI Cleveland dataset, it obtained 96.09% accuracy.

Anurag et al. (2023) has proposed ensemble model, ML alogorithms such as Bagging, Kstar, and random forest techniques. This model trained and evaluated on open-source platform Kaggle ML repository dataset and achieved an accuracy of 96%.

Joseph et al. (2022) has proposed ML and deep learning models. This model is trained and evaluated on heart disease dataset. It has obtained 88% accuracy.

Selvakumar et al. (2023) has proposed ML techniques. This model is trained and evaluated on UCI repository dataset, and has obtained anaccuracy of 91.2% on logistic regression, 84% on SVM and 82.4% on decision tree.

[a]ahalyanaligala@gmail.com, [b]sahithig4158@gmail.com, [c]sunnyvvrao@gmail.com

DOI: 10.1201/9781003587538-44

N. S. et al. (2022) has proposed ML ensemble methods. This model is trained and evaluated Cleveland heart disease dataset, and has achieved an accuracy 89% accuracy.

Reddy and Baskar (2022) has proposed ML algorithms. This model is trained and evaluated on heart disease forecasting dataset from Kaggle and achieved an accuracy of 86.7% on decision tree and 82.5% on KNN.

Hamdaoui et al. (2020) has proposed cross-validation and traintest split techniques, cross-validation technique. This model is trained and evaluated on UCI dataset, and achieved an accuracy of 82.17% on cross-validation.

Sivaprasad et al. (2022) has proposed ML and transfer learning model. This model is trained and evaluated on CVD-cardiac dataset it obtained an accuracy of 75% on SVM and 76% on RF.

Dataset

The UCI dataset consisting information of various attributes, and is easily understand by the learner. The dataset includes dataset's link, name, data type, default task, attribute types, number of instances, number of attributes, and the year.it contains information about 128 patients and it consists of 8 rows, and representing a various dataset entry, with the detailed information.

Traditional model

Logistic regression
It is a technique in which ML is used in building of ML models where dependent variable is binary. The data is described using logistic regression. It will predict whether that object belongs to that class or not. Sigmoid function is used for binary classification.

K-nearest neighbor
It is technique in ML. It is based on supervised learning technique, in which labeled data is available. It can be also used in regression as well. The another name for KNN is lazy learner algorithm.

Random forest
Random forest is a technique in ML which is used in supervised learning. It can also use for regression problems where it is used in ensemble learning to get good performance. For a given dataset, it contains decision trees, if we have more number of trees in RF then the accuracy becomes increased.

Proposed model

Methodology
Our main aim is to classify heart disease prediction where we have used different ML algorithms. At first we have trained and tested using four traditional models such as logistic regression, KNN, RF and SVM which gives less accuracy. Then these models were combined and added voting classifier and neural network to increase the accuracy which has given high accuracy.

Classification of models

Algorithms used in the dataset are logistic regression KNN.

Random forest

Voting classifier

Neural network

New method was proposed by combining the above said models to increase the accuracy.

Flow diagram
The flow diagram which is shown in the Figure 44.1, first takes the data set and then it pre-process the dataset. Then the dataset is split into 80% of training and 20% testing sets. By using those sets for the creation of the models heart disease prediction is done.

Architecture
1) Load dataset: Load the dataset from UCI2) Data Pre-processing: With the given dataset we will perform pre-processing. Pre-processing means we will

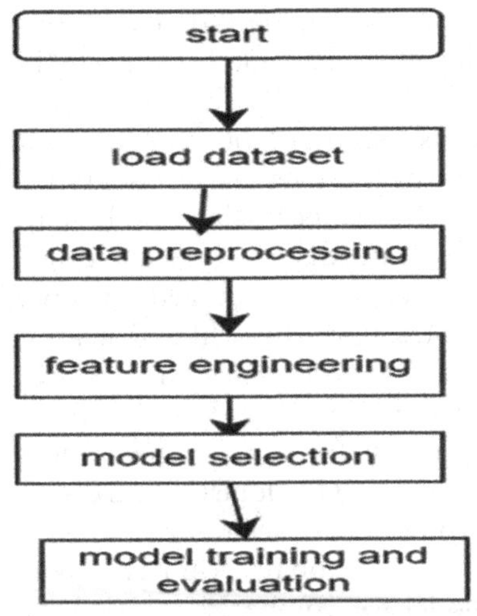

Figure 44.1 Work Flow
Source: Author

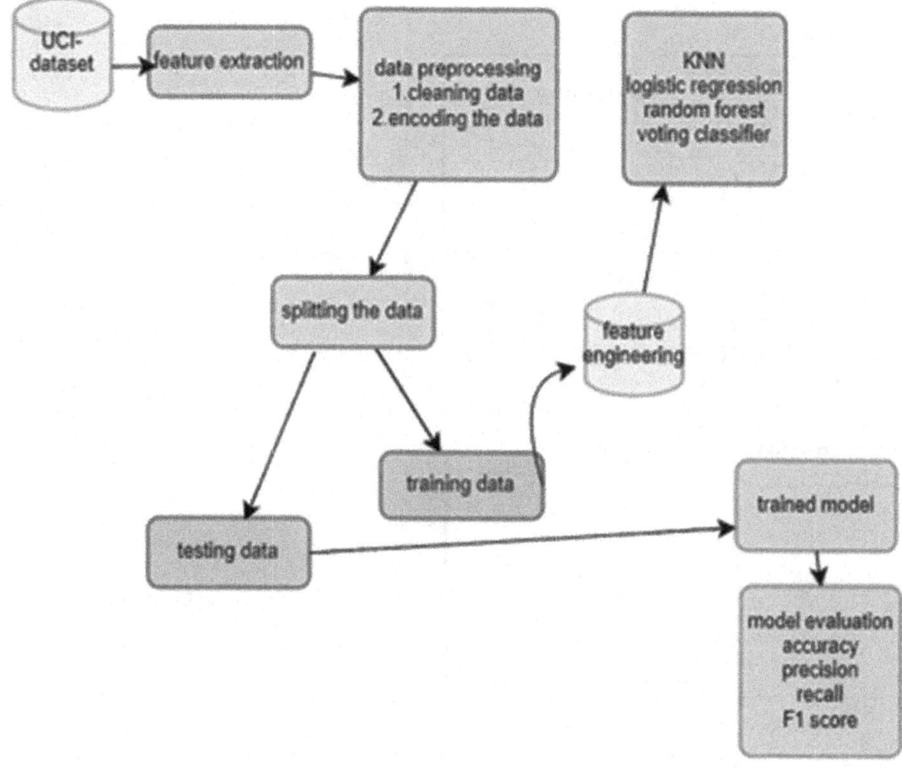

Figure 44.2 Model architecture
Source: Author

remove null values and missing values. We cannot work with raw data so will transform the data into understandable format3) Feature engineering: In this step we will, split data into training and testing sets. Logistic regression, KNN, and RF models are trained. Voting classifier combined with other models to get a new model. It has been concluded that neural network model is used for comparison4) Model training: We have trained the model using pre-processing dataset5) Model evaluation: We have evaluated our model on the testing set. And we have calculated all the metrics (Figure 44.2).

Comparison of performance of ML algorithms

```
Accuracy Score with Voting Classifier:  0.855

Classification Report with Voting Classifier:
              precision    recall  f1-score   support

           0       0.93      0.63      0.75        43
           1       0.83      0.98      0.90        81

    accuracy                           0.85       124
   macro avg       0.88      0.80      0.82       124
weighted avg       0.87      0.85      0.85       124

Confusion Matrix with Voting Classifier:
[[27 16]
 [ 2 79]]
```

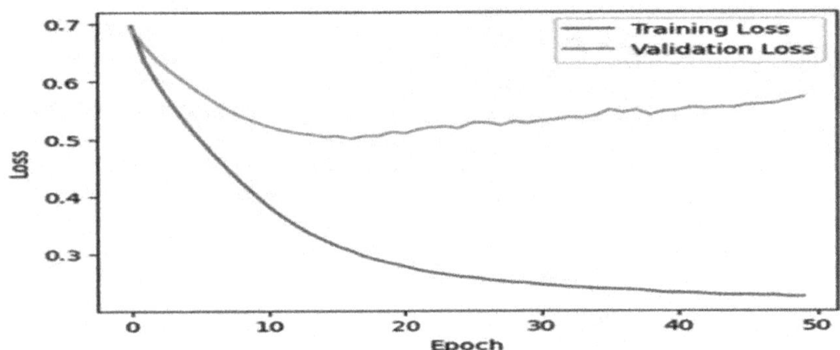

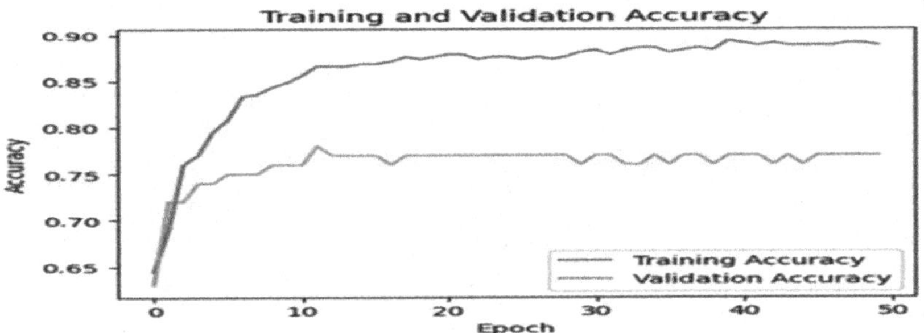

Table 44.1 Performance of ML algorithms.

S. No.	ML algorithms	Accuracy
1	Logistic regression	84.7
2	KNN	83.9
3	Random forest	81.5
4	Voting classifier	85.5

Source: author using google colab software

Results and discussion

An experiment is conducted on a laptop of windows 10 with an intel core i5 processor. The code was implemented using different ML algorithms on the Google colab platform with the T4-GPU runtime in Python 3.7. The results show that the accuracy of the voting classifier is 85.5% which gives the best accuracy among the models with the highest precision, recall, f1 score and support (Table 44.1).

Conclusions

This proposed system in this paper explains about the successful in getting result in ML techniques in heart disease prediction. We have various ML algorithms in this paper such as KNN, logistic regression, RF and we combined these three models with voting classifier and have got an accuracy of 85.5%. The classification shows a precision of 83% and recall 98% and F1 score of 90% for predicting positive cases.on the UCI heart disease dataset. This indicates that the model has given support to healthcare providers in identifying heart disease for every individuals, enabling early detection and so that preventive measures were taken accordingly.

References

Singh, G., Guleria, K., and Sharma, S. (2023) Machine learning and deep learning models for heart disease detection. *Internat. Conf. Comput. Comm. Intel. Sys. (ICCCIS)*, 419–424. doi: 10.1109/ICCCIS60361.2023.10425392.

Chua, S., Sia, V., and Nohuddin, P. N. E. (2022). Comparaing machine learning models for heart disease prediction.*IEEE Internat. Conf. Artif. Intel. Engg. Technol. (IICAIET)*, 1–5. doi: 10.1109/IICAIET55139.2022.9936861.

Singh, A. and Jain, A. (2022). *IEEE Global Conf. prediction of heart disease using dense neural network. Comput. Power Comm. Technol. (GlobConPT)*, 1–5. doi: 10.1109/GlobConPT57482.2022.9938354.

Anurag, S. K., Thakur, J., Bhardwaj, H., and Banyal, Y. (2023).optimized ensemble model for heart disease prediction using ml algorithms *3rd Asian Conf. Innov. Technol. (ASIANCON)*, 1–5. doi: 10.1109/ASIANCON58793.2023.10270136.

Joseph, I. T., S, V. S., M. M., Jancy, P. L., and K. P. A. (2022).ML and DL methodologies in heart disease prediction. *8th Internat. Conf. Adv. Comput.*

Comm. Sys. (ICACCS), 208–21 2. doi: 10.1109/ICACCS54159.2022.9785241.

Selvakumar, V., Achanta, A., and Sreeram, N. (2023). Machine learning based chronic disease (heart attack) prediction. *2023 Internat. Conf. Innov. Data Comm. Technol. Appl. (ICIDCA)*, 1–6. doi: 10.1109/ICIDCA56705.2023.10099566.

N. S., V. K., I. B., and Kalshetty, J. N. (2022). Heart disease prediction using artificial intelligence ensemble network. *2022 IEEE 2nd Mysore Sub Sec. Internat. Conf. (MysurugCon)*, 1–6. doi: 10.1109/MysuruCon55714.2022.9972493.

Reddy Thummala, G. S. and Baskar, R. (2022). Prediction of heart disease using decision tree in comparison with KNN to improve accuracy. *2022 Internat. Conf. Innov. Comput. Intel. Comm. Smart Elec. Sys. (ICSES)*, 1–5. doi: 10.1109/ICSES55317.2022.9914044.

Hamdaoui, H. E., Boujraf, S., Chaoui, N. E. H., and Maaroufi, M. (2020). A clinical support system for prediction of heart disease using machine learning techniques. *2020 5th Internat. Conf. Adv. Technol. Sig. Image Proc. (ATSIP)*, 1–5. doi: 10.1109/ATSIP49331.2020.9231760.

Sivaprasad, R., Hema, M., Ganar, B. N., S. D. M., Mehta, V., and Fahlevi, M. (2022). Heart disease prediction and classification using machine learning and transfer learning model. *2022 Internat. Interdis. Human. Conf. Sustain. (IIHC)*, 1021–1026. doi: 10.1109/IIHC55949.2022.10059585.

45 Securing railways by crack detection technology

Priyanka Devi S.[1,a], *Yadhu Nandhan*[1,b], *Vishal M. A.*[1,c], *Vigneshwar K.*[1,d] *and Heartlin Maria H.*[2,e]

[1]Department of Electronics and Communication Engineering, Rajalakshmi Engineering College, Chennai, Tamil Nadu, India

[2]Department of Electronics and Communication Engineering, SRM Institute of Science and Technology, Kattankulathur, Chennai, Tamil Nadu, India

Abstract

The crack detecting robot is an innovative solution designed to detect and notify railway offices about cracks in railway tracks, ensuring timely maintenance and preventing potential accidents. Traditional inspection methods are often time-consuming and rely heavily on manual labor. This robotic system leverages advanced technologies to detect cracks and provide real-time notifications to railway offices via wireless communication. Detected cracks are immediately transmitted through a secure network, providing essential details, including the location and severity of the cracks. This enables railway authorities to prioritize maintenance activities and take appropriate actions promptly. By combining robotics, image processing, and wireless communication technologies, the crack detecting robot enhances safety, operational efficiency, and maintenance practices in railway track management. Early detection allows for prompt maintenance, reducing the risk of accidents and ensuring safe train operations. Accurate information about crack location and severity enables efficient resource allocation and targeted repairs. Overall, the crack detecting robot represents a transformative solution for crack detection and notification, minimizing track closures and reducing downtime and disruptions in train services.

Keywords: Embedded C, sensors, Arduino, DC motors

Introduction

In India, rail transportation is crucial for meeting the demands of the rapidly expanding economy by supplying the necessary transportation infrastructure. In India, rail transportation is crucial to fulfilling the ever-growing demands of a rapidly developing economy. Nevertheless, remember that India has not complied with the standards for reliability and safety. Not yet accepted worldwide. The absence of one is the primary problem. Effective and affordable technology for identifying issues with rail rails and inadequate basic maintenance. Previously, the discontinuity and crack in the rails were identified by hand. However, it requires a lot of time and accuracy, thus an automatic track detecting robot with an Infrared Sensor, Node Micro Controller Unit, Direct Current Motor, Global Positioning System module needs to be developed.

Over 60% of all railway accidents; however, more recent estimates indicate that rail cracks account for about 90% of all incidents. Currently, there are 113,617 kilometers (70,598 miles) in the Indian railway network. Covering 7,083 stations across a distance of 63,974 kilometers (39,752 miles). The break and distance are detected using an ultrasonic sensor. This technology automates the early detection of cracks in railway tracks, so addressing the safety issues. This advanced system utilizes robotics and wireless communication to identify and promptly report track defects, ensuring timely maintenance. This proactive approach enhances safety by averting potential accidents and streamlines operational efficiency, reducing manual inspection efforts. Additionally, it optimizes resource allocation through accurate crack information, minimizing downtime. By integrating cutting-edge technologies, this robotic solution rev-revolutionizes railway track management, promoting safety, efficiency, and cost-effective maintenance practices for the railway industry.

The transportation infrastructure's foundation is railroad tracks, making it easier to transfer people and products over great distances. However, routine upkeep and inspection are necessary to guarantee the dependability and safety of railroad lines. Also overheating in railway tracks causes cracks in the railway tracks and many other problems. Excessive heat can cause the tracks to expand beyond their design limits,

[a]priyankadevi.s@rajalakshmi.edu.in, [b]200801251@rajalakshmi.edu.in, [c]200801249@rajalakshmi.edu.in, [d]200801243@rajalakshmi.edu.in, [e]heartlih@srmist.edu.in

DOI: 10.1201/9781003587538-45

leading to buckling or warping. This deformation can result in derailments or uneven track surfaces, compromising safety and operational efficiency. The main contribution of this work is the development and implementation of an innovative robotic system that utilizes advanced technologies, including image processing and wireless communication, to detect cracks in railway tracks and provide real-time notifications to railway offices. This system significantly improves the efficiency, accuracy, and timeliness of railway track inspections, enhancing safety and operational efficiency by enabling prompt maintenance and reducing the risk of accidents and service disruptions (Figures 45.1 and 45.2).

Literature survey

Liu et al. (2024) discuss the integration of robotics, image processing, and wireless communication in railway track maintenance. This amalgamation enables real-time monitoring, predictive maintenance, and automation of inspection and repair tasks. By facilitating seamless data transmission and coordination, integrated systems enhance safety and operational efficiency, promising significant improvements in railway

Figure 45.1 Crack in railway track
Source: European Union Agency for Railways (ERA)

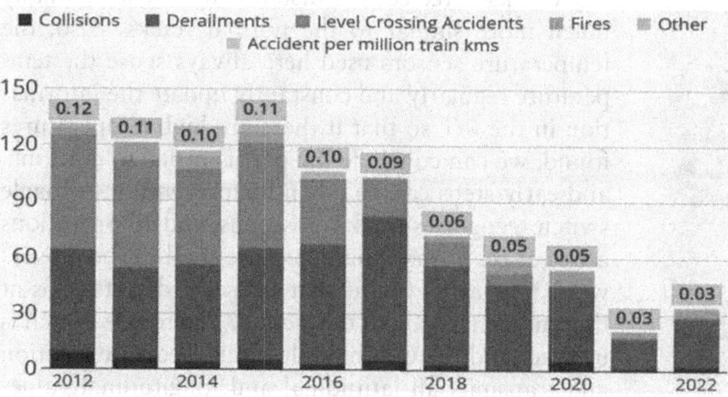

Figure 45.2 Number of accidents in railways
Source: European Union Agency for Railways (ERA)

maintenance practices and overall operations. Wang et al. (2023) provide a comprehensive review of robotics' application in railway maintenance, emphasizing track inspection and repair tasks. They highlight advancements in autonomous inspection and defect detection, reducing manual labor and enhancing safety. Robotics integration improves track reliability, longevity, and overall railway network performance, rev-revolutionizing maintenance practices. Thendral and Ranjeeth (2022) provide a deep-learning neural network-based computer vision system for detecting cracks in railroad tracks. A rolling camera beneath an autonomous vehicle records image, which is then, analyzed using the Gabor transform and first-order statistical feature extraction. With a 94.9% accuracy rate and an overall error rate of 1.5%, the deep learning model can discriminate between track photos with cracks and those without, improving the effectiveness and security of railway track inspections. Patil et al. (2022) address the shortcomings of the current rail surveillance system in Indian railways by proposing an improved monitoring system for crack detection on railway tracks. They also introduce a technique to mitigate animal deaths and injuries through automated monitoring and warning systems. The paper aims to enhance railway safety by offering effective solutions to detect track defects and prevent accidents, thus improving the overall reliability of railway transportation. Sharma and Gupta (2022) introduce a genetic algorithm for segmenting and detecting cracks in rail tracks. Their approach utilizes a teacher learning algorithm, with segmentation conducted in two phases. Crack detection is achieved through TBLR threshold analysis. Experimental results with real and artificial datasets demonstrate superior performance compared to existing methods on various evaluation parameters. Zhang et al. (2022) review robotics' role in railway maintenance, emphasizing track inspection and repair. They highlight advancements in autonomous inspection, defect detection, and maintenance, reducing manual labor and enhancing safety. Robotics integration streamlines processes, improves track reliability, and advances railway operations towards greater efficiency and safety. Zhang et al. (2021) discuss problems with the conventional artificial track slab fracture detection method for ballastless tracks of the CRTSII type and provide an enhanced YOLOv3 algorithm. For effective feature extraction, they incorporate SENet in the residual module and deep separable convolution with inverted residual structure. Path fusion improves the identification of minor cracks, while the Mish activation function increases stability and accuracy. The experimental findings show a 26% speed boost and 5.3% accuracy increase over the original YOLOv3. Compared to traditional methods, their approach exhibits superior detection effectiveness and robustness, offering promising prospects for track slab crack detection in high-speed railway maintenance and compared with the already stored features in the database and producing the output depending on the highest resemblance. Vongservewattana et al. (2021) validate acoustic emission for railway track crack analysis using Mel-Frequency Cepstral Coefficients (MFCC). Acoustic emission is commonly used for crack detection in aerospace materials and bearing defects. They adapt MFCC for acoustic feature extraction, suitable for the 100–400 kHz range relevant to railway track crack detection. Experimental results demonstrate the efficacy of this technique in extracting and classifying railway track crack features, highlighting its potential for enhancing railway track monitoring systems.

Problem identification

The problem addressed by the crack detecting robot is the timely detection and notification of cracks in railway tracks to prevent accidents and operational disruptions. Traditional inspection methods are labor-intensive and often slow, making early crack detection challenging. This poses a significant safety risk and leads to operational inefficiencies. The robot's objective is to leverage advanced technologies for precise crack detection and wireless communication for real-time notifications. By doing so, it aims to enhance safety, operational efficiency, and maintenance practices by automating the detection process, reducing downtime, and enabling efficient resource allocation. The problem definition encompasses the need for a transformative solution in railway track management (Figure 45.3).

Figure 45.3 The traditional crack detection method
Source: European Union Agency for Railways (ERA)

Proposed system

The proposed crack detecting robot integrates an Arduino Uno microcontroller with a sensor data reading and location tracking module, with the IoT data transmission module and the motor driver module. The crack detecting robot autonomously traverses infrastructure surfaces, detecting cracks through IR sensor readings. When a crack is detected, its GPS coordinates are logged and transmitted via Wi-Fi to an open-source website which will be supervised by the railway officers. The temperature sensor ensures the optimal condition of the railway track. The motor driver facilitates movement, while the toggle switch enables manual control when necessary. This simple yet effective system offers early crack detection, reducing the risk of structural failure and maintenance costs, while IoT integration and GPS tracking enhance efficiency and accuracy in infrastructure maintenance and planning. The automated crack-detecting robot uses an Arduino microcontroller, temperature sensors, IR sensors, a motor driver controller, DC motors, GPS, and IoT, enabling efficient track navigation and crack detection. The IR sensor detects cracks; it just operates using its infrared radiation, which hits the tracks and bounces back. This sensor calculates its distance and checks if it is in its normal condition; if there are any cracks detected, the distance will be abnormally much more similar to the normal tracks. Also, the temperature sensors used here always sense the temperature regularly and constantly update the information in the IoT so that if there are high temperatures found, we can conclude the part is prone to cracking, and early steps can be taken for prevention. A toggle switch recognizes track transitions, and all operations are shut off at that phase because if all of the sensors work regularly even at that phase, it detects gaps at that moment to avoid this action. The toggle switch is utilized, and the GPS module tracks accurate location and transmits all latitudinal and longitudinal values through the IoT so that manual labor can only come to the correct spot. Two DC motors are employed by the robot, and their direction is controlled by the motor driver controller (Figure 45.4).

The respected railroad officers receive all of the data collected by the Arduino from sensors and processed data that is sent to them via an internet protocol module. All the information about cracks, temperatures, and GPS locations is displayed on the internet (Figure 45.5).

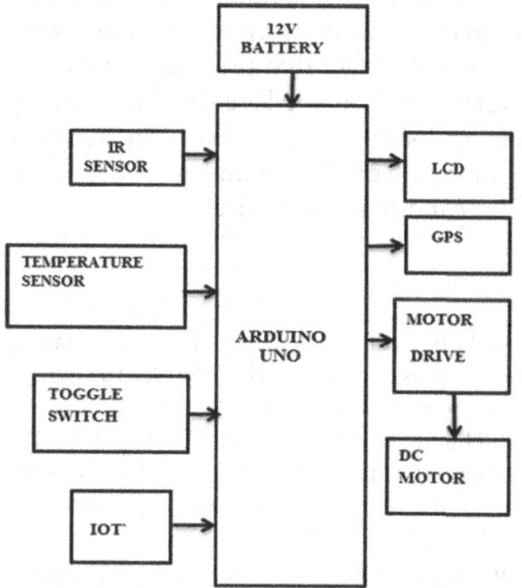

Figure 45.4 Block diagram of the proposed system
Source: European Union Agency for Railways (ERA)

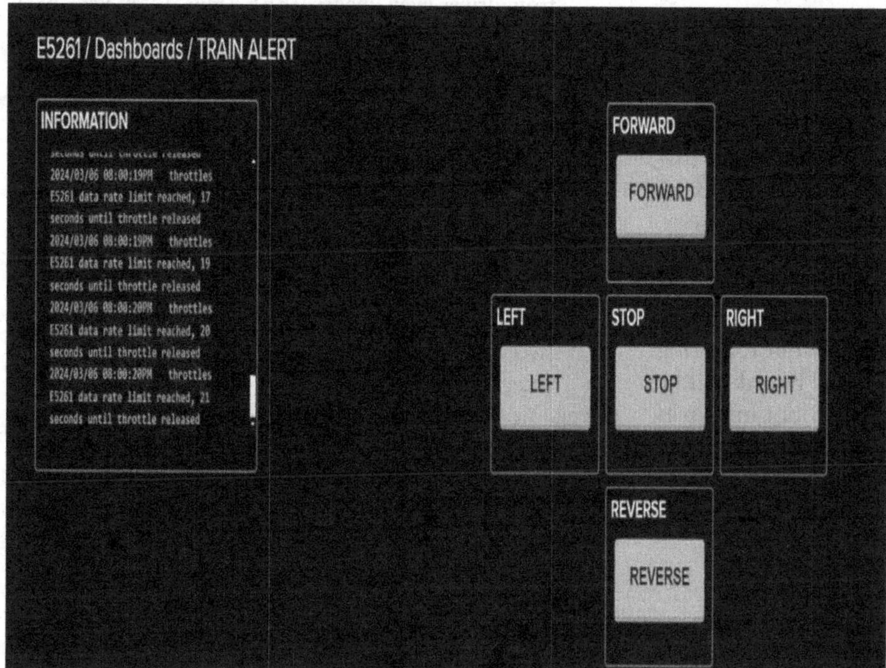

Figure 45.5 Open-source website to control the robot
Source: European Union Agency for Railways (ERA)

Our website's dashboard, which has controls for moving left, right, forward, and reverse, truly allows us to operate our robot. By making the most use of all of its parts, this integrated system improves the efficiency and safety of railway maintenance by managing temperature, identifying tracks, detecting fractures, and providing precise location tracking.

Results and discussion

The crack detecting robot project aims to enhance railway track safety and maintenance practices by utilizing advanced technologies for crack detection and notification. The project's results and discussion focus on the effectiveness of the robot in detecting cracks, and its impact on railway operations. For every sensor to function, a unique set of codes must be supplied; these codes must be written under a single program for every sensor to operate simultaneously. The robot demonstrates high accuracy in detecting cracks on railway tracks. It can identify cracks of varying sizes and severity levels, ensuring that even minor cracks are detected and reported (Figure 45.6).

We obtain data from temperature, infrared sensor, and GPS readings. The necessary prior steps can be immediately made based on these details. When a temperature reading exceeds a pre-determined threshold, it is evident that the area is susceptible to future cracking. Railway inspectors can take effective prompt

Figure 45.6 The working model
Source: European Union Agency for Railways (ERA)

action by using infrared sensors to gather pertinent information. The GPS position of the sites where the cracks are located is transmitted, allowing the proper steps to be conducted at the precise locations and saving a significant amount of time. It sends the track information detecting cracks and if any are found, it notifies in real time so that previous actions can be taken.

Conclusions and future scope

We conclude that by identifying track cracks, the crack-detecting robot aids in overcoming track discontinuity. We can lower the labor cost for inspecting railway track cracks by utilizing this technology. This robot may be operated in the control room via the Adafruit.com open-source website. This system serves as an example of how the railway department can use IoT. Other than this, it runs without requiring any power. The sensors and node MCU board will be used to collect the necessary data. The robot improves operating efficiency by cutting down on track maintenance downtime by optimizing the fracture detection procedure. By giving railway authorities precise information regarding the location and extent of cracks, the robot helps them to prioritize maintenance tasks. This guarantees that maintenance efforts are focused where they are most needed and aids in the optimization of resource allocation. This robot is reasonably priced and represents a one-time investment. The long-term advantages in terms of increased safety and lower maintenance expenses can offset the robot's initial cost, even though it may need a sizable initial investment. All that's needed for upkeep, compared to using human power, is a rechargeable battery. Another option for self-recharging is to build the robot with solar panels on top. The fact that this technique takes less time is very significant. More significantly, by identifying the cracks early on, it lowers the accident rate. This is a more effective mechanism that the railway department can use in the future.

References

Colin, M., Palhol, F., and Leuxe, A. (2016). Adaptation of transport infrastructures and networks to climate change. *Transport. Res. Proc.*, 14, 86–95.

Wan, Y., Cao, J., Huang, W., Guo, J., and Wei, Y. (2020). Perimeter control of multi region urban traffic networks with time-varying delays. *IEEE Trans. Sys. Man. Cybern. Sys.*, 50(8), 2795–2803.

Gopalakrishnan, K. (2018). Deep learning in data-driven pavement image analysis and automated distress detection: A review. *Data*, 3(3), 28.

Levering, A., Tomko, M., Tuia, D., and Khoshelham, K. (2021). Detecting unsigned physical road incidents from driver-view images. *IEEE Trans. Intel. Veh.*, 6(1), 24–33.

Han, C., Ma, T., Xu, G., Chen, S., and Huang, R. (2022). Intelligent decision model of road maintenance based on improved weight random forest algorithm. *Int. J. Pavement Engg.*, 23(4), 985–997.

Yuan, Y., Islam, M. S., Yuan, Y., Wang, S., Baker, T., and Kolbe, L. M. (2021). EcRD: Edge-cloud computing framework for smart road damage detection and warning. *IEEE Internet Things J.*, 8(16), 12734–12747.

Peng, S., Su, G., Chen, J., and Du, P. (2017). Design of an iot-bim-gis based risk management system for hospital basic operation. *Proc. IEEE Symp. Service-Orient. Sys. Engg. (SOSE)*, 69–74.

Deng, Y., Cheng, J. C., and Anumba, C. (2016). Mapping between BIM and 3D GIS in different levels of detail using schema mediation and instance comparison. *Autom. Const.*, 67, 1–21.

46 A comparative analysis for air quality prediction by AQI calculation using different machine learning algorithms

Rohit Kumar[a], *V. Krishna Likitha*[b], *Md. Harshida*[c], *Sk. Afreen*[d], *Guduru Manideep*[e] *and Maridu Bhargavi*[f]

Department of CSE, Vignan University, Guntur, Andhra Pradesh, India

Abstract

Air pollution contributes to lung cancer, heart problems, respiratory diseases, inflammation, organ damage, and depletion of ozone, acid rain, and other environmental problems. It also has a significant negative influence on human health. So it's important to know the exact air quality index (AQI) which shows how unhealthy the ambient air is. Some pollutant concentrations (PM10, PM2.5, NO, SO_2, CO, O_3) and AQI can be used to predict the air quality. We have taken city air quality dataset of 16 attribute and then used logistic regression and got 78.44% accuracy. Additionally, we used linear regression with an Root mean square error (RMSE) value of 14.19 and R'2 value of 0.96 and gradient boosting regression with 99.97% accuracy, and random forest with an R'2 value of 0.34 and Mean absolute error (MAE) value 82.78 and then we tried ensemble model and got 96.75% accuracy. Three underlying regression models—linear regression, random forest, and gradient boosting are used to build a voting regressor ensemble model. The model' performance is being assessed by Mean squared error (MSE) or R-squared score. This analysis demonstrates that the optimal algorithm is gradient boosting or random forest but if we want to give custom values then our proposed ensemble method outperform it.

Keywords: Air quality index, machine learning algorithms, ensemble model

Introduction

Air quality prediction plays a vital role in health protection, environment impact assessment, climate consideration, and economic implications. Accurate forecasting of the air quality index (AQI) helps in understanding pollutants in the atmosphere, which is vital for making informed decisions related to health and environmental policies. In this project, we're comparing different machine learning (ML) methods to see which one is best at predicting air quality. We want to find out which method gives us the most accurate and dependable forecasts.

Utilizing ML regression techniques, the prediction of air quality has seen a rise in application. The dataset employed for this purpose incorporates atmospheric factors including carbon monoxide (CO), nitrogen dioxide (NO_2), ozone (O_3), sulfur dioxide (SO_2), particulate matter smaller than 2.5 μm in diameter (PM2.5), particulate matter smaller than 10 μm in diameter (PM10), and ammonia (NH_3). The AQI is subsequently calculated based on these parameters, offering a comprehensive assessment of air quality conditions. Employing a voting regressor, which acts as an ensemble meta-estimator, enhances prediction accuracy by merging three base regression models: Linear regression, random forest, and gradient boosting. This ensemble approach is designed to elevate both prediction accuracy and model robustness.

Literature review

Mihirani et al., utilized ML model-style algorithms including random forest regression, K-nearest neighbor regression, linear regression, and Lasso regression for air quality forecasting. Random forest regression showed the highest accuracy (99%) and lowest RMSE error.

Reddy et al., employed random forest algorithm and Naive Bayes algorithms for air pollution forecasting, with random forest outperforming Naive Bayes in accuracy and error.

Srivastava et al., used random forest algorithm, support vector machine (SVM), and logistic regression to predict air pollution, with random forest regression achieving 93.5% accuracy.

Kasetty et al., employed K-means clustering unsupervised ML for air pollution analysis, with k=2 showing optimum accuracy.

Mohammad et al., utilized SMOTE-Tomek method and random forest algorithms to handle uneven data effectively, achieving 95% precision with random forest classifier.

[a]raajrohit806@gmail.com, [b]krishnalikithavegulla@gmail.com, [c]harshida136@gmail.com, [d]afreenshaik2693@gmail.com, [e]manideepguduru76@gmail.com, [f]bhargaviformal@gmail.com

DOI: 10.1201/9781003587538-46

Muljana et al., investigated the impact of reviews and ratings on educating consumers about app quality using multivariate analysis, achieving low prediction error despite dataset restrictions.

Kulkarni et al., suggested using ML methods to forecast app ratings on Google play store, with support vector regression and random forest achieving accuracies of 76.49% and 73.55%, respectively.

Akanksha et al., employed random forest, decision tree, and linear regression methods to generate air quality indices for Indian cities, emphasizing the importance of lower RMSE for higher accuracy.

Kekulanadara et al., used random forest, SVM, and decision tree techniques for classification, highlighting decision tree's superiority for large datasets.

Vyas et al., utilized linear regression models for air quality prediction, achieving 75% accuracy and suggesting it as a preferred model based on RMSE.

Shafi et al., employed K-means clustering for air pollution analysis, emphasizing its simplicity, scalability, and effectiveness for large datasets.

Data and variables

The dataset consists of 16 attributes, featuring data such as "city", "date", and various pollutant levels like "PM2.5", "PM10", "NOx", "NH_3", "CO", "NO", "NO_2", "SO_2", "O_3", "Xylene", "Benzene", and "Toluene". Additionally, it includes attributes related to air quality such as "AQI" and "AQIBucket", which categorizes AQI into different ranges or buckets.

Dependent variable: AQI and AQI Bucket can be considered dependent variables as they are measures of air quality derived from the levels of various pollutants.

Independent variable: "PM2.5", "PM10", "NOx", "NH_3", "CO", "NO", "NO_2", "SO_2", "O_3", "Xylene", "Benzene", and "Toluene" are independent variables as they represent different pollutant levels that may affect air.

Control variables: "city" and "date" could potentially serve as control variables. "city" may control for geographical variations, while "date" could control for temporal variations in air quality measurements. Additionally, other factors like weather conditions, geographical features, and population density could also be considered control variables if included in the dataset.

Methodology and model specifications

Proposed architecture

The air pollution dataset serves as the training data for our model. The pre-processing steps involve cleaning and preparing the data to make it suitable for use in the model. The dataset is splitted into training to train the model and testing for model evaluation. Our model comprises an ensemble of regression algorithms, including linear regression, random forest regression, and gradient-boosting regression using voting regressor. Finally, model evaluation is conducted by calculating metrics such as R-squared, relative absolute error (RAE), accuracy score and root mean square (RMS) values.

Model evaluation

Our model evaluation results indicate strong performance across various algorithms for air quality

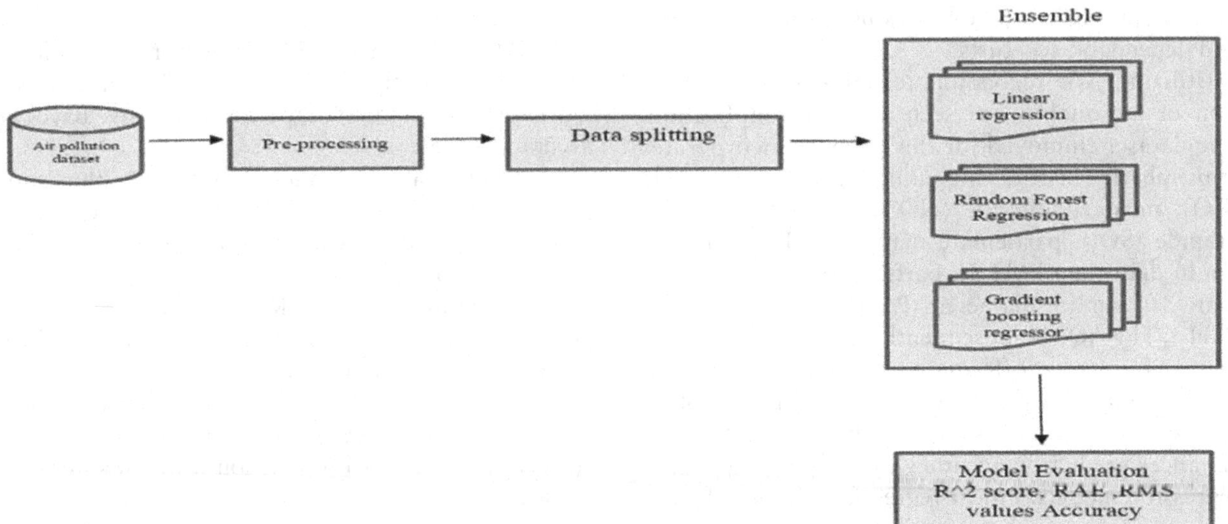

Figure 46.1 Architecture of the ensemble model
Source: https://ieeexplore.ieee.org/document/10205615

Table 46.1 Accuracy table.

Algorithm	Accuracy (%)
Logistic regression	78.44
Linear regression	81.29
Random forest	99.93
Gradient boosting	99.97
Ensemble model	96.75

Source: Author

Table 46.2 Accuracy table for composite dataset.

Algorithm	Accuracy (%)
Gradient boosting	91.4
Proposed ensemble model	96.3

Source: Author

detection. Notably, random forest and gradient boosting models achieved near-perfect accuracy scores, showcasing their efficacy in predicting air quality levels. Linear regression also demonstrated solid performance. The ensemble model yielded a commendable accuracy of 96.75%, validating its effectiveness in aggregating predictions from diverse algorithms. Each model was evaluated based on the mean squared error (MSE), with lower values indicating better predictive accuracy. Among the models tested, the hybrid neural network exhibited the lowest MSE of 0.215, indicating superior performance compared to other approaches. It outperforms the random forest and gradient boosting algorithm on custom dataset which makes our model effective.

$$MSE = \frac{1}{n}\sum_{i=1}^{n}(Y_i - \hat{Y}_i)\text{^}2$$

Empirical results

Random forest, with RMSE values of 0.84 and 1.34, is the most accurate algorithm for predicting air quality. On the other hand, with an accuracy of 99.93% and 99.97%, respectively, our research demonstrates that random forest and gradient boosting are the optimal algorithms. These results provide insights into the performance of each model in predicting air quality levels. Notably, the random forest and gradient boosting models achieved exceptionally high accuracy scores, indicating their effectiveness in air quality detection.

Conclusions

The group method combining gradient boosting, random forest, and linear regression with a voting regressor. The use of algorithms has shown to be a successful method for forecasting air pollution levels. The model's potential for practical uses in air quality management and monitoring was demonstrated by its high R^2 score, low MSE, and where applicable satisfactory accuracy. Better outcomes could be achieved by further optimizing and fine-tuning the ensemble model, which would open the door to improved environmental forecasting and decision-making. Notable results come from evaluating several ML models for the detection of quality of air. Random forest and gradient boosting models perform exceptionally well, with test accuracies above 99.9%, while logistic regression and linear regression offer modest levels of accuracy. The ensemble model stands out in particular because it achieves an amazing accuracy of 96.75% by combining predictions from different algorithms. It is more appropriate for the composite dataset as it combines features from all models.

References

Mihirani, M., Yasakethu, L., and Balasooriya, S. (2023). Machine learning-based air pollution prediction model. *2023 IEEE IAS Global Conf. Emerg. Technol. (GlobConET)*, 1–6. doi: 10.1109/GlobConET56651.2023.10150203.

Reddy, P. D. and Parvathy, L. R. (2022). Prediction analysis using random forest algorithms to forecast the air pollution level in a particular location. *2022 3rd Internat. Conf. Smart Elec. Comm. (ICOSEC)*, 1585–1589. doi:10.1109/ICOSEC54921.2022.9952138.

Srivastava, H., Sahoo. G. K., Das, S. K., and Singh, P. (2022). Performance analysis of machine learning models for air pollution prediction. *2022 Internat. Conf. Smart Gen. Comput. Comm. Netw (SMART GENCON)*, 1–6. doi: 10.1109/SMARTGENCON56628.2022.10084037.

Kasetty, S. B. and Nagini, S. (2022). A survey paper on an IoT-based machine learning model to predict air pollution levels. *2022 4th Internat. Conf. Adv. Comput. Comm. Control Netw. (ICAC3N)*, 1408–1412. doi: 10.1109/ICAC3N56670.2022.10074555.

Mohammad, J. and Kashem, M. A. (2022). Air pollution comparison RFM model using machine learning approach. *2022 IEEE 7th Internat. Conf. Converg. Technol. (I2CT)*, 1–5. doi: 10.1109/I2CT54291.2022.9824248.

Muljana, R., Ayuningtyas. L. D., Daksa, R. P., Djamhari, S. F., Fiezayyan, M. A., and Sagala, N. T. M. (2023). Air pollution prediction using random forest classifier: A case study of DKI Jakarta. *2023 Internat. Conf. Comp. Sci. Inform. Technol. Engg. (ICCoSITE)*, 428–433. doi:10.1109/ICCoSITE57641.2023.10127759.

Kulkarni, M., Raut, A., Chavan, S., Rajule, N., and Pawar, S. (2022). Air quality monitoring and prediction using SVM. *2022 6th Internat. Conf. Comput. Comm. Control Automat. (ICCUBEA)*, 1–4. doi: 10.1109/ICCUBEA54992.2022.10010942.

Akanksha, A., Maurya, N., Jain, M., and Arya, S. (2023) . Prediction and analysis of air pollution using machine learning algorithms. *2023 3rd Internat. Conf. Intel. Technol. (CONIT)*, Hubli, India, 2023, pp. 1–6.

Chattopadhyay, M., et al. (2021). Subjective probabilistic expectations, household air pollution, and health: Evidence from cooking fuel use patterns in West Bengal, India. *Res. Ener. Econ.*, 66, 101262.

Agrawal, G., Mohan, D., and Rahman, H. (2021). Ambient air pollution in selected small cities in India: Observed trends and future challenges. *IATSS Res.*, 45(1), 19–30.

Shree, R. N., Santhiya, G., and Bhargavi, R. (2020). Assessment of spatial hazard and impact of PM10 using machine learning. *2020 4th Internat. Conf. Comp. Comm. Sig. Proc. (ICCCSP)*, IEEE, 2020, pp.1–4.

Nandini, K. and Fathima, G. (2019). Urban air quality analysis and prediction using machine learning. *2019 1st Internat. Conf. Adv. Technol. Intel. Control Environ. Comput. Comm. Engg. (ICATIECE)*, Bangalore, India, 2019, pp.98–102.

Punugoti, R., Dutt, V., Kumar, A., and Bhati, N. (2023). Boosting the accuracy of cardiovascular disease prediction through SMOTE. *2023 Internat. Conf. IoT Comm. Autom. Technol. (ICICAT)*, 1–6.

47 Optimization of 6T SRAM cell delay using Sleepy Keeper in Cadence Virtuoso 90 nm technology

Madiraju Joshitha[a], *Mavillapati Nicharitha*[b] *and K. Mahendran*[c]

Departmant of ECE, Saveetha Engineering College, Thandalam, Chennai-602105, Tamil Nadu, India

Abstract

This study investigates how to use Cadence Virtuoso's Sleepy Keeper to optimize the performance of static random access memory (SRAM) in a 90 nm technology node. The basis is a thorough analysis of Sleepy Keeper principles, Cadence Virtuoso tools, and SRAM architecture. To evaluate Sleepy Keeper's effect on SRAM latency and power consumption, circuit-level simulations and power assessments are routinely used. Layout improvements are also used to reduce parasitic components and improve cell performance metrics. To implement power management within the SRAM array, extra transistors are strategically included together with control logic. Through an iterative process driven by repeated simulations and layout improvements, SRAM latency, power efficiency, and overall performance are all balancedly improved. With regard to the incorporation of Sleepy Keeper for SRAM optimization, this work provides useful insights and important advances in low-power memory design for modern semiconductor technologies. The procedure is fully documented at every stage of the design process and can be used as a guide for future work in the ever-evolving field of semiconductor memory optimization.

Keywords: Static Random Access Memory (SRAM), Sleepy Keeper, SRAM latency

Introduction

A mainstay of contemporary semiconductor design, static random access memory (SRAM) is essential for a wide range of applications, from high-performance computing to embedded systems. The need for effective and low-power memory solutions is growing as technology nodes get smaller Yeap, G. K. et al. (1998), Hina, M. and Nayar, S. (2013). In order to meet this requirement, this study examines how to best utilize the power-saving method Sleepy Keeper to maximize SRAM performance within the framework of a 90 nm technology node. The foundation for a thorough examination of Sleepy Keeper's use and effects is laid by exploring the complexities of SRAM architecture and the powers of the Cadence Virtuoso platform Park, J. (2014). The work utilizes circuit-level simulations, power assessments, and layout modifications to achieve a synergistic improvement in SRAM functionality, with a focus on minimizing delay and boosting power efficiency Taehui, et al. (2013). The goal of the research is to contribute significantly to the changing field of semiconductor technology by delivering useful methods and insights for improving low-power memory architecture. Memory optimization is at the forefront of technological progress due to semiconductor devices' constant quest for more computing power and energy efficiency. Static random access memory (SRAM) is a key component in this environment and frequently serves as a major bottleneck for reaching targeted performance metrics Narender, H. (2011), Andrea, C. et al. (2012). This work explores the challenging problem of enhancing SRAM efficiency in the constraints of a 90 nm technology node by utilizing the novel Sleepy Keeper method. Sleepy Keeper offers a potential solution for cutting power usage without sacrificing functionality because of its capacity to specifically deactivate specific memory cells during idle times. The intricacies of SRAM architecture, the powerful powers of Cadence Virtuoso tools, and the theoretical foundations of Sleepy Keeper are all thoroughly explored in the following sections, provides the foundation for a sophisticated comprehension and utilization of this power-saving approach. The need for memory solutions that achieve the best possible balance between performance and power efficiency is growing as semiconductor technologies develop. A viable path to accomplishing these goals is provided by the combination of Sleepy Keeper integration, SRAM optimization, and Cadence Virtuoso's advanced features.

Design of SRAM cell

The core component of SRAM cell is what makes it possible to store binary data in a static, non-volatile way. The two cross-coupled inverters at the heart of the SRAM cell are made up of complementary metal-oxide-semiconductor (CMOS) transistors, or NMOS and PMOS, respectively. This configuration creates a

[a]Joshithamadiraju28@gmail.com, [b]nicharithamavillapati@gmail.com, [c]mahendrank@saveetha.ac.in

DOI: 10.1201/9781003587538-47

bistable circuit that can store logic 0 or logic 1, or one of two states Geetha Priya, M. et al. (2012). Each SRAM cell is coupled to two bit lines (BL and BL_bar) that enable read and write operations. The SRAM cells are arranged into arrays. Wordlines provide access to particular rows of SRAM cells by being selectively triggered during operations. The bit lines are set to the appropriate logic levels and the wordline corresponding to the target row is active during a write operation Anu, T. and Goyal, S. (2015). The supplied data is then latch-loaded onto the cross-coupled inverters, which store it in the SRAM cell. On the other hand, during a read operation, the bit lines' levels are used to determine the logic states that are stored in the cell, and the wordline's activation links the cell to them. SRAM cells are essential parts of many systems where quick and effective data storage is crucial because of their simplity and speed (Figure 47.1).

Operation of SRAM cell

Electrical signals interact dynamically in SRAM cell during write and read operations. During a write operation, the associated wordline activates the targeted SRAM cell and precharges the bit lines. The cross-coupled inverters in the cell are identified by the differential voltage that is produced when the input data is applied to the bit lines. The input data is stored in a stable logic state thanks to this bistable latch mechanism. On the other hand, in a read operation, the chosen wordline opens the cell and connects it to the bit lines without changing the data that has been stored. Sense amplifiers are used to measure and amplify the voltage difference across the bit lines, which allow one to determine the logic state that is stored in the SRAM cell. The wordline is then turned-off, isolating the cell and ending the read cycle Catthoor, F. et al. (1998). The SRAM cell's vital function in offering quick and non-volatile memory storage is highlighted by the complex dance of electrical signals that occurs within it (Table 47.1).

Characteristics of SRAM cell

The characteristics of SRAM are speed, volatility, and stability. SRAM provides quick access to stored data because it doesn't require frequent refreshing like dynamic RAM (DRAM) does. Six transistors arranged in a flip-flop format make up the conventional SRAM cell structure, which increases stability but also makes it larger and less dense than DRAM. SRAM is appropriate for applications needing high-speed access, like

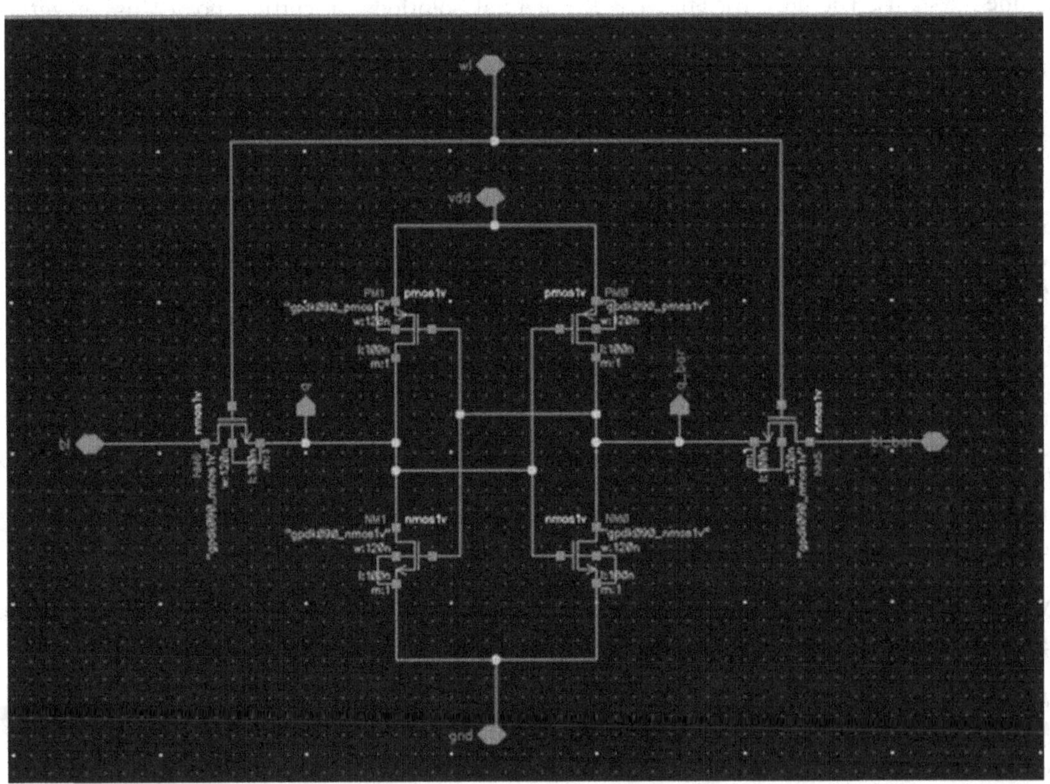

Figure 47.1 Design of SRAM cell
Source: Author

Table 47.1 Read and write operation of SRAM cell.

Read operation		Write operation	
Read "1"	Pre-charge both the bit lines word line=1 bit bar=0 (discharges to 0) bit=1	Write "1"	Q=1, Qbar=0 Word line=1 bit bar=0 bit =1
Read "0"	Pre-charge both the bit lines word line=1 bit bar=1 bit =0 (discharges to 0)	Write "0"	Q=0, Qbar=1 word line=1 bit bar=1 bit =0

Source: Author

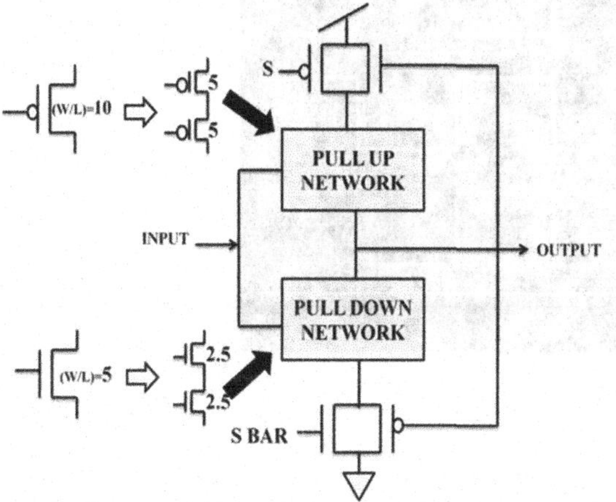

Figure 47.2 SRAM cell using sleepy keeper technique
Source: Author

processor cache memory, since it supports both read and write operations. In contrast to DRAM, this speed is accompanied by a higher power consumption. The complexity, greater number of transistors, and larger cell size of SRAM all lead to higher production costs Rabaey, J. M. and Chandrakasan, A. P. (2003). SRAM proves its importance in computer memory systems by finding a niche where speed and responsiveness are critical, despite its higher cost and lesser density.

Power gating techniques

In SRAM design, power gating techniques are essential to the goal of energy efficiency. The Sleepy Keeper technique is one such technique that incorporates sleep transistors into SRAM cells in a deliberate manner. By selectively cutting-off power to dormant cells during idle times, these sleep transistors minimize leakage current and hence lower total power usage. These sleep transistors are activated and deactivated using control logic, which guarantees a smooth transition between powered and power-gated states. Although power gating in SRAM offers a potentially effective way to reduce energy consumption, timing issues and the intricacy of the control logic provide practical hurdles. Notwithstanding these difficulties, power gating is an essential component of energy-efficient memory system design since it minimizes leakage power during idle times, which is especially useful in situations where power consumption is a key consideration.

Sleepy Keeper technique for high speed SRAM cell

Using power gating, the Sleepy Keeper approach selectively deactivates inactive memory cells while preserving their data in retention registers, so saving power for high-speed SRAM. By turning-off inactive cells, this method lowers total power consumption, however, reactivating the cells requires restoring data from retention registers, which adds latency. Power gating and wakeup are managed by control logic, which maximizes power savings and minimizes performance impact. Sleepy Keeper increases power efficiency in SRAM arrays, which are essential for battery-operated systems and mobile devices, even though it adds complexity. The secret of Sleepy Keeper's success is its ability to strike a balance between latency overhead and power savings, guaranteeing maximum energy economy in high-speed SRAM arrays without sacrificing performance. In order to accomplish the needed power reduction while preserving data integrity and responsiveness, implementation requires careful design and optimization of control logic, retention registers, and power gating methods. All things considered, Sleepy Keeper is an essential strategy for increasing battery life and enhancing energy efficiency in contemporary electronic gadgets and systems that depend on fast SRAM (Figure 47.2).

Results and discussion

Using the Sleepy Keeper technique and a 90 nm technology process, Cadence Virtuoso IC design has created a low-power SRAM cell. The analysis of delay characteristics is the main goal of the simulation. Data "1" is written into the SRAM cell when the word line (WL) is set to "1" and the bit lines (BL) stay high at "1" while the complementary bit line (BLB) stays low at "0" in the Sleepy Keeper technique-based SRAM cell simulation. On the other hand, data "0" is stored into the SRAM cell when WL = 1 and BLB = 1 (remaining

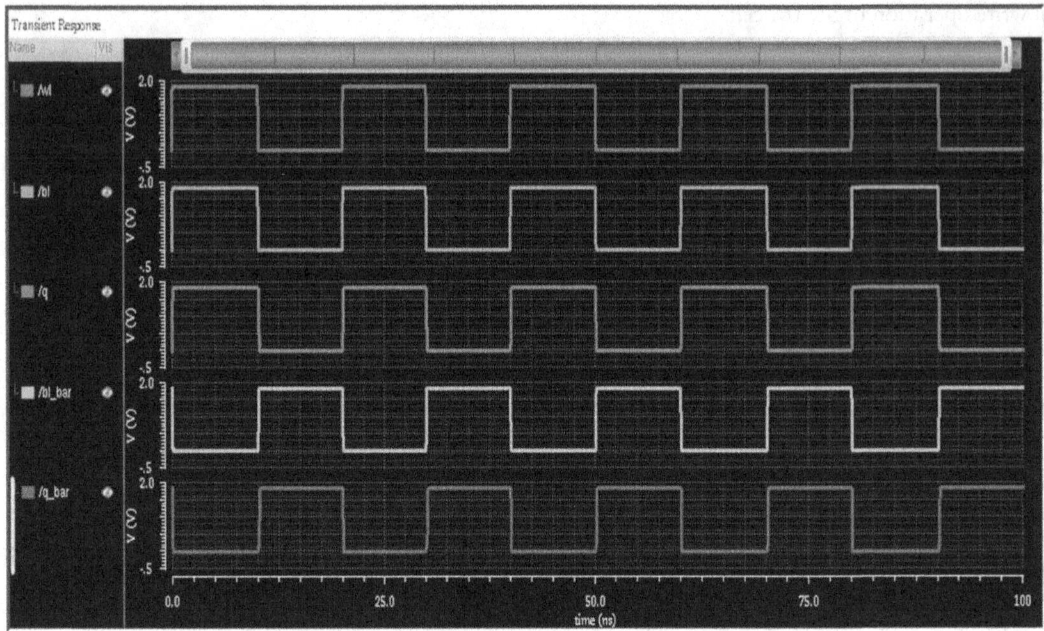

Figure 47.3 Waveform for SRAM cell using sleepy keeper technique
Source: Author

Without power gating	35.44 ps
Sleep technique	27.30 ps
Stack technique	38.38 ps
Sleepy stack technique	37.22 ps
Sleepy keeper technique	23.36 ps
Lector technique	25.98 ps

high) while BL = 0 (remaining low). Figure 47.3 shows the waveform of the SRAM cell simulation using the Sleepy Keeper technique.

Conclusions

By using the Sleepy Keeper technique, the SRAM cell delay was decreased from 35.44 ps to 23.36 ps. This indicates a drop of about 12.08 ps. As a result, the Sleepy Keeper technique's delay improvement comes out to roughly 34.1% of the initial delay value. This significant decrease highlights how power-saving techniques such as Sleepy Keeper can optimize SRAM performance and lead to increased efficiency in electronic systems and integrated circuits.

References

Yeap, G. K. et al. (1998). Practical low power digital VLSI design. Kluwer Academic Publishers. Norwell, MA, ISBN: 0792380096, 233.

Hina, M. and Nayar, S. (2013). A new approach for leakage power reduction techniques in deep submicron technologies in CMOS circuit for VLSI application. *IEEE Int. Solid State Circ. Conf. (ISSCC) Dig. Tech.*, 3, 11–16.

Park, J. (2014). Sleepy stack: A new approach to low power VLSI and memory. Ph.D. dissertation. *School Elec. Comp. Engg,* http://hdl.handle.net/1853/7283

Taehui, Na, Woo, S.-H., Kimg, J., Jeon, H., and Jung, S.-O. (2013). Comparative study of various latch-type sense amplifiers. *IEEE Trans. Very Large Scale Integ. (VLSI) Sys.*, 12(6), 1–5.

Narender, H. (2011). LECTOR: A technique for leakage reduction in CMOS circuits. *IEEE Trans. Very Large Scale Integ. (VLSI) Sys.*, 12(2), 198–205.

Andrea, C., Macii, A., Macii, E., and Poncino, M. (2012). Design techniques and architectures for low-leakage SRAMs. *IEEE Trans. Circ. Sys.*, 59(9), 1992–2007.

Geetha Priya, M., Baskaran, K., and Krishnaveni, D. (2012). Leakage power reduction technique in deep submicron technologies for VLSI applications. *Elsevier-Sciverse Sci. Direct Proc. Engg.*, 30, 1163–1170.

Anu, T. and Goyal, S. (2015). A literature review on leakage and power reduction techniques in CMOS VLSI design. *Internat. J. Adv. Res. Comp. Comm. Engg.*, 3(2), 554–558.

Catthoor, F., Wuytack, S., and DeGreef, E. (1998). Custom memory management methodology exploration of memory organization for embedded multimedia system design. Kluer Acadamic Publishers, Boston.

Rabaey, J. M. and Chandrakasan, A. P. (2003). Digital integrated circuits: A design perspective. Upper Saddle River, NJ, USA: Prentice-Hall.

48 Energy management system for hybrid AC/DC microgrids and electric vehicles-based on clouded leopard optimization control

S. Sruthi[1,a], *K. Karthikumar*[2,b] *and P. ChandraSekar*[3,c]

[1]Research Scholar, Dept. of EEE, Veltech Rangarajan Sagunthala R and D Institute of Science and Technology, Avadi, Chennai, India

[2]Associate Professor, Dept. of EEE, Veltech Rangarajan Dr. Sagunthala R and D Institute of Science and Technology, Avadi, Chennai, India

[3]Professor, Dept. of EEE, Veltech Rangarajan Dr. Sagunthala R and D Institute of Science and Technology, Avadi, Chennai, India

Abstract

This research introduces an efficient energy management methodology using hybrid renewable power in AC-DC microgrids, including electric vehicle (EV) utilizing sources such as solar and wind energy. These systems offer promising solutions for enhancing security, reliability, and efficiency in power systems, with the added benefit of reducing greenhouse gas emissions. This approach addresses challenges faced by existing control methods, such as instability and complexity, by simplifying control through clouded leopard optimization (CLO) and facilitating efficient power sharing. By filling in current research gaps and resolving issues with distributed generation power, energy storage AC grid power integration, system state of charge (SOC) and load demand the novel approach helps to enhance energy management in hybrids microgrids. The proposed approach achieves an efficiency of about 94%.

Keywords: State of charge, power management, hybrid microgrids, AC grid power integration, electric vehicle power references, varying power profiles

Introduction

Hybrid AC/DC microgrids present promising opportunities for distributed generation (DG) systems by combining both AC and DC technologies without the need for converters (Nourollahi etal., 2020). This allows for direct energy harvesting from various renewable sources such as fuel cells (DC power), solar units (DC power), dispatchable units like microturbines and wind turbines (AC power), offering high power quality and reliability at a lower capital cost. Mixed AC/DC microgrids have several benefits over just-AC or just-DC microgrids, including reduced startup costs, no need for converters, improved power quality, and on-site direct power delivery (Wang et al., 2020; Gholami et al., 2022).

On-going analysis endeavors have tended to these difficulties through imaginative energy the board methodologies (EMS). A counterfeit brain organization (ANN)-based EMS for dealing with a DC microgrid utilizing a HESS was proposed by Ramu et al. (2024). An hourly power stream (PF) study for AC/DC mixture micro grids associated by an interlinkage converter (IC) within both islanded and lattice associated structures was introduced by Moradi et al. (2024) inside an EMS. Beluga Whale Streamlining (BWO) was utilized by Alharbi et al. (2024) to make a proficient EMS for disseminating load in a DC microgrid. Ghasemi et al. (2024) focused on overseeing generator units, setting up a versatile burden the board plan, and charging and releasing electric vehicles. The concept of hybrid AC/DC microgrids has garnered significant attention from researchers due to the combined benefits of DC and AC structures. This study looks into how best to operate and manage hybrid AC/DC micro grids using renewable energy sources when there are electric vehicles (EVs). This is a difficult undertaking because EV charging demand is unpredictable and complicated. Four charging options with varying market share and capacity are taken into consideration while modeling and analyzing plug-in hybrid EVs at two distinct penetration levels (80% and 40%) within the hybrid microgrid. This will reduce the adverse effects of EV charging demand on the grid. In order to handle the stochastic nature of solar and wind energy sources and the requirement for EV charging, a point estimate approach-based stochastic technique is proposed.

[a]sruthi.seasapram91@gmail.com, [b]drkarthikumark@veltech.edu.in, [c]drchandrasekar@veltech.edu.in

DOI: 10.1201/9781003587538-48

Overview of system to establish energy management in hybrid microgrid

Figure 48.1 illustrates the optimal management using the proposed system architecture. For the HRES and utility to manage energy quality challenges, stable real and reactive power regulation is crucial. This section outlines the control method for real power and reactive power operation modes connected to the grid.

Active and reactive control strategies in hybrid microgrid

This section analyzes improvements in energy management for hybrid microgrid and load modes. Power from the main grid and hybrid microgrid must both adapt as load demands alter. To meet load needs, photovoltaic array, power from the fuel cell, and main grid must be integrated. The power balance is displayed in Equation (1) and should be fulfilled at the DC-connection and PCC.

$$P_{HRES}(t) = \sum P_{EV}(t) + P_{PV}(t) + P_{FC}(t) + P_{Bat}(t) \tag{1}$$

Acquirable renewable energy resources over time t, with a representation of total power production $P_{HRES}(t)$. The detailed modeling process has been referred from Praveen Kumar et al., (2019).

Proposed technique-based optimal energy management in hybrid microgrid

The design of clouded leopard optimization (CLO) involves mathematical simulations of natural behaviors exhibited by clouded leopards (Trojovská and Dehghani, 2022). Equation 2 is first used to randomly initialize the locations of the leopards with clouds in the area of search at the beginning of the procedure.

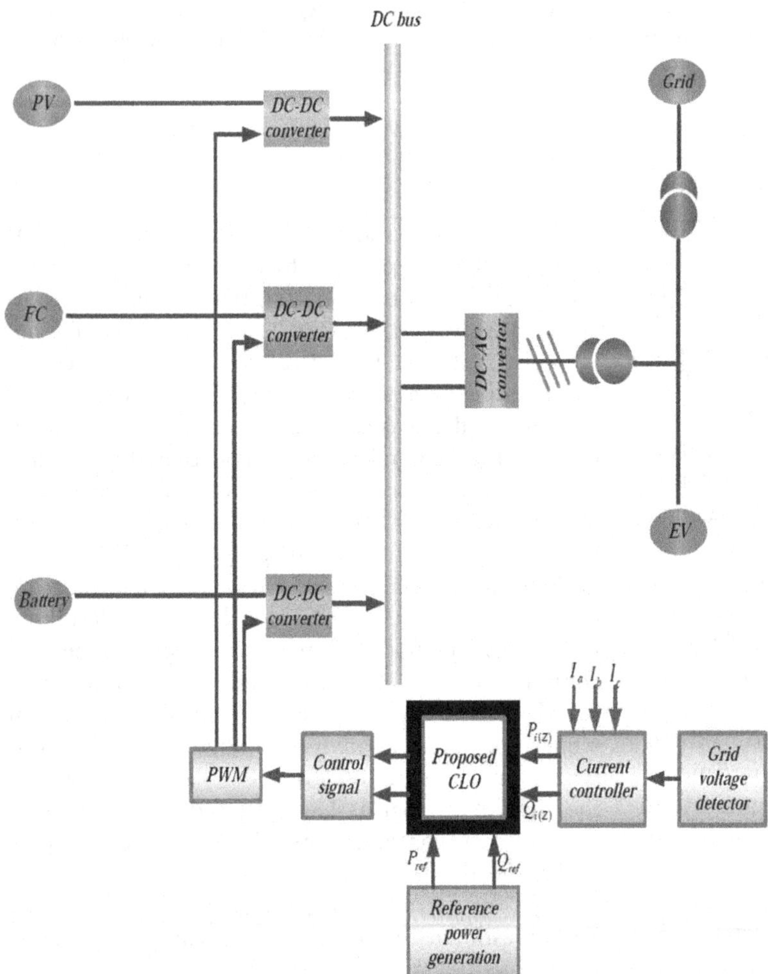

Figure 48.1 proposed optimal management system architecture
Source: Author

$$X_i : x_{i,j} = [l_j + r_{i,j}(u_j - l_j)],$$
$$i = 1,2,\text{L},N;\ j = 1,2,\text{L},m \quad (2)$$

where, i^{th} cloudy leopard (i.e., candidate solution) is expressed as X_i, decision variable is expressed as $x_{i,j}$, population size of clouded leopard is expressed as "N", number of decision variables is specified as "m", randomly selected numbers in [0, 1] is specified as $r_{i,j}$, lower and upper bound are specified as u_j,l_j. The CLO population, which is made up of all cloudy leopards, may be expressed numerically by a matrix as shown in Equation 3.

$$X = \begin{bmatrix} X_1 \\ X_i \\ X_N \end{bmatrix}_{N \times m} \quad (3)$$

where, population matrix of CLO is expressed as "X". The following vector expression represents the computed values for the goal of the function.

$$F = \begin{bmatrix} F_1 \\ F_i \\ F_N \end{bmatrix}_{N \times 1} = \begin{bmatrix} F(X_1) \\ F(X_i) \\ F(X_N) \end{bmatrix}_{N \times 1} \quad (4)$$

where, vector of the objective function is expressed as "F", the value of the aimed function derived from the clouded leopard is denoted as F_i. Equation 5 is used to produce a new scenario for the corresponding clouded leopard according to the intended prey's position. Equation 6 states that this new location of the related cloudy leopard takes precedence over the prior position if it produces a better value for the goal function.

$$x_{i,j}^{P1} = \begin{cases} x_{i,j} + r_{i,j}(p_{i,j} - l_{i,j}x_{i,j}), & F_i^P < F_i; \\ x_{i,j} + r_{i,j}(x_{i,j} - l_{i,j}p_{i,j}), & else, \end{cases} \quad (5)$$

$$X_i = \begin{cases} X_i^{P1}, & F_i^{P1} < F_i; \\ X_i, & else, \end{cases} \quad (6)$$

whereas, according to the first phase of CLO, X_i^{P1} describes the innovative position proposed for the *i*th cloudy leopard, F_i^{P1} is the desired function value, $x_{i,j}^{P1}$ is its *j*th dimension, F_i^{P1} is the target function value of the prey p_i, $p_{i,j}$ is the *j*th dimension of the target, $l_{i,j}$ are random values chosen from the set {1, 2}. Equation 7 is used to generate a random position close to each clouded leopard in order to mimic their activity.

$$x_{i,j}^{P2} = x_{i,j} + \frac{l_{i,j} + r_{i,j}(u_j - l_j)}{t}(2r_{i,j} - 1). \quad (7)$$

$$X_i = \begin{cases} X_i^{P2}, & F_i^{P2} < F_i; \\ X_i, & else, \end{cases} \quad (8)$$

where X_i^{P2} is the new position that is recommended for the *i*th cloudy leopard based on the CLO's first phase; X_i^{P2} the function's objective value is its *j*th dimension and the random integers are chosen from the range {0, 1}.

Outcomes and analysis

This particular section exhibits the execution of the suggested procedure on the MATLAB/Simulink stage, pointed toward upgrading power quality in hybrid microgrids. Abrupt variations in voltage or current signals under typical operating settings might result in an imbalanced supply, which can lead to problems with power quality, especially with solar PV systems. The simulation results depict the variations in solar PV irradiance levels (400 W/m^2, 800 W/m^2, 100 W/m^2 and 600 W/m^2) during system operation. Charging and discharging activities of the battery are indicated by negative and positive power values, respectively. As the irradiance fluctuates, Figure 48.2 demonstrates how this combination control strategy successfully maintains the DC bus voltage at 400 V, shifting to 800 W/m^2 at T = 1 s.

When the battery is in CV mode, it charges continuously until the current flow crosses a predefined threshold, signifying that the battery is fully charged. Until the voltage reaches a cut-off threshold, the charging current in CC mode remains constant. On the other hand, in CV mode, the current that charges drops but the voltage stays constant. A system for managing batteries built into the car's battery system controls the entire charged process. Usually, the EV battery has a low voltage when it first starts to charge. For the battery and charger to last as long as possible, a steady charging current must be maintained.

Fortunately, changes in irradiance during different periods (1, 2, 3, and 4) do not affect the battery charging process. With battery state of charge (SOC) setting to 9, the voltage and current profiles of the battery, as depicted in Figure 48.3, confirm that the battery is indeed charging despite fluctuations in irradiance. The control algorithm successfully keeps the DC connection voltage at 400 V during the variation in irradiance. As seen in Figure 48.4, there is a significant shift in the SOC value at T = 2 s. Consequently, the battery runs out and the voltage drops when the fuel

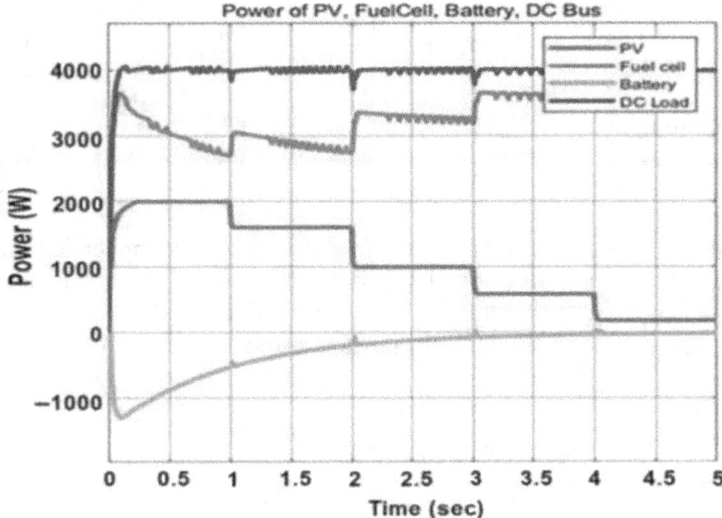

Figure 48.2 Analysis of total system power using proposed technique
Source: Author

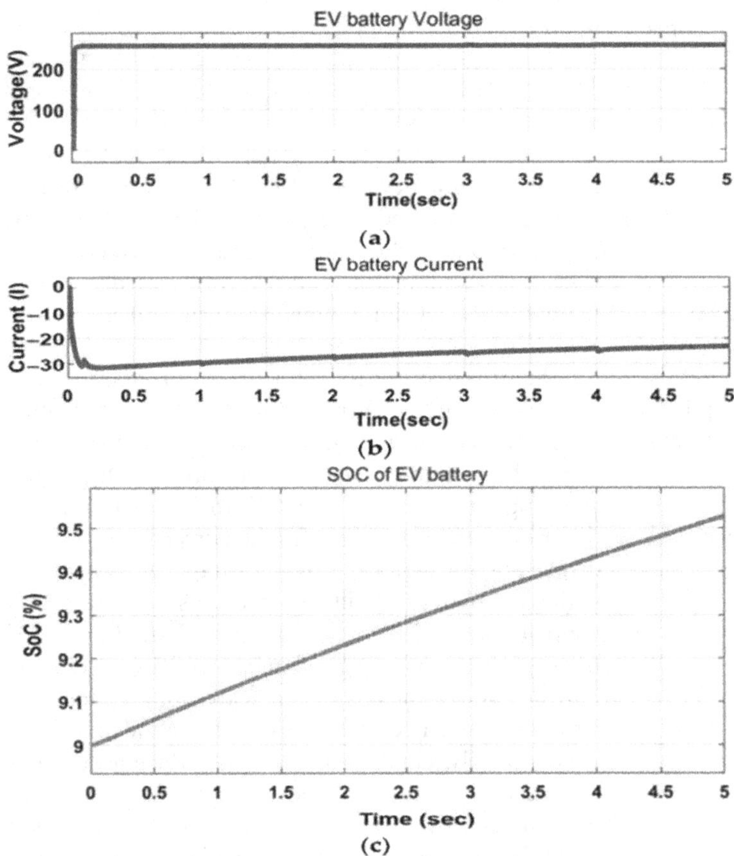

Figure 48.3 Investigation of EV (a) Voltage (b) Current (c) SOC
Source: Author

cell begins to deliver power to the grid. Later in the timeline, at T = 4 s, the fuel cell's output of electricity as the battery's discharging curves both increase as the SOC value continues to decline and the battery voltage lowers. All control techniques become efficient when a recommended controller optimizes the performance of DC-DC converters with better outcomes. Efficiency is evaluated and compared and shown in Table 48.1

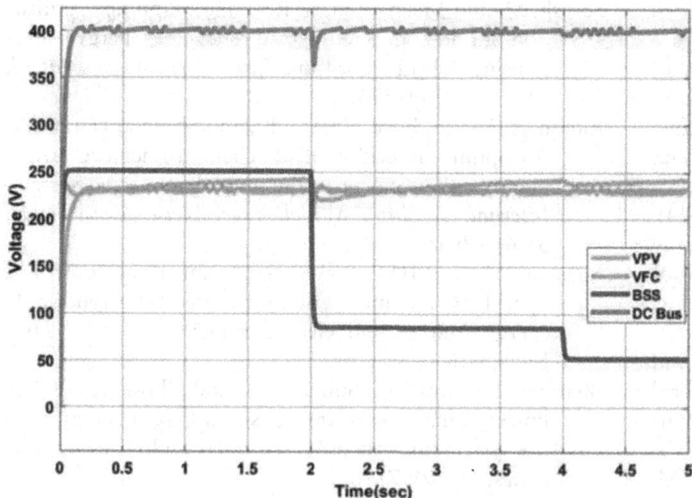

Figure 48.4 Analysis of voltage at various sources and DC bus
Source: Author

Table 48.1 Comparison of source power efficiency in two scenarios.

Efficiency							
Bases	Clarification methods						
	GA	PSO	WOA	AWOTS	GSA	FA	Developed technique
PV	81.136	74.26588	81.2357	80.0967	81.3473	82.4384	90.9375
EV	69.8746	70.9835	81.7112	82.0069	82.1038	85.7384	94.387
Battery	55.907	65.7123	88.1177	89.9911	90.3564	91.4356	93.7560

Source: Author

under balanced and random distribution scenarios, taking into account variables including power from the hybrid microgrid, grid power, load power, photovoltaic voltage and inverter voltage.

Conclusions

This study investigates the combination battery energy sources, solar PV systems of fuel cells, and to produce a hybrid microgrid that can power EV charging and consumer loads using the CLO approach. A resilient EMS is established to addresses the difficulties presented by erratic and dynamic events. The suggested hybrid intelligence algorithm is used to examine the microgrid system at different phases under two situations. Based on experimental results, the recommended hybrid EMS guarantees a balanced power distribution among the storage system and sources guaranteeing a continuous power supply. It quickly adjusts the DC bus voltage when there are variations and efficiently controls the terminal power balance of the system's components.

References

Nourollahi, R., Zare, K., and Nojavan, S. (2020). Energy management of hybrid AC-DC microgrid under demand response programs: Real-time pricing versus time-of-use pricing. *Demand Res. Appl. Smart Grids Oper. Iss.*, 2, 75–93.

Wang, P., Wang, D., Zhu, C., Yang, Y., Abdullah, H. M., and Mohamed, M. A. (2020). Stochastic management of hybrid AC/DC microgrids considering electric vehicles charging demands. *Ener. Reports*, 6, 1338–1352.

Gholami, K., Azizivahed, A., and Arefi, A. (2022). Risk-oriented energy management strategy for electric vehicle fleets in hybrid AC-DC microgrids. *J. Ener. Storage*, 50, 104258.

Ramu, S. K., Vairavasundaram, I., Palaniyappan, B., Bragadeshwaran, A., and Aljafari, B. (2024). Enhanced energy management of DC microgrid: Artificial neural networks-driven hybrid energy storage system with integration of bidirectional DC-DC converter. *J. Ener. Storage*, 88, 111562.

Moradi, S., Tinajero, G. D. A., Vasquez, J. C., Zizzo, G., Guerrero, J. M., and Sanseverino, E. R. (2024). Hierarchical-power-flow-based energy management for alternative/direct current hybrid microgrids. *Sustain. Ener. Grids Netw.*, 101384.

Alharbi, A. G., Olabi, A. G., Rezk, H., Fathy, A., and Abdelkareem, M. A. (2024). Optimized energy management and control strategy of photovoltaic/PEM fuel cell/batteries/supercapacitors DC microgrid system. *Energy*, 290, 130121.

Ghasemi, E., Ranjbaran, A., and Pourhossein, J. (2024). Designing multi-objective electric and thermal energy management system of microgrid in the presence of controllable loads and electric vehicles. *Elec. Engg.*, 106(2), 1519–1532.

Belkhier, Y. and Oubelaid, A. (2024). Novel design and adaptive coordinated energy management of hybrid fuel-cells/tidal/wind/PV array energy systems with battery storage for microgrids. *Ener. Storage*, e556.

Praveen Kumar, T., Subrahmanyam, N., and Sydulu, M. (2019). Power flow management of the grid-connected hybrid renewable energy system: A PLSANN control approach. *IETE J. Res.*, 1–16.

Trojovská, E. and Dehghani, M. (2022). Clouded leopard optimization: a new nature-inspired optimization algorithm. *IEEE Acc.*, 10, 102876–102906.

Suresh, M. and Meenakumari, R. (2020). Optimum utilization of grid connected hybrid renewable energy sources using hybrid algorithm. *Trans. Inst. Meas. Control*, 014233122091374.

Rajesh, P., Shajin, F. H., Rajani, B., and Sharma, D. (2022). An optimal hybrid control scheme to achieve power quality enhancement in micro grid connected system. *Internat. J. Num. Model. Elec. Netw. Dev. Fields*, 35(6), e3019.

Rahman, M. S. (2018) . Power management and control of hybrid AC/DC microgrids integrated with renewable energy sources and electric vehicles. *Doctoral Dis.*, Macquarie University.

Bharatee, A., Ray, P. K., Subudhi, B., and Ghosh, A. (2022). Power management strategies in a hybrid energy storage system integrated AC/DC microgrid: A review. *Energies*, 15(19), 7176.

Sarwar, S., Kirli, D., Merlin, M. M., and Kiprakis, A. E. (2022). Major challenges towards energy management and power sharing in a hybrid AC/DC microgrid: A review. *Energies*, 15(23), 8851.

49 Aquatic health guardian: A water quality monitoring system for fish farming

Pandiyan P.[a], *Aakash M.*[b], *Dhinagaran M.*[c], *Dharanish Raja K.*[d] *and Mukilan M.*[e]

Department of EEE, KPR Institute of Engineering and Technology, Coimbatore, Tamilnadu, India

Abstract

This paper emphasizes the importance of automated water quality monitoring systems in fish farming to tackle hurdles in upholding ideal circumstances for fish health and productivity. It presents the planning and implementation of an integrated monitoring system for water quality with the goal of guaranteeing optimal conditions for the health and development of fish. The system focuses on monitoring vital parameters including temperature, pH, total dissolved solids (TDS), and turbidity. Utilizing tools such as the Blynk app, the Naive Bayes algorithm, Internet of Things (IoT) technology, advanced sensors, and Anaconda software, the system delivers real-time data and alerts, enabling farmers to make informed decisions and apply effective farm management strategies.

Keywords: Water quality monitoring, fish farming, aquatic health guardian system, IoT technologies, Blynk app, Naive Bayes algorithm

Introduction

Fish farming, also known as aquaculture, is becoming more and more important in supplying the rising demand for seafood worldwide. However, preserving ideal water quality is crucial for the well-being and productivity of farmed fish. Variations in parameters such as pH, temperature, total dissolved solids (TDS), and turbidity can have significant impacts on fish health, growth rates, and overall farm profitability. Traditional water quality monitoring techniques often rely on sporadic sampling by hand as well as laboratory analysis, which can be time- and labor-consuming, and may not provide timely feedback to farmers. Furthermore, these methods may leave farmers vulnerable to risks such as disease outbreaks or reduced productivity if they fail to detect sudden changes or localized variations in water quality (Wu et al., 2022). To tackle these challenges, there's a growing need for automated and integrated monitoring systems, custom-made specifically for fish farming operations. With such devices, real-time data on important aspects related to water quality allows farmers to be proactive and make well-informed decisions to maintain ideal circumstances for fish growth and health (Gao et al., 2019).

Conventional water quality standards might not always be enough to decide if fish farm wastewater is okay to release into other waters. To deal with this, people suggest using advanced technology like Internet of Things (IoT)-based systems for better monitoring in fish farms (El-Atab et al., 2020). For instance, in the Mekong Delta, they've set up an IoT-based monitoring system for fish farming to tackle worsening water quality and more diseases specifically for Pangasius fish (Danh et al., 2020). Using IoT technology in fish farming and aquaculture could make things more efficient and easier to manage overall (Parri, et al., 2020). IoT-powered real-time monitoring systems play a crucial role in providing vital information on water quality parameters within fish cages and enable the remote management of offshore sea farms (Rabeea et al., 2020). Additionally, there's a proposal to enhance water quality surveillance in aquaculture, including fish farming, by designing hybrid aerial underwater robotic systems. These systems would utilize water quality sensors integrated onto robotic sensing platforms (Ouyang et al., 2021). Moreover, research has explored machine learning (ML) methods for predicting water quality metrics in fish farming environments. By employing ML models, it becomes feasible to forecast parameters such as pond temperature, pH and dissolved oxygen, all of which are essential for maintaining optimal fish farming conditions (Zambrano et al., 2021). This study introduces the design and implementation of an integrated pH, temperature, TDS, and turbidity sensor-based fish farming water quality

[a]pandyya@gmail.com, [b]pardeepaakash007@gmail.com, [c]dhinagaranmuthusamy@gmail.com, [d]dharanish25selvi@gmail.com, [e]mukilanmuki2611@gmail.com

DOI: 10.1201/9781003587538-49

monitoring system. The system aims to enhance farm management practices and increase overall output by providing farmers with real-time data visualization, customizable alert systems, and continuous monitoring capabilities.

Proposed framework

Figure 49.1 depicts the block diagram of the proposed work. They illustrate the suggested fish farming water quality monitoring system, encompassing all essential components for effective monitoring in aquatic environments. In this study, the selection of appropriate sensors capable of measuring crucial factors essential for maintaining ideal conditions for fish growth and health is meticulously examined. To ensure precise and reliable data collection, parameters including pH, temperature, TDS, and turbidity are thoughtfully chosen. By integrating these sensors with advanced technology such as the ESP8266 module and Arduino UNO board, seamless wireless connectivity is achieved, enabling smooth data transmission throughout the monitoring system.

Data collection is managed with accuracy and dependability using Arduino sketches, effectively gathering sensor data to maintain precision that reflects the aquatic surroundings. Incorporating the Blynk app elevates performance, delivering instantaneous monitoring and presentation of sensor readings to enable quick decision-making and action. Python scripts integrated into the Anaconda software platform allow for the graphical depiction of gathered data, enriching comprehension and supporting knowledgeable decision-making in fish farming operations management. The Naive Bayes classifier algorithm is among the ML methods employed to enhance the system's effectiveness. Leveraging large datasets of water quality measurements, the model identifies patterns and anomalies in these metrics, aiding proactive decision-making. Integration with IoT technologies enables fish producers to access crucial information from anywhere, optimizing conditions for fish health and growth through remote monitoring and control. Table 49.1 displays the water quality parameters relevant to fish farming. Additionally, the comparison table outlines the main characteristics of temperature,

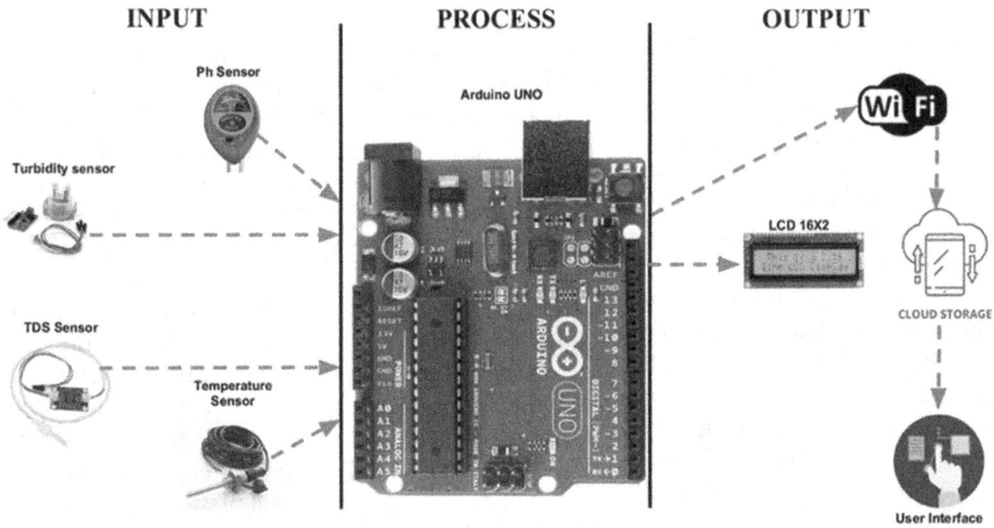

Figure 49.1 Proposed work block diagram
Source: Author

Table 49.1 Water quality parameters for fish farming.

Sensor	Data 1	Status	Data 2	Status
pH sensor	5	Normal	10	Abnormal
TDS sensor	108	Normal	401	Abnormal
Temperature sensor	30	Normal	40	Abnormal
Turbidity sensor	45	Normal	95	Abnormal

Source: Author

turbidity, pH, and TDS sensors utilized in monitoring fish farming water quality. pH sensors provide acidity or alkalinity readings (within a range of 0–14 pH), with an accuracy of ±0.1 to ±0.2 pH. TDS sensors vary in accuracy and measuring range, requiring periodic cleaning and calibration, the costs of which depend on the technology used. Temperature sensors, with a moderate cost, offer accuracy ranging from ±0.5 to ±1°C. Turbidity sensors assess water clarity, with prices and accuracy varying. Integrating all sensors with IoT platforms enables remote monitoring and control, enhancing the efficiency of fish farming practices. Regular calibration and maintenance procedures ensure the system's accuracy and reliability over time.

Statistical analysis and ML predictions enable timely identification of issues and anomalies in water quality parameters. The Blynk app triggers notifications if parameters deviate from acceptable levels, alerting farmers to take corrective action. Field testing and validation in authentic fish farming environments confirm system performance and reliability. Scalability and adaptability accommodate future expansions, ensuring the system evolves alongside industry needs. Comprehensive documentation and training materials empower users to deploy and utilize the system effectively, fostering sustainable and efficient fish farming practices for long-term success.

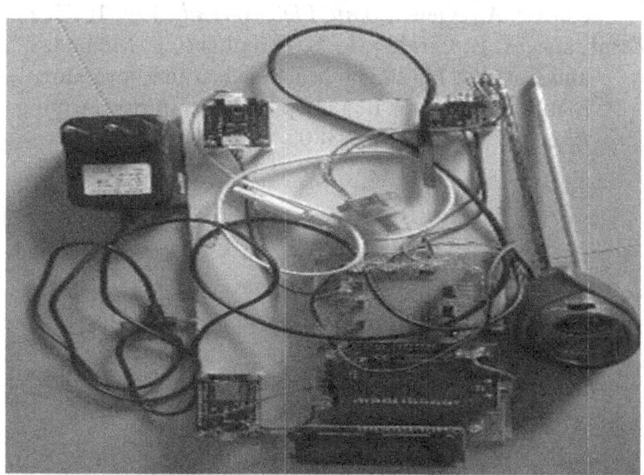

Figure 49.2 Hardware setup of the proposed system
Source: Author

Results and discussion

The hardware setup of the proposed system is depicted in Figure 49.2. This system undergoes testing with water and outputs data to the Blynk app. The dataset used in this study has 1000 whereas 500 data are normal and remaining 500 data are abnormal for testing the water quality parameters.

The final output of the fish farming water quality monitoring system is depicted in Figure 49.3. The hardware setup is tested with fish tank. If all parameters including pH, temperature, TDS, and turbidity are within the normal range, it shows the measured value as shown in Figure 49.3 (a). The water appears clear, indicating favorable conditions for fish farming. With the temperature also in the normal range, the overall assessment suggests that the water is suitable for aquaculture purposes. If any one of the parameters exceeds the abnormal condition, the shows the measured readings as well as notifies the fish farmers for the recommendations (Figure 49.3b). Factors such as acid rain, pollution, and runoff from agricultural land may contribute to abnormal pH levels. If concerned, testing the water with a home testing kit is advisable. Additional details from the image include water temperature: 30°C, TDS: 29 mg/L, phosphate level (P): 5 mg/L and battery level (Blynk): 73%. The proposed system offers a concise overview of the system's ability to monitor and maintain optimal water quality conditions essential for successful fish farming operations.

(a)

(b)

Figure 49.3 Water quality parameter monitoring in fish tank. (a) Normal (b) Abnormal
Source: Author

Conclusions

The proposed work has demonstrated the effectiveness of an integrated fish farming water quality monitoring system in addressing the critical need for actual monitoring and optimization of conditions for fish health and growth. By successfully integrating pH, temperature, TDS, and turbidity sensors with IoT technologies, the aquatic health guardian system provides valuable insights and alerts to fish farmers, enabling informed decision-making and proactive measures to maintain optimal water quality. The study's findings have not only addressed the challenges faced by fish farmers in traditional monitoring methods but have also contributed to advancing knowledge in aquaculture practices. While addressing the limitations in the study, such as scalability and system adaptability, future research directions could focus on enhancing system capabilities and expanding applications in diverse aquaculture settings.

References

Rabeea, A. M., Al-Rawi, A. S., Mohammad, O. J., and Hussien, M. B. (2020). The residual effect of fish farms on the water quality of the Euphrates River, Iraq. *Egypt. J. Aquat. Biol. Fisher.*, 24(4), 549–561.

Danh, L. V. Q., Dung, D. V. M., Danh, T. H., and Ngon, N. C. (2020). Design and deployment of an IoT-based water quality monitoring system for aquaculture in Mekong Delta. *Internat. J. Mech. Engg. Robot. Res.*, 9(8), 1170–1175.

El-Atab, N., Almansour, R., Alhazzany, A., Suwaidan, R., Alghamdi, Y., Babatain, W., and Hussain, M. M. (2020). Heterogeneous cubic multidimensional integrated circuit for water and food security in fish farming ponds. *Small*, 16(4), 1905399.

Gao, G., Xiao, K., and Chen, M. (2019). An intelligent IoT-based control and traceability system to forecast and maintain water quality in freshwater fish farms. *Comp. Elec. Agricul.*, 166, 105013.

Ouyang, B., Wills, P. S., Tang, Y., Hallstrom, J. O., Su, T. C., Namuduri, K., and Den Ouden, C. J. (2021). Initial development of the hybrid aerial underwater robotic system (haucs): Internet of things (iot) for aquaculture farms. *IEEE Internet Things J.*, 8(18), 14013–14027.

Parri, L., Parrino, S., Peruzzi, G., and Pozzebon, A. (2020, May). A LoRaWAN network infrastructure for the remote monitoring of offshore sea farms. *2020 IEEE Internat. Instrumen. Meas. Technol. Conf. (I2MTC)*, 1–6.

Wu, Y., Duan, Y., Wei, Y., An, D., and Liu, J. (2022). Application of intelligent and unmanned equipment in aquaculture: A review. *Comp. Elec. Agricul.*, 199, 107201.

Zambrano, A. F., Giraldo, L. F., Quimbayo, J., Medina, B., and Castillo, E. (2021). Machine learning for manually-measured water quality prediction in fish farming. *Plos One*, 16(8), e0256380.

50 Comparative analysis of facial expression recognition system for age, gender, and emotions

Sheeba Joice C.[1,a], *Jenisha C.*[1,b], *Hitha Shanthini S.*[1,c] *and R. Vedhapriyavadhana*[2]

[1]Department of ECE, Saveetha Engineering College, Chennai, Tamilnadu, India

[2]Department of School of Computing Engineering and Physical Sciences, University of the West of Scotland, London

Abstract

Facial expression recognition systems have advanced quickly, allowing for accurate real-time detection of age, gender, and emotion. Facial features are retrieved effectively for exact identification using deep learning models such as MobileNetV2, ResNet, and DenseNet, as well as approaches like the Haar cascade classifier. The use of datasets such as FER 2013 promotes robust model training, while pre-processing procedures provide peak performance. Models such as the Caffe model achieve great accuracy in detectingage and gender in real time through transfer learning and fine-tuning. These improvements highlight the potential of facial recognition systems in a variety ofapplications, including enhanced security, healthcare, and human-computer interaction.

Keywords: Facial expression recognition (FER), deep learning, age, gender, emotions, Haar cascade classifier, Caffe model, MobileNetV2

Introduction

Facial expression recognition (FER) combines computer vision, artificial intelligence, and psychology to interpret human emotions from facial cues, with applications in human-computer interaction and mental health monitoring. Groundbreaking work by psychologist Paul Ekman laid the foundation for FER (Allaert et al., 2022). Recent advancements in deep learning have revolutionized facial emotion identification, enabling real-time recognition in various domains like gaming and surveillance (Abu Ghrban and Abbadi, 2023). FER relies on state-of-the-art algorithms and datasets like FER 2013, Adience, and IMDB-WIKI for robust training, employing technologies such as MobileNetV2 (Sandler et al., 2018) and the Haar cascade classifier for real-time emotion recognition and rapid face detection. The Caffe model underscores the significance of transfer learning for accurate age and gender prediction (Jia et al., 2014).

Giri et al. (2022) underscores the importance of age and gender considerations in emotion detection, advocating for inclusive systems. Their proposed real-time emotion detection approach, utilizing CNNs and OpenCV, shows promise, but further validation across diverse demographics is needed. Barahate and Saturwar, (2010) introduces a robust face recognition method employing PCA and BPNN, achieving an average recognition rate of 87.5%, albeit facing challenges with fast-moving objects and varying lighting conditions. Reddy et al., (2021) presents a hybrid face mask detection model blending deep learning and traditional methods, achieving peak accuracies of 96.3% with InceptionV3 and logistic regression, demonstrating superior performance. Joice C. Sheeba et al. (2024) propose a smart security system integrating Viola-Jones algorithm and CNNs for face detection and recognition, enhanced by MATLAB's image processing and GSM module for real-time alerts, offering a reliable security solution. This research aims to inspire further innovation in facial expression recognition methodologies and applications.

Methodology

Data collection and data pre-processing

The facial expression recognition 2013 dataset (FER, 2013) is highly valued for emotion recognition due to its extensive features. It includes over 35,000 labeled facial images with diverse expressions and high resolution, facilitating robust model training and improving accuracy compared to earlier datasets. A large dataset is crucial for testing deep convolutional neural network (DCNN) models. Irrelevant features can hinder performance, so data cleaning through image processing techniques like angle adjustment, zooming, and color correction is essential. Key techniques like rescaling and preprocessing standardize pixel values and enhance image quality, reducing distractions and highlighting important details for the model.

[a]sheebajoice@saveetha.ac.in, [b]jenishac@saveetha.ac.in, [c]hithashanthini@saveetha.ac.in

DOI: 10.1201/9781003587538-50

Emotion model

Emotion recognition utilizes a specialized two-channel architecture to extract emotional features from the FER2013 dataset, feeding them into the inception layer of a CNN during training. MobileNetV2, optimized for mobile and embedded devices, employs depth-wise separable convolution and linear bottleneck layers for real-time emotion recognition. Training involves optimizing network parameters with an algorithm like Adam and labeled facial image batches, adjusting weights to minimize emotion label differences. Post-training, MobileNetV2 pre-processes inputs, extracts features, and predicts emotion labels through fully connected layers. Figure 50.1 illustrates the different types of emotions.

Figure 50.1 Different types of emotions
Source: Author

Transfer learning models like ResNet, Inception, VGG16, VGG19, and DenseNet are also employed. In real-time face mask detection, VGG16 and VGG19 identify masks during feature extraction and facial recognition. MobileNetV2's lightweight architecture and balanced accuracy and efficiency make it ideal for real-time processing on devices with limited resources. Table 50.1 highlights that MobileNet V2 is better suited for real-time emotion recognition.

Table 50.1 Highlights of MobileNetV2.

Model	Architecture	Parameters	Computational efficiency	Accuracy	Real-time performance
MobileNetV2	Depth-wise separable convolution, linear bottleneck layers	Low	High	Moderate	Excellent
ResNet	Deep residual network with skip connections	High	Moderate	High	Moderate
Inception	Inception modules with multiple convolutional paths	Moderate	Moderate	High	Poor
VGG	Deep convolutional network with repeated blocks.	High	Low	Moderate	Poor
DenseNet	Dense Connectivity between layers	Very high	Moderate	High	Poor

Source: Author

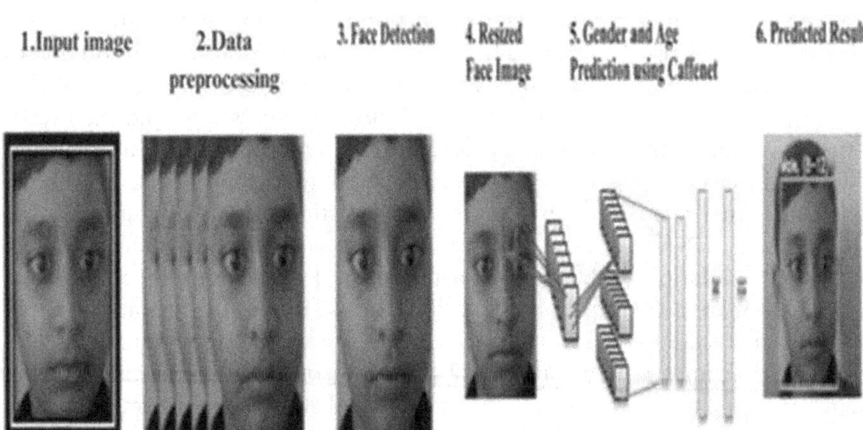

Figure 50.2 Caffe model flow diagram
Source: Author

Age and gender model

The Caffe model provides a robust platform for age and gender detection, leveraging its facial recognition-tailored architecture to extract detailed facial features essential for precise detection. Figure 50.2 illustrates the Caffe model flow diagram and Figure 50.3 depicts the gender specification. Utilizing deep learning technology, likely based on CNNs, the Caffe model excels at capturing facial characteristics. Through transfer learning from datasets like Adience or IMDB-WIKI, the model refines its architecture and weights, fine-tuning age and gender datasets. Training involves parameter adjustments and monitoring for optimal performance, leading to real-time implementation with webcam integration via libraries like OpenCV for instant age and gender detection of multiple faces within a single image. Table 50.2 presents a comparison of prominent deep learning frameworks across different models, facilitating the assessment of age and gender prediction models developed with the Caffe framework.

Results

While the Haar cascade classifier excels in face detection, its performance suffers in dynamic scenarios and varying lighting conditions, highlighting its limitations compared to CNNs. Emotion recognition models leveraging MobileNetV2 offer a promising balance of accuracy and real-time processing. Similarly, age and gender prediction models based on frameworks like Caffe show significant advancements in demographic attribute prediction. The proposed model demonstrated remarkable results during experimentation with volunteers of varying ages as depicted in Figure 50.4.

Contrasting the approach of using OpenCV and CNN for emotion detection in facial feature recognition with MobileNetV2, disparities arise in methodologies and performance. While effective, the emotion detection and recognition (EDR) software's reliance on OpenCV and CNN (Giri et al., 2022) may face limitations in accuracy and speed. In contrast, MobileNetV2's specialized architecture ensures

Figure 50.3 Gender specification
Source: Author

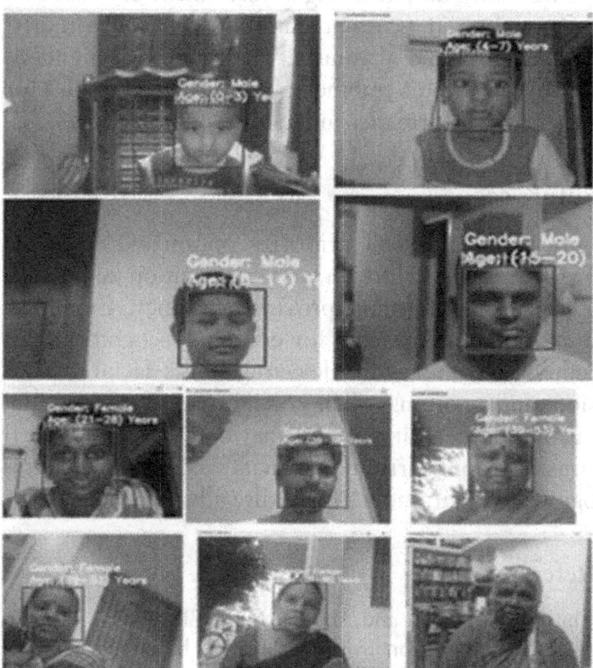

Figure 50.4 Prediction of age, gender and emotions in real-time from a webcam
Source: Author

Table 50.2 Comparison of different deep learning techniques.

Framework	Core language	CPU	GPU	Open source	Trained model	Pre-trained model
Caffe	C++	✓	✓	✓	✓	✓
Cuda-convent	C++	----	✓	✓	✓	---
Decaf	Python	✓	----	✓	✓	✓
Over feat	Lua	✓	✓	----	----	✓
Theano/Pyle arn2	Python	✓	✓	✓	✓	----
Torch	Lua	✓	✓	✓	✓	----

Source: Author

swift and efficient operation, particularly on mobile devices, comprehensively capturing emotions like smiles or frowns. Thus, MobileNetV2 emerges as a superior choice for real-time and precise emotion detection tasks. MobileNetV2's results surpass existing methodologies, with added enhancement through age and gender detection, broadening its capabilities and usability.

Conclusions

While the Haar cascade classifier remains efficient, its limitations in dynamic scenarios necessitate more sophisticated approaches such as CNNs. MobileNetV2-based emotion recognition models provide a compelling solution for real-time applications with their balanced accuracy and efficiency. Additionally, deep learning models like those based on Caffe show significant potential in accurately predicting demographic attributes. These findings highlight the importance of embracing advanced techniques for enhanced performance in facial recognition tasks.

Future scope

Future research will prioritize optimizing real-time processing capabilities in facial recognition tasks, with a specific focus on improving efficiency. Exploring multimodal approaches for emotion recognition and refining techniques for detecting subtle facial expressions and emotions will be crucial areas of exploration. Furthermore, addressing privacy concerns and ethical considerations surrounding facial recognition technology will be paramount for future advancements.

References

Abu Ghrban, Z. S. and Abbadi, N. K. EL. (2023). Human age predication from face images based on combining deep wavelet network and machine learning algorithms. *J. Comp. Sci.*, 19(5), 654–666. https://doi.org/10.3844/jcssp.2023.654.666.

Allaert, B., Bilasco, I. M., and Djeraba, C. (2022). Micro and macro facial expression recognition using advanced local motion patterns. *IEEE Trans. Affec. Comput.*, 13(1), 147–158. https://doi.org/10.1109/TAFFC.2019.2949559.

Barahate, S. R. and Saturwar, J. (2010). Face recognition using PCA based algorithm and neural network. *Proc. Internat. Conf. Workshop Emerg. Trends Technol.*, 249–252. https://doi.org/10.1145/1741906.1741963.

Giri, S., Singh, G., Kumar, B., Singh, M., Vashisht, D., Sharma, S., and Jain, P. (2022). Emotion detection with facial feature recognition using CNN & OpenCV. *2022 2nd Internat. Conf. Adv. Comput. Innov. Technol. Engg. (ICACITE)*, 230–232. https://doi.org/10.1109/ICACITE53722.2022.9823786.

Jia, Y., Shelhamer, E., Donahue, J., Karayev, S., Long, J., Girshick, R., Guadarrama, S., and Darrell, T. (2014). Caffe. *Proc. 22nd ACM Internat. Conf. Multim.*, 675–678. https://doi.org/10.1145/2647868.2654889.

Joice, C. S., Shanthini, H., Dedeepya, M., Roshini, M. M., and Laya, N. S. (2024). *Smart Door Design Based on Face Detection and Recognition Using CNN and Viola Jones Algorithm* (1st Edition). Taylor and Francis Group, 68–72.

Reddy, S., Goel, S., and Nijhawan, R. (2021). Real-time face mask detection using machine learning/deep feature-based classifiers for face mask recognition. *2021 IEEE Bombay Sec. Sig. Conf. (IBSSC)*, 1–6. https://doi.org/10.1109/IBSSC53889.2021.9673170.

Sandler, M., Howard, A., Zhu, M., Zhmoginov, A., and Chen, L.-C. (2018). MobileNetV2: Inverted residuals and linear bottlenecks. *2018 IEEE/CVF Conf. Comp. Vis. Patt. Recogn.*, 4510–4520. https://doi.org/10.1109/CVPR.2018.00474.

51 Unlocking the learning potential: Exploring the outcome of jigsaw-based cooperative learning on the scholastic achievement

S. Sharmila Queenthy[a] *and M. Thirumavalavan*[b]
Department of Chemistry, Saveetha Engineering College, Chennai, Tamilnadu, India

Abstract

This task delved into the efficiency of the "Jigsaw-based supportive learnedness method" in which students equally led, prepared, and participated in a group, learning for themselves as well as for their group. This study revealed that the enhanced student-friendliness of teaching and learning practices have shown beneficial in holding student's interest and motivating them to study and perform better. **Goal:** The intention of this work was to examine how well the scholastic achievement and satisfaction of the first year under graduate engineering chemistry students were affected by the Jigsaw-based teaching method. **Sampling:** The task was conducted in two distinct classes, with one serving as the control group (non-jigsaw) and the other as the experimental group (jigsaw). Within the class of 30 students, the students were separated into 5 groups of six students each for the jigsaw approach. The jigsaw group learners were further divided into five "home groups," each consisting of six students, aligning with the division of the electrochemistry topic into six subtopics labeled A, B, C, D, E, and F. The subjects covered were oxidation and reduction reaction (A), electrochemical cell (B), electrode potential (C), electrochemical series (D), differentiation of electrolytic and electrochemical cell (E) and Nernst equation (F). **Instruments:** Through assessments of academic achievements and feedback, the study sought to elucidate the potential benefits and challenges of implementing the jigsaw-based cooperative learning in educational settings. **Conclusion:** The observed results suggested that the jigsaw-based tactic was an excellent method to analyze the learning ability of the learners as well as for the students to enhance their scores.

Keywords: Jigsaw-based approach, scholastic achievement, learning ability, electrochemistry

Introduction

Studies into the implementation of the jigsaw method demonstrated benefits when jigsaw tasks were organized and included several components that were essential for successful cooperative learning. The core components include positive student interdependence, encouraging personal responsibility; emphasize face-to-face interactions, prioritizing interpersonal and small-group abilities, and facilitating group reflection (Johnson et al., 1991). It was stated if the students were evenly distributed among the groups; the jigsaw method could also be helpful when various types of students were present in the same class (dominant, slow, bright, or competitive individuals, etc.). While lectures remain prevalent in various fields, there is increasing recognition of the advantages of active learning methods. These approaches are noted for fostering deeper understanding (Prince, 2004), better retention of information over time (Laal and Laal, 2012), upgraded ability to apply knowledge to different situations (Roehl et al., 2013), increased engagement and social interaction (Minifie and Davis, 2013), and the creation of inclusive learning environments (Johnson, 2019).

The jigsaw technique typically involves two stages: firstly, students collaborate in "expert groups," focusing on mastering a particular topic, and secondly, these groups are rearranged into "jigsaw groups," ensuring each new group contains an expert from each area who will then educate their peers (Aronson, 1978).

Notably, when compared to traditional lecturing, the performance would be greatly improved by employing the active learning methodologies. For instance, in a meta-study of undergraduate courses viz. science, technology, engineering, and mathematics (STEM), dynamic learning methods vary greatly in their formats and durations, including activities such as group problem-solving, in-class worksheet or tutorial completion, utilization of personal response systems, and implementation of studio or workshop course structures (Freeman et al., 2014). The study indicates that the students in dynamic learning settings are 1.5 times less likely to fail compared to those in traditional lecture-based formats. Furthermore, exam scores for courses utilizing active learning approaches

[a]sharmialwin@gmail.com, [b]mtvala@gmail.com

DOI: 10.1201/9781003587538-51

increased by 6% compared to those using traditional lecture formats.

Literature review

The results from the reviewed studies on "jigsaw cooperative learning" generally substantiated the discoveries of dynamic learning investigation. The scholars responded positively to this teaching method and demonstrated improved performance. Most of the literature assessed focused on STEM education. Consequently, it was anticipated that the outcomes of this study would enhance the body of knowledge concerning active learning and jigsaw cooperative learning.

Multiple studies investigating the impact of "jigsaw cooperative learning" on students' academic achievement consistently demonstrate positive effects (Azmin, 2016; Doymus, 2008; Gömleksiz, 2007; Johnson and Stanne, 2000; Karacop and Doymus, 2013; Kiliç, 2008; Morgan et al., 2008; Tran and Lewis, 2012). Overall, it has been observed that students' performance improves, and they also express enjoyment in engaging with various activities. Gömleksiz (2007) specifically determined that "jigsaw cooperative learning" could lead to the students for retaining the information for longer periods when compared to teacher-centered instruction.

Methodology

To effectively implement the jigsaw activity, this work included a sample of 60 undergraduate students from two separate classes enrolled in an engineering chemistry course during the academic years 2022–2023. One class, comprising 30 students, served as the non-jigsaw (control) group, where the electrochemistry topic was taught using traditional lecture methods. The other class, also consisting of 30 students, served as the jigsaw (experimental) group and received instruction through cooperative learning (jigsaw). The primary tool for data collection was the electrochemistry quiz, administered to both groups.

The activity investigated the cooperative learning of the students understanding on electrochemistry topic. The teachings in both groups were done by the same educator. As depicted in Figure 51.1, the jigsaw group members were organized into five "home groups", each comprising six students. This grouping was based on the subdivision of the electrochemistry topic into six subtopics labeled as A, B, C, D, E, and F. The subjects covered were oxidation and reduction reaction (A), electrochemical cell (B), electrode potential (C), electrochemical series (D), differentiation of electrolytic and electrochemical cell (E) and Nernst equation (F).

In the Jigsaw group, each student was assigned one specific subtopic of content to study. These groups were responsible for coaching their allotted subtopic to the other scholars within their learning group. Once each student had individually studied their subtopic, they convened with others who had been assigned the same subtopic, forming "expert group". Within these "expert groups", students shared their ideas and collaborated to prepare the presentations for their "Jigsaw groups" and presented as shown in Figure 51.2. This collaborative process allowed for gaps in individual knowledge to be addressed, misconceptions to be corrected and key concepts to be reinforced. Each student in the "expert groups" took turns presenting

Figure 51.2 Student presenting the topic to jigsaw group
Source: Author

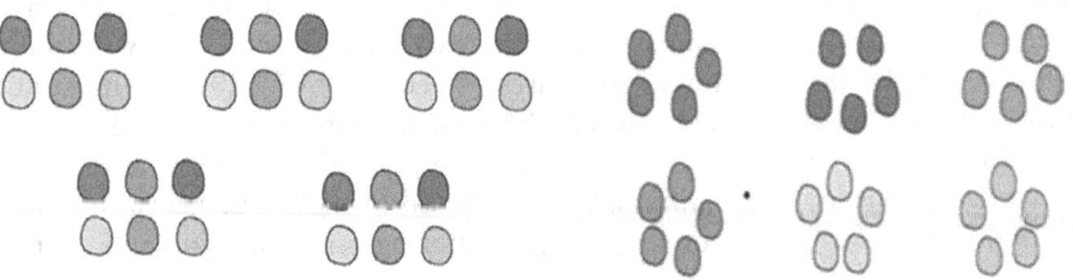

Figure 51.1 Concept of jigsaw group
Source: The Jigsaw Method www.cultofpedagogy.com © 2015 by Jennifer Gonzalez

their subtopic, while the remaining students listened attentively, took notes, and asked questions to ensure their understanding of the material.

The evaluation method could involve a straightforward quiz to ensure that all students had grasped a fundamental understanding of the entire material covered. It was ensured that the quiz encompassed all the content from the subtopics.

The questionnaire survey consisted of ten questions, which were designed to measure the student's views towards the jigsaw method. This questionnaire was given to the participants at the end of the final lesson.

Results and discussion

The "jigsaw cooperative learning group" and the "non-jigsaw lecture-style learning group" equally participated in the evaluation quiz together. The results of the quiz, completed by students after covering the six subtopics of the electrochemistry learning component, were analyzed. Table 51.1 presents the outcomes of the quiz.

Based on the data, the jigsaw group demonstrated a higher mean score (89.6) compared to the non-jigsaw group (80.9), suggesting that participants in the jigsaw group achieved higher scores on average. The substantial T-value of 4.2 indicates a significant difference, while the very low p-value of 0.0001 underscores high confidence in these findings.

The comparison of quiz scores between the two groups clearly showed that the "jigsaw activity approach" resulted in higher scores among students compared to the traditional non-jigsaw lecture-style learning group. Figure 51.3, clearly depict that the students who were capable of scoring 90% and above 90% showed no significant variations in the score analysis. But, however, it was suggested that this jigsaw activity approach was more helpful for the students who could secure the average score (60–85%). In these cases, 90% scores 81–100 which would be definitely enhanced with the help of jigsaw activity approach. In the case of jigsaw activity approach, the score did not decrease below 80%. Moreover, the difference between the individual score in the non-jigsaw activity approach was significantly more whereas there was a little difference only in the scores of the students involved in the jigsaw activity. Thus, it was confirmed that the jigsaw approach was an effective method for the students to enhance their scores as well as an efficient method to analyze the learning ability of the learners.

Table 51.1 Independent sample T-test analysis of quiz score.

Groups	N	Mean	SD	t	p
Non-jigsaw	30	80.9	10.3	4.2	0.0001
Jigsaw	30	89.6	4.9		

Source: Author

Challenges

- Ensuring all students remain actively engaged and motivated throughout the jigsaw activity.
- Managing the time needed for students to thoroughly understand their assigned topics and effectively convey that knowledge to their peers.
- A small number of students perceived that their peers didn't approach the activity seriously enough.
- Some students were hesitant about the responsibility of teaching their peers.
- Students expressed dissatisfaction regarding the inability to choose their own groups.

Feedback through survey

To analyze the influence of this active learning activity on students, feedback was gathered through a combination of questions and ratings.

Q 1. The activity helped me to understand the subject matter.

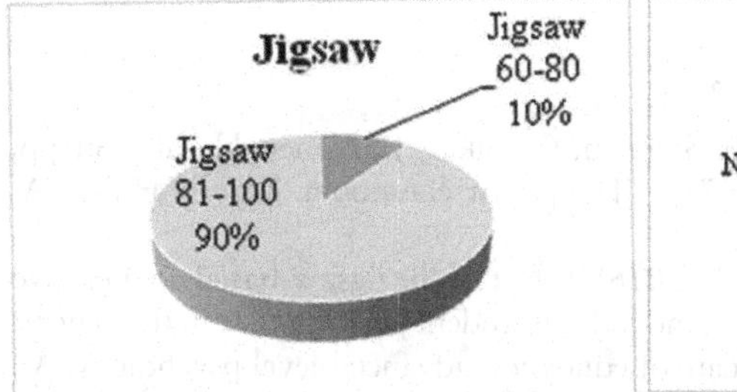

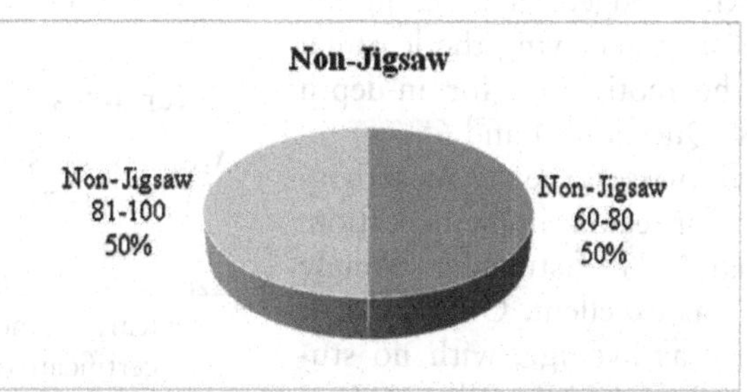

Figure 51.3 Comparison of the quiz score of the students of non-jigsaw and jigsaw group
Source: original data

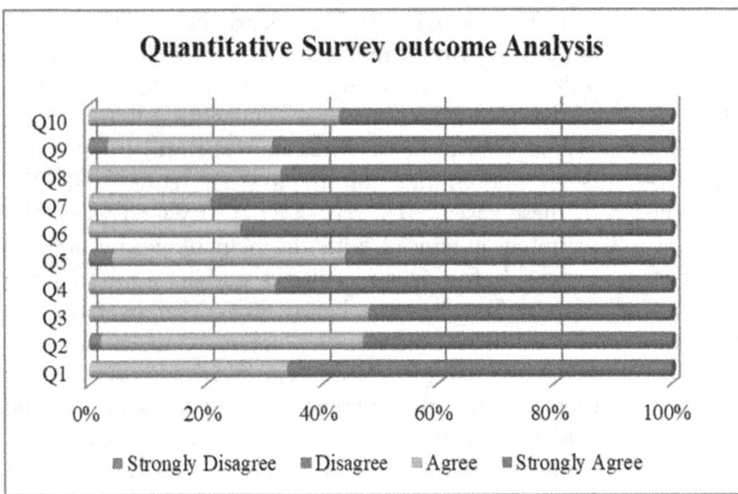

Figure 51.4 Feedback rating given by students to jigsaw activity
Source: Author

Q 2. The activity made it possible to cover the subject in-depth.
Q 3. The exercise improved the participant's communication abilities.
Q 4. A detailed examination of the subjects improved analytical ability.
Q 5. The activity assisted in mitigating shyness and hesitation.
Q 6. The activity should be integrated for all chemistry topics.
Q 7. You are assured that this knowledge could be implemented in an engineering context.
Q 8. The exercise was enjoyable.
Q 9. This represents a highly efficient method of learning.
Q 10. The activity is good enough to perform on regular basis

The conclusions were made from the preceding questions where the participants were prompted to rate their level of agreement as shown in Figure 51.3.

According to the data provided in the feedback report, students also acknowledged that the jigsaw exercise did not only assist in achieving the learning aim but also provided the motivation for in-depth study of the same themes (Questions 4 and 6).

Students were requested to assess the jigsaw activity to determine its suitability for regular implementation. Figure 51.4 illustrates that 63% of students strongly agreed that this activity was excellent. Only 11% of students rated the activity as average, with no students providing a poor rating. Additionally, 26% of students deemed the activity good enough to be conducted regularly. These statistics not only indicate the effectiveness of the jigsaw method but also highlight students' willingness to embrace this active learning approach on a consistent basis.

Conclusions

In this study a significant difference in student scores between "jigsaw cooperative learning" and "traditional teaching-based" pedagogy was observed. Nonetheless, "jigsaw cooperative learning", 90% scores 81–100 and remains an effective teaching strategy as the students participated actively in knowledge assimilation and it adapts classroom procedures without detrimentally affecting the grades. This study highlights several advantages of employing jigsaw cooperative learning to enhance the educational experience for students. The suggested jigsaw active learning technique could promote proactive, creative, confident, and flexible learners who would excel in critical thinking, problem-solving, effective communication, and teamwork. Additionally, there is a noticeable rise in the students' bonding, which is a crucial component needed to motivate the learners to engage themselves in extracurricular activity.

References

Aronson, E., Stephen, C., Sikes, J., Blaney, N., and Snapp, M. (1978). The jigsaw classroom. Beverly Hills, CA: Sage.

Azmin, N. H. (2016). Effect of the jigsaw-based cooperative learning method on student performance in the general certificate of education advanced-level psychology: An exploratory Brunei case study. *Internat. Edu. Stud.*, 9(1), 91–106. Retrieved from http://www.ccsenet.org/journal/index.php/ies/article/view/49895.

Doymus, K. (2008). Teaching chemical bonding through jigsaw cooperative learning. *Res. Sci. Technol. Edu.*, 26(1), 47–57. https://doi.org/10.1080/02635140701847470.

Freeman, S., Eddy, S. L., McDonough, M., Smith, M. K., Okoroafor, N., Jordt, H., and Wenderoth, M. P. (2014). Active learning increases student performance in science, engineering, and mathematics. *Proc. Nat. Acad. Sci.*, 111(23), 8410–8415. https://doi.org/10.1073/pnas.1319030111.

Gömleksiz, M. N. (2007). Effectiveness of cooperative learning (jigsaw II) method in teaching English as a foreign language to engineering students (case of Firat University, Turkey). *Eur. J. Engg. Edu.*, 32(5), 613–625. https://doi.org/10.1080/03043790701433343.

Johnson, D. W., Johnson, R. T., and Smith, K. A. (1991). Cooperative learning: Increasing college faculty instructional productivity. *ASHE-ERIC Reports Higher Edu.* Report No.4

Jones, T. N., Graham, K. J., and Schaller, C. P., (2012). A jigsaw classroom activity for learning IR analysis in organic chemistry. *J. Chem. Educ.*, 89(10), 1293–1294.

Karacop, A. and Doymus, K. (2013). Effects of jigsaw cooperative learning and animation techniques on students' understanding of chemical bonding and their conceptions of the particulate nature of matter. *J. Sci. Edu. Technol.*, 22(2), 186–203. https://doi.org/10.1007/s10956-012-9385-9.

Kiliç, D. (2008). The effects of the jigsaw technique on learning the concepts of the principles and methods of teaching. *World Appl. Sci. J.*, 4(supp. 1), 109–114. Retrieved from http://www.idosi.org/wasj/wasj4(s1)/18.pdf.

Laal, M. and Laal, M. (2012). Collaborative learning: What is it? *Proc.-Soc. Behav. Sci.*, 31, 491–495. https://doi.org/10.1016/j.sbspro.2011.12.092.

Minifie, J. R. and Davis, K. (2013). Ensuring Gen Y students come prepared for class; then leveraging active learning techniques to most effectively engage them. *Am. J. Busin. Manag.*, 2(1), 13–19.

Morgan, B. M., Rodriguez, A. D., and Rosenberg, G. P. (2008). Cooperative learning, jigsaw strategies, and reflections of graduate and undergraduate education students. *College Teach. Methods Styles J.*, 4(2), 1–6. https://doi.org/10.19030/ctms.v4i2.5519.

Prince, M. (2004). Does active learning work? A review of the research. *J. Engg. Edu.*, 93(3), 223–231. https://doi.org/10.1002/j.2168-9830.2004.tb00809.x.

Roehl, A., Reddy, S. L., and Shannon, G. J. (2013). The flipped classroom: An opportunity to engage millennial students through active learning strategies. *J. Family Cons. Sci.*, 105(2), 44–49.

Tran, V. and Lewis, R. (2012). Effects of cooperative learning on students at An Giang University in Vietnam. *Internat. Edu. Stud.*, 5(1), 86–99. Retrieved from http://www.ccsenet.org/journal/index.php/ies/article/view/12121.

Printed in the United States
by Baker & Taylor Publisher Services